ARCHAEOLOGY OF EAST ASIA

Bronze Weapons of the Qin Terracotta Warriors

Standardisation, Craft Specialisation and Labour Organisation

XIUZHEN LI

BAR INTERNATIONAL SERIES 2992 VOLUME 3

BAR PUBLISHING

Published in 2020 by
BAR Publishing, Oxford

BAR International Series 2992

Archaeology of East Asia, 3
Bronze Weapons of the Qin Terracotta Warriors

ISBN 978 1 4073 1690 1 paperback
ISBN 978 1 4073 5602 0 e-format

DOI https://doi.org/10.30861/9781407316901

A catalogue record for this book is available from the British Library

COVER IMAGE *'Bronze Arrows for the Terracotta Crossbowman', Courtesy of Zhao Zhen.*

BAR
PUBLISHING

BAR titles are available from:

BAR Publishing
122 Banbury Rd, Oxford, OX2 7BP, UK
EMAIL info@barpublishing.com
PHONE +44 (0)1865 310431
FAX +44 (0)1865 316916
www.barpublishing.com

ARCHAEOLOGY OF EAST ASIA

Series Editor: Anke Hein, University of Oxford, UK

This series provides a platform for data-rich studies on a variety of topics and materials from all over East Asia as well as conference proceedings reflecting the newest research insights and trends. The series takes a regional approach to East Asia, encouraging projects that cross national borders, even into adjoining regions, and/or cover areas usually overlooked in mainstream research. This includes all parts of China, Japan, Korea, Mongolia, the Russian Far East, the Tibetan Plateau as a whole and the northern reaches of Southeast Asia. Especially encouraged are submissions proposing and conducting new approaches and methods in all aspects of archaeology including scientific techniques, spatial analysis, various digital methods, but also theory and model-based or traditional chronology-focused studies.

If you are interested in publishing in the Archaeology of East Asia series, please contact editor@barpublishing.com.

A Zooarchaeological Study of the Haimenkou Site, Yunnan Province, China
Juan Wang

Oxford, BAR Publishing, 2018 BAR International Series 2902

The Origin of Cattle in China from the Neolithic to the Early Bronze Age
Chong Yu

Oxford, BAR Publishing, 2020 BAR International Series 2959

Of Related Interest

Contacts Between the Shang and the South c. 1300–1045 BC
Celine Y. Y. Lai

Oxford, BAR Publishing, 2019 BAR International Series **2915**

The Technology of Large-Scale Zinc Production in Chongqing in Ming and Qing China
Wenli Zhou

Oxford, BAR Publishing, 2016 BAR International Series **2835**

Chariots in Early China
Origins, cultural interaction, and identity
Hsiao-yun Wu

Oxford, BAR Publishing, 2013 BAR International Series **2457**

Ancient Society and Metallurgy
A comparative study of Bronze Age societies in Central Eurasia and North China
Liangren Zhang

Oxford, BAR Publishing, 2012 BAR International Series **2328**

The Distribution of Bronze Drums in Early Southeast Asia
Trade routes and cultural spheres
Ambra Calò

Oxford, BAR Publishing, 2009 BAR International Series **1913**

*To Guo Baofa*郭宝发 *and Guo Yin*郭音

Acknowledgements

This research was based on my PhD thesis submitted to University College London (UCL) in 2011, and funded by the Sun Hung Kai Kwok Foundation in Hong Kong (2007–2010). Financial assistant was also kindly provided by the Museum of the Terracotta Army, the Rio Tinto Group via the Institute for Archaeo-Metallurgical Studies, British Academy, and UCL Graduate School.

My research, as part of the wider cooperative project, greatly benefits two cooperative parties, the Emperor Qin Shihuang's Mausoleum Site Museum, China, and the UCL Institute of Archaeology, UK. My first and deepest thanks go to the late Professor Peter Ucko, who took the effort to establish this Sino-British cooperation between the Museum and the Institute. Without his encouragement and support, none of the research for this book would have taken place. Many thanks also go to Director Wu Yongqi and Professor Thilo Rehren for having provided me with a stable cooperative environment and for their constructive ideas and suggestions on the project.

My sincere gratitude to both Deputy Director Song Xinchao and Director Zhao Rong for their assistance and support of my research project; further, I am grateful for the international cooperation from the State Administration of Cultural Heritage and the Shaanxi Provincial Administration of Cultural Heritage.

This book could not have been written without the invaluable critiques, suggestions, comments, and feedback from my joint supervisors Professor Marcos Martinón-Torres, Professor Andrew Bevan, and Professor Thilo Rehren. Their stimulating and insightful comments on each draft were especially helpful in clarifying ideas and arguments in my research. Their patient encouragement in some critical moments strengthened me to carry on.

I am also grateful to Liu Zhancheng, Zhao Kun, Xia Yin, and my other colleagues involved in the cooperative project between the two institutions, and for assisting me in data collection in the Department of Conservation. I would also like to thank Chang Qiuyue, Yan Hongxia, Zhang Min, Liu Jing, and Wang Ting for their time in taking photographs of these weapons in the storage at the Emperor Qin Shihuang's Mausoleum Site Museum.

Many heart-felt thanks go to Shiona Airlie, a long-time friend and the first person to read each chapter's draft. Glenn Hughes also deserves the warmest thanks for his understanding and support over those years for my research. My niece Jingjing also deserves my deepest thanks for making my writing-up time much less burdensome.

This book is dedicated to my late husband Guo Baofa and daughter Guo Yin. The suffering of their deaths is my whole life pain, and has inspired me to be stronger to fulfil this research and find justice for them.

Table of Contents

List of figures xiii
List of tables xvii
Abstract xviii

1. Introduction 1
1.1. Introduction 1
1.2 Research review and defining research questions 1
1.2.1 Previous research on the bronze weapons 1
1.2.2 Defining research questions 3
1.3 The Emperor and his tomb complex – defining the temporal and spatial context 3
1.3.1 The Qin Dynasty and Emperor Qin Shihuang 3
1.3.2 The tomb complex, the terracotta warriors and the bronze weapons 5
1.4 Chapter summaries 8

2. Approaches to Standardisation and Labour Organisation 9
2.1 Introduction 9
2.2 Standardisation 9
2.2.1 A definition of standardisation 9
2.2.2 Parameters affecting standardisation 9
2.2.3 How to define and measure standardisation from archaeological data? 9
2.3 Standardisation and technological organisation 11
2.3.1 Casting and assembling 11
2.3.2 Technological models 11
2.4 Standardisation and specialisation 12
2.4.1 Specialisation 12
2.4.2 The hypothesis of standardisation and specialisation 13
2.5 Standardisation versus variation: copying error 14
2.6 Spatial signatures of workshop activity and labour organisation 14
2.6.1 Point pattern analysis 15
2.6.2 Spatial randomness, regularity, and clustering 15
2.6.3 First-order effect and second-order effect 15
2.7 Summary 16

3. Methodology 17
3.1 Introduction 17
3.2 Qualitative classification 17
3.2.1 Functional classification of the weapons 17
3.2.2 Micro-features 21
3.2.3 Inscriptions 21
3.3 Sampling 22
3.3.1 Principles of sampling 22
3.3.2 Sampling strategy for this project 22
3.4 Dimensional measurements and quantitative analysis 22
3.4.1 Measuring techniques 22
3.4.2 Metric variables 23
3.4.3 Accuracy and precision assessment 23
3.4.4 Statistical analysis 24
3.4.4.1 Cluster analysis 24
3.4.4.2 Principal Component Analysis 25
3.4.4.3 Frequency histogram 25
3.4.4.4 Coefficient of Variation 25
3.5 Establishing a GIS database and spatial analysis 26
3.5.1 Original map 26

3.5.2 Digital map 26
3.5.3 Using a local grid 26
3.5.4 Spatial statistics – pair correlation functions 26
3.6 Integration with archaeometric data 27
3.7 Summary 27

4. Inscriptions on the Weapons 29
4.1 Introduction 29
4.2 Types of inscriptions 29
4.2.1 Cast inscriptions 29
4.2.2 Carved or chiselled inscriptions 30
4.2.3 Ink inscriptions 32
4.3 The content of the inscriptions and the types of weapons 32
4.3.1 Inscriptions on dagger-axes, spears, and halberds 32
4.3.2 Inscriptions on lances 34
4.3.3 Inscriptions on swords 37
4.3.4 Inscriptions on triggers 37
4.3.5 Inscriptions on ferrules 39
4.4 Spatial patterns of the inscribed weapons 42
4.4.1 Spatial patterns of inscribed halberds 42
4.4.2 Spatial patterns of inscribed lances 42
4.4.3 Spatial patterns of inscribed triggers 42
4.4.4 Spatial patterns of inscribed ferrules 46
4.5 Function of the inscriptions 47
4.5.1 Quality control 47
4.5.2 Assembling triggers 48
4.5.3 Other functions 48
4.6 Summary 48

5. Bronze Triggers 49
5.1 Introduction 49
5.2 Research review 49
5.3 Measurements of the bronze triggers 50
5.3.1 Sources of data 50
5.3.2 Triggers' function and manipulation 50
5.3.3 Method of measuring 51
5.3.4 Patterns in trigger measurements 53
5.3.4.1 Trigger Part A 53
5.3.4.2 Trigger Part B 55
5.3.4.3 Trigger Part C 56
5.3.5 Analysing the patterns of the triggers 56
5.4 Coefficients of variation and the degree of standardisation 58
5.5 Trigger assembly and technological organisation 61
5.6 Spatial analysis 65
5.6.1 Parameters affecting the spatial patterns 65
5.6.2 A pair correlation function 71
5.6.3 Results obtained on assembled triggers 71
5.6.4 Activity areas in the pit 76
5.7 Summary 76

6. Bronze Arrows 79
6.1 Introduction 79
6.2 Research review 79
6.3 Arrow data 79
6.3.1 The archaeological origin of the bronze arrows 79
6.3.2 Bronze arrows from the five easternmost trenches of Pit 1 80
6.3.2.1 An enumeration of arrows in each corridor 81
6.3.2.2 The size of the arrow bundles 81
6.3.2.3 Arrow bundles and quivers 81
6.3.2.4 Arrow types 85

6.3.3 A comparison between the arrows from Pit 1 and those from Pit 2 and other sites 87
6.4 Measurements of the bronze arrows 88
6.4.1 Sampling strategy 88
6.4.2 Methodology of the measurements 89
6.4.3 Patterns in the dimensions of the arrows 89
6.4.3.1 Measurements of the arrow tangs 89
6.4.3.2 Measurements of the arrowheads 89
6.4.4 Coefficient of variation values and degree of standardisation 92
6.4.4.1 CV values for the arrow tangs 92
6.4.4.2 CV values for the arrowheads 93
6.5 Arrow production, assembly and labour organisation 94
6.5.1 Consistency and variability of arrow bundles 96
6.5.2 Batch or cell production organisation 97
6.6 Spatial data 98
6.6.1 Parameters affecting the spatial patterns 98
6.6.2 A pair correlation function 102
6.7 Summary 103

7. Bronze Ferrules and Long Weapons 105
7.1 Introduction 105
7.2 Research review 106
7.2.1 Ferrules 106
7.2.2 Bronze blades of long weapons 108
7.3 Measurements of the bronze ferrules 108
7.3.1 Sources of data 108
7.3.2 Methodology for the measurements 109
7.3.3 Measuring patterns in the ferrules 109
7.3.4 Analysis of the patterns of each type of ferrules 110
7.3.4.1 Subgroups within Type I ferrules 113
7.3.4.2 Subgroups within type III ferrules 115
7.3.5 Coefficients of variation (CV) and degrees of standardisation 117
7.4 The bronze long weapons 118
7.4.1 Sources of data 118
7.4.2 Measurements of the long weapons 120
7.5 Spatial data 121
7.5.1 Parameters affecting the spatial patterns 121
7.5.1.1 The ferrules, long weapons, and the battle formation 122
7.5.1.2 Workshop practice and labour organisation 127
7.5.1.2.1 The ferrules 127
7.5.1.2.2 The long weapons 127
7.5.2 Pair correlation functions on the ferrules 127
7.6 Summary and discussion 131

8. Discussion and Conclusions 135
8.1 Standardisation 135
8.1.1 Projectile Weapons 135
8.1.2 Long Weapons 135
8.1.3 Parameters affecting standardisation 136
8.1.3.1 Models, Moulds, and Finishing Practices 136
8.1.3.2 Behaviour of the craftspeople 136
8.1.3.3 Efficiency and quality control? 137
8.2 Organisation of production 137
8.2.1 Model of production organisation 137
8.2.2 Organisational structure and practices 139
8.2.3 Craft specialisation and attached production 139
8.3 Spatial statistics and activity areas 140
8.3.1 Spatial patterns of triggers and activity areas 140
8.3.2 Placement of arrows into the pit 141
8.3.3 Ferrules and long weapons 141
8.4 Directions for future research 141

8.4.1 Research model for other research within the tomb complex 141
8.4.2 Contemporary weapons at other sites 141
8.4.3 Bronze versus iron weapons 142

Bibliography 143

Appendix 1: Spatial and metric data on triggers: Part A 151
Appendix 2: Spatial and metric data on triggers: Part B 159
Appendix 3: Spatial and metric data on triggers: Part C 167
Appendix 4: Spatial and metric data on triggers: Part D and E 175
Appendix 5: Correspondence between trigger part groups 183
Appendix 6: Inscriptions on triggers 189
Appendix 7: Count and metric data on arrow bundles 195
Appendix 8: Spatial and metric data on ferrules 219

List of figures

Fig. 1.1. Map of the Qin Empire ... 3

Fig. 1.2. Tomb complex of the Emperor Qin Shihuang ... 5

Fig. 1.3. The layout of three pits containing the terracotta warriors ... 6

Fig. 1.4. Pit 1: (a) the five easternmost trenches are hatched (1, 2, 10, 19, and 20), and (b) the corridor numbering for the five easternmost trenches ... 6

Fig. 1.5. (a) A selection of bronze weapons; (b) terracotta warriors discovered in Pit 1; (c) spatial distribution of the terracotta warriors; (d) spatial distribution of the bronze weapons ... 7

Fig. 3.1. The distribution of swords and hooks in Pit 1, shown on top of the distribution of robed and armoured terracotta warriors ... 18

Fig. 3.2. The distribution of ferrules and long weapons in Pit 1, shown on top of the distribution of robed and armoured terracotta warriors ... 19

Fig. 3.3. The distribution of triggers and arrows in Pit 1, shown on top of the distribution of robed and armoured terracotta warriors ... 20

Fig. 3.4. Sampling strategy for the bronze weapons ... 22

Fig. 4.1. A cast *Sigong* and the carved inscriptions on a dagger-axe ... 30

Fig. 4.2. *Sigong* inscription on a lance showing the use of both carving and chiselling techniques ... 30

Fig. 4.3. The carved inscription on a lance (ID00860) ... 31

Fig. 4.4. A trigger part showing both ink and chiselled inscriptions (ID00976B), with details of the chisel marks obtained through the examination of silicone rubber impressions observed under the SEM ... 31

Fig. 4.5. Ink inscription on a bronze trigger (ID 002388) ... 32

Fig. 4.6. Detail of an ink inscription and a chiselled inscription on a trigger part ... 33

Fig. 4.7. Organisational structure of the production of bronze halberds based on their inscriptions ... 34

Fig. 4.8. Organisational structure of the production of bronze lances based on their inscriptions mentioned in Table 4.2 ... 36

Fig. 4.9. Chronology of the bronze halberds and lances, based on the regnal years recorded in their inscriptions mentioned above ... 36

Fig. 4.10. The earth branches indicate the animal year and every two-hour each day ... 38

Fig. 4.11. Trigger parts A, B, C, D, E and the locations of the inscriptions ... 40

Fig. 4.12. One of the stem-branch characters (you酉) on trigger location Ba ... 40

Fig. 4.13. A *Sigong* inscription on a ferrule ... 42

Fig. 4.14. Spatial distribution of the halberds noting their dates of manufacture as recorded in the inscriptions ... 42

Fig. 4.15. Blade width, craftspeople and years of production for the lances ... 43

Fig. 4.16. Spatial pattern of the lances, discriminated by their width ... 43

Fig. 4.17. Spatial patterns of the lances according to the (a) year of production and (b) workers involved ... 44

Fig. 4.18. Spatial patterns of the trigger inscriptions on the Ba location ... 45

Fig. 4.19. Spatial patterns of the triggers with recurrent inscriptions ... 46

Fig. 4.20. Spatial patterns of the inscribed and uninscribed ferrules ... 47

Fig. 5.1. Triggers with (a) and without (b) a case ... 50

Fig. 5.2. The five parts of a trigger with the labels used to refer to them in this study 51
Fig. 5.3. Manipulation of the trigger in the crossbow (image courtesy of Zhao Zhen) 52
Fig. 5.4. Dimensions measured in each of the trigger parts 52
Fig. 5.5. Triggers labelled with their museum IDs 54
Fig. 5.6. A plot of the first two principal components of trigger part A pieces with the 4 approximate trigger subgroups 55
Fig. 5.7. Digital reproduction of the four subtypes of trigger part A 55
Fig. 5.8. Typological variations in trigger part B 56
Fig. 5.9. A plot of the first two principal components of the trigger part B pieces; the green dots show the notched and the blue dots the unnotched triggers 57
Fig. 5.10. Notches on otherwise different groups, Bg1 and Bg2. 58
Fig. 5.11. Typological variations in trigger part C. 58
Fig. 5.12. The four groups resulting from the measurement of trigger part C 59
Fig. 5.13. CV results for trigger part A 60
Fig. 5.14. CV results for trigger part B 60
Fig. 5.15. CV results for trigger part B pieces with '*Gong*' inscription versus those without 60
Fig. 5.16. CV results for trigger part C 60
Fig. 5.17. Assembly patterns of trigger parts based on micro-features 62
Fig. 5.18. Trigger assembly, part A to B and C to B 63
Fig. 5.19. Trigger assembly, part A to C and B to C 64
Fig. 5.20. Trigger assembly, part B to A and C to A 66
Fig. 5.21. Assembly of trigger parts A, B, and C 67
Fig. 5.22. Concentrated patterns of assembled trigger groups 68
Fig. 5.23. The association between inscriptions at location Ba and assembled trigger groups 69
Fig. 5.24. Left: spatial distribution of crossbowmen in battle formation and triggers in Pit 1, showing the generally good correspondence between the two. Right: drawings of a crossbowman carrying a crossbow and an infantryman holding a halberd 69
Fig. 5.25. Triggers and arrows in the easternmost five trenches of Pit 1 70
Fig. 5.26 The spatial distribution of Assembly Group 2 triggers 72
Fig. 5.27. The spatial distribution of Assembly Group 5 triggers 72
Fig. 5.28. The spatial distribution of Assembly Group 6 triggers 73
Fig. 5.29. The spatial distribution of Assembly Group 7 triggers 73
Fig. 5.30. The spatial distribution of Assembly Group 11 triggers 74
Fig. 5.31. The spatial distribution of Assembly Group 12 triggers 74
Fig. 5.32. The spatial distribution of Assembly Group 8 triggers 75
Fig. 5.33. The spatial distribution of Assembly Group 9 triggers 75
Fig. 5.34. Activity areas in Pit 1 77
Fig. 6.1. Numbered foot plates of the terracotta warriors and arrow bundles 80
Fig. 6.2. Count of the arrows (a) versus triggers (b) in each corridor, and number of arrows per trigger (c) 82
Fig. 6.3 Frequency histogram of the number of arrows and arrow bundles 83
Fig. 6.4 Frequency histogram of the number of loose arrows or small groups with up to 10 arrows 83

Fig. 6.5. Frequency histogram of the number of arrows in bundles containing between 70 and 120 arrows 84
Fig. 6.6. Traces of a quiver with a bundle of arrows in Pit 1 84
Fig. 6.7. Reconstruction of a Qin arrow quiver 84
Fig. 6.8 Bronze arrows from the Qin bronze chariots 85
Fig. 6.9. Bronze arrows with linen strings coiled around them 86
Fig. 6.10. Bronze arrows showing bamboo remains of shaft 86
Fig. 6.11. Linking between an arrowhead and tang from the bronze chariot 86
Fig. 6.12. The neck between the arrowhead and tang 86
Fig. 6.13. Shorter and longer arrow tangs 87
Fig. 6.14. Bronze arrowhead with iron tang 87
Fig. 6.15. A special arrow from Pit 1 87
Fig. 6.16. Type A of the bronze bolts from Pit 2 88
Fig. 6.17. Type B of the bronze bolts from Pit 2 88
Fig. 6.18. (a) An arrow from another Qin site, and (b) one from Pit 1 of the terracotta warriors 88
Fig. 6.19. Sampling strategy used for the bronze arrows 88
Fig. 6.20. The arrow bundle (ID001017) and the six sampled arrows 89
Fig. 6.21. Histogram of the frequency distribution observed on the arrow tang length measurements 90
Fig. 6.22. Images of examples of the two subgroups of arrow tangs: (a) short and (b) long 91
Fig. 6.23. Histogram of the frequency distribution observed on the arrow tang length measurements 91
Fig. 6.24. Histogram of the frequency distribution of six length measurements of the arrowheads 92
Fig. 6.25. Histogram of the frequency distribution of 6 width measurements of the arrowheads 93
Fig. 6.26. Mean length of arrow tangs versus their CV values from each bundle 93
Fig. 6.27. Examples of the longer and shorter tangs in an arrow bundle 94
Fig. 6.28. Arrows labelled with their museum IDs 95
Fig. 6.29. Mean length versus mean width of the six sampled arrowheads from each bundle 96
Fig. 6.30. CV values of arrowheads length and width in six series 96
Fig. 6.31. Two arrow bundles selected for comparison 97
Fig. 6.32. Chemical composition of arrows in bundles ID 1152 and ID 1985 98
Fig. 6.33. Spatial distribution of arrows (a) and arrow bundles (b) 99
Fig. 6.34. Location of the arrows showing iron tangs assembled with bronze arrowheads 100
Fig. 6.35. Location of the special bronze arrows 100
Fig. 6.36. The location of bronze arrows matched to the crossbowmen in the pit 101
Fig. 6.37. The spatial distribution of short tang arrows within the five easternmost trenches of Pit 1 (a), and (b) a pair correlation function considering the spatial pattern over the first 10 metres 102
Fig. 6.38. The spatial distribution of the longer tang arrows within the five easternmost trenches of Pit 1 (a), and (b) a pair correlation function considering the spatial pattern over the first 10 metres 103
Fig. 7.1. Examples of ferrules unearthed from the pits of the terracotta warriors 105
Fig. 7.2. Ferrule, wooden shaft and halberd unearthed from Pit 1 106
Fig. 7.3. Reconstruction drawing of a warrior holding a long weapon 106
Fig. 7.4. Ferrule with inscription referring to *Shang Yang* 107
Fig. 7.5. Dimensions measured for the ferrules 109

Fig. 7.6. Ferrules from the five easternmost trenches of pit 1, labelled with their museum IDs 110
Fig. 7.7. Scatterplot of the ferrule measurements: length versus T1-diameter 113
Fig. 7.8. A plot of the first two PCA components of the ferrules' measurements including all variables recorded. 114
Fig. 7.9. Type I, type II, and type III ferrules 114
Fig. 7.10. Proportions of the three types of ferrules within the assemblage 115
Fig. 7.11. A small hole and a pin shown on type I ferrules 116
Fig. 7.12. Two inside pins and a surface trace on a type III ferrule 116
Fig. 7.13. Association between the ferrules and the blade weapons 116
Fig. 7.14. A plot of the first two principal components of ferrule type I 117
Fig. 7.15. A plot of the first two principal components of ferrule type III 118
Fig. 7.16. '*Sigong*' inscriptions shown on types III-A and III-B, but not on the III-C ferrule 119
Fig. 7.17. Inscribed and uninscribed type III ferrules 119
Fig. 7.18. CV values for ferrules belonging to subgroups I-A, III-A, and III-B 120
Fig. 7.19. CV values for the inscribed (III-B-1) and uninscribed (III-B-0) ferrules 120
Fig. 7.20. Bronze spear, dagger-axe, halberd, and lance 120
Fig. 7.21. A plot of the first two principal components of the spears 122
Fig. 7.22. Dimensions measured for the lances 123
Fig. 7.23. A plot of the first two principal components of the lances coded by the dates of production 123
Fig. 7.24. A plot of the first two principal components of the lances coded by the craftspeople involved in their production 124
Fig. 7.25. Spatial distribution of ferrules (a) and with warriors (b) in Pit 1 125
Fig. 7.26. Type I ferrules, halberds, and bronze and iron spears in Pit 1 126
Fig. 7.27. Type III ferrules and bronze lances in Pit 1 128
Fig. 7.28. Spatial patterns of subgroups I-A, I-B, and I-C 128
Fig. 7.29. Spatial patterns of the inscribed and uninscribed type III ferrules 129
Fig. 7.30. Spatial patterns of type II ferrules 129
Fig. 7.31. Spatial patterns of the lance subtypes 130
Fig. 7.32. The spatial distribution of type I-A ferrules within the five easternmost trenches of Pit 1 (a), and (b) a pair correlation function considering the spatial pattern over the first 10 metres 131
Fig. 7.33. The spatial distribution of ferrules III-A with *Sigong* inscription within the five easternmost trenches of Pit 1 (a), and (b) a pair correlation function considering the spatial pattern over the first 10 metres 132
Fig. 7.34. The spatial distribution of ferrules III-B within the five easternmost trenches of Pit 1 (a), and (b) a pair correlation function considering the spatial pattern over the first 10 metres 133
Fig. 7.35. The spatial distribution of ferrules III-B with *Sigong* inscription within the five easternmost trenches of Pit 1 (a), and (b) a pair correlation function considering the spatial pattern over the first 10 metres 133
Fig. 7.36. The spatial distribution of ferrules III-B without *Sigong* inscription within the five easternmost trenches of Pit 1 (a), and (b) a pair correlation function considering the spatial pattern over the first 10 metres 134

List of tables

Table 3.1. Test measurements on three random selected triggers ... 24
Table 3.2. Test measurements on three random selected triggers ... 24
Table 3.3. Standard deviation and coefficient variation of the above 14 measurements for each trigger dimension ... 25
Table 4.1. Inscriptions on the halberds, dagger-axes, and spears ... 33
Table 4.2. Inscriptions on the lances and their fittings ... 35
Table 4.3. Ten heaven stems ... 37
Table 4.4. Twelve earth branches ... 37
Table 4.5. The sexagenary cycle based on the stem and branch combination ... 38
Table 4.6. The variety of characters recorded on the trigger inscriptions ... 39
Table 4.7. Details of triggers bearing the same inscription in at least three parts ... 41
Table 4.8. Inscribed versus uninscribed ferrules in each corridor ... 47
Table 5.1. The correlations between the three original variables of trigger part A and the three principal components ... 53
Table 5.2. Variation in the different trigger part A groups and assembled parts ... 56
Table 5.3. The correlations between the three original variables of trigger part B and the three principal components ... 57
Table 5.4. The correlations between the three original variables of trigger part C and the three principal components ... 59
Table 5.5. Assembled triggers and their count ... 67
Table 5.6. Pair correlation results on assembled trigger groups ... 76
Table 6.1. The archaeological origin of the bronze arrows ... 80
Table 6.2. The number of arrows and bundles (≥90) in each corridor of Pit 1 ... 81
Table 6.3. Pair correlation results on shorter and longer tang arrows ... 94
Table 7.1. The types of ferrules, their locations, and the research methods employed ... 108
Table 7.2. Measurements taken on type I ferrules ... 111
Table 7.3. Measurements taken on type II ferrules ... 111
Table 7.4. Measurements taken on type III ferrules ... 112
Table 7.5. The correlations between the six original variables of ferrules and the five principal components ... 114
Table 7.6. Type I, type II, and type III ferrules ... 115
Table 7.7. The correlations between the six original variables of ferrule type I and the five principal components ... 117
Table 7.8. The three subgroups within type I ferrules ... 117
Table 7.9. The correlations between the six original variables of ferrule type III and the five principal components ... 118
Table 7.10. Subgroups of type III ferrules and their inscriptions ... 120
Table 7.11. Counts of inscribed and uninscribed ferrules against their types ... 120
Table 7.12. Bronze long weapons and their archaeological origins ... 121
Table 7.13. Origins and inscriptions of the halberds or dagger-axes ... 121
Table 7.14. The correlations between the seven original variables of the spears and the five principal components ... 122
Table 7.15. The correlations between the seven original variables of lances and the five principal components ... 124
Table 7.16. Pair correlation results on the subgroups of ferrules ... 134

Abstract

Alongside the thousands of terracotta warriors discovered in the tomb complex of the first emperor of China were tens of thousands of bronze weapons, including arrowheads and crossbow triggers, lances, spears, halberds (and the ferrules associated with them), swords and a few other special types. This quantity and quality of bronze weaponry provides an extremely rare opportunity to investigate patterns of standardisation and labour organisation within a single, very large and intentional assemblage as well as to consider the role of bronze production during the Qin period (325–206 BC), which marks perhaps the most crucial early stage in Chinese political unification.

This book draws upon extensive measurements, typological analysis and related statistical treatment, as well as a study of the spatial distribution of those bronze weapons found in the most extensively excavated part of the tomb complex (the five easternmost trenches in Pit 1). Metric data and statistical assessment of inter- and intra-group variation (e.g. coefficients of variation) suggest interesting patterns with regard to relative degrees of standardisation. A combination of geographic information systems (GIS) and point pattern analysis is used to assess formally any spatial patterning in the weapons and their analytical attributes, which then also provides further information about the labour organisation behind the production, transportation and placement of weapons as they were moved from the workshop and/or arsenal to the funeral pits. Combining these insights with those obtained from inscriptions found on some of the weapons and from ancient documents, this project investigates what technologies and crafting behaviour affected weapons production and labour organisation in a centralised imperial system.

This research project fills a gap in the study of mass production, the behaviour of craftspeople, and related logistical organisation in ancient China and, for the first time, provides empirical data by analysing systematically the types, dimensions and spatial patterns of Qin bronze weapons for the Terracotta Army in the Emperor Qin Shihuang's tomb complex.

1

Introduction

1.1. Introduction

In the 1970s, three pits containing thousands of terracotta warriors and horses were discovered in the tomb complex of Emperor Qin Shihuang (秦始皇 259–210 BC) near Xian, China. Qin Shihuang was the first emperor of what can be considered as the first unified Chinese state. In 221 BC, he established the Qin Empire (221–206 BC) following a series of military campaigns in central China. Considered by many as a ruthless autocrat, he was nevertheless a political, economic and military innovator and sought immortality by building a vast tomb for his afterlife that he modelled after an underground empire, guarded by an army of terracotta warriors (Institute and Team, 1988; Yuan, 1990, 2014; Lindesay and Guo, 1998; Portal, 2007; Loewe, 2007; Rawson, 2007).

Together with the life-sized terracotta warriors, a large number of bronze weapons, such as swords (剑), spears (矛), lances (铍), dagger-axes (戈), halberds (戟), hooks (钩), ceremonial weapons *Su* (殳), triggers (弩机), arrows (镞), and ferrules (镦), was discovered in the three pits located to the east of the tomb mound. For example, within the 2,000 square metres fully excavated to date in Pit 1 (the five easternmost trenches), over 40,000 individual arrows were discovered (some found in bundles and some loose). Besides these, 486 other bronze weapons were unearthed (Institute and Team, 1988:249). Although a considerable amount of research has been conducted on the terracotta warriors themselves in terms of manufacture, detailed sculpturing, polychrome paints, and military battle formation (Yuan, 1990; Wang 1994; Ledderose, 2000; Blansdorf et al., 2001; Portal, 2007; Khayutina, 2013), as well as birth of the Qin Empire (Pines et al., 2014), comparatively less attention has been paid to the weapons assemblage.

This comparative lack of attention is surprising given the fact that the quantity and quality of bronze weapons discovered in the tomb complex provides a great deal of information about Qin mass production methods, metallurgical know-how, labour organisation and even the behaviour of individual craftspeople. These bronze weapons were all cast using models and moulds, and finished by filing, grinding and polishing of their surfaces (Li et al., 2011). Most of the bronze weapons were well preserved after being buried for more than 2,000 years; some of the blades, particularly those of swords and lances, appear to have undergone an anti-rust treatment, because they still had metallic lustre when found in the pit of the terracotta warriors. But this still poses a mystery for arguments raised in recent decades (Han et al., 1981; Yuan, 1990; He, 1996; Zhang et al., 2011; Martinón-Torres et al., 2019), and it seems neither chromium nor soil have played an important role in well preserving such weapons. Some of the Qin bronze weapons were discovered in other archaeological contexts, but those finds remain pieces of evidence isolated from the contexts of tombs and battlefields. The availability of such a large quantity of Qin bronze weapons from one depositional context makes it possible to conduct a systematic study into the standardisation and labour organisation of bronze production in a very specific political and ideological setting. The spatial arrangement of the weapons, the inscriptions on some of them, the metrical and typological variation they exhibit, their manufacture traces, and their chemical composition are all very important dimensions offering complementary information.

This book is based on my PhD research, initiating a broader project carried out between the Emperor Qin Shihuang's Mausoleum Site Museum and the UCL Institute of Archaeology, which aims to reconstruct the Qin bronze industry that produced these weapons. The research addresses issues such as the supply of raw materials, casting technology, production processes, quality control and monitoring, as well as offering insights into the large amount of labour and organisation needed to produce and place the bronzes in the pits within a limited period of time (246–210 BC). The goal of this book is to develop the study of bronze specialisation and standardisation in Chinese archaeology by carrying out the systematic typological and spatial analysis of this large assemblage of bronze weaponry from Pit 1 of the Terracotta Army within the Qin tomb complex and to analyse the archaeological, metrical and spatial data alongside other relevant patterns. This study provided theoretical and methodological vision for continuing archaeometric research carried out by the Sino-British cooperative team.

1.2 Research review and defining research questions

1.2.1 Previous research on the bronze weapons

Over the past three decades, in addition to basic typological and historical research on this large quantity of bronze weapons (Institute and Team, 1988; Yuan 1990; Wang 1994), Chinese archaeologists and scientists have also undertaken research on three other main aspects. These are: **a**) the interpretation of the inscriptions and their implications for the organisation of weapons production during the Qin period (Yuan, 1984, 1990; Huang, 1983, 1990; Liu and Jiang, 2006); **b**) the military functions of the weapons, their placement in an army battle formation in the pit and the military strategy they imply (Qin, 1975; Qin and Zhang, 1983; Liu, 1986; Dang, 1987; Bai, 1994;

Yates, 2007; Yang, 2017); **c**) the casting technology and chemical composition (Wang, 1980a, 1987; Yuan et al, 1981a, 1981b; Han et al., 1983; Yuan and Cheng, 1986). Previous research has particularly addressed the specific types of Qin bronze weapons from a historical perspective or has carried out scientific analysis on a limited number of samples. However, relatively little research has been undertaken on the standardisation of production and its relationship with craft organisation (for exceptions, see Wang, 1980a; Yuan et al., 1981a, 1981b).

Thus, a few previous publications address to the topic of standardisation, but the subject has not benefited from a systematic and comprehensive approach so far. Yuan Weihua et al. (1981a) investigated the bronze casting and processing technology, and based on a relatively small number of analytical samples, suggested that the similarity of the weapons in terms of chemical composition was a reflection of high levels of specialisation and standardisation. Wang Xueli (1980a) offered discussion on these topics and noted that the moulds used for the mass production of bronze weapons in the Qin Dynasty were normally made from two pieces fixed together, but that the casting methods were slightly different for different kinds of weapons. For example, the sprue on a sword, spear or hook was at the end of the handle, but the bronze arrows were made in an overlapping mould where the tangs were cast in a first stage, then the arrowheads cast onto them in a second stage. The chemical composition of the bronzes also varied according to the function of the weapons. For example, the swords needed sharpness and resilience, while the arrows had to be sharp but hard-pointed; consequently, the tin content of swords appears lower than that of the arrows. In another paper, Wang Xueli (1987) also presented a detailed list of the standardised aspects of the bronze weapons:

- the length of the arrowheads, arrow tangs (bronze) and arrow shafts (wood or bamboo) was consistent;
- all parts of the crossbow triggers were assumed interchangeable;
- the bronze swords were exactly symmetrical from the spine to the blades, which indicates standardisation in mould production.

The supervision, quality control, and system of rewards and punishments behind the production of Qin bronze weapons have also been discussed, using the inscriptions on the weapons as well as bamboo slips and ancient written documents as sources (Yuan et al., 1981b; Wang, 1987; Li and Gao, 2010).

Overall, previous research has offered a general overview of the standardisation and labour organisation prevalent in the production of bronze weapons during the Qin period, based on a relatively limited number of samples. However, to date there has been a lack of robust data collection and systematic, large-scale analysis for assessing the degree of standardisation, a dearth of sound archaeological theory, and few if any attempts to interpret any spatial pattern of technological variation in the pit. Overall, there has been very little discussion as to what kind of systematic research is most appropriate for a large assemblage of weaponry, integrated with theoretical models of specialised production, in an early complex society.

Craft specialisation and standardisation have been widely discussed from theoretical perspectives in the Western archaeological literature (for details see Chapter 2). They have been considered in relation to such issues as the origins of food surplus, leisure time, and population growth (Boas, 1940) and to the rise of social complexity (Rice, 1981; Clark and Parry, 1990). Other researchers have concentrated on defining the term 'specialisation' (Muller, 1984; Tosi, 1984; Costin, 1986; Stark, 1991); establishing parameters to identify types and degrees of specialisation (Earle, 1981; Costin, 1991); exploring the relationship between standardisation and specialisation (Clark, 1986; Torrence, 1986; Rice, 1991; Costin and Hagstrum, 1995; Roux, 2003); and proposing techniques for assessing the degree of standardisation (Eerkens and Bettinger, 2001). However, these Western publications rarely consider Chinese materials. In recent years, some scholars from China and the West have begun to study craft specialisation in the production of Shang and Zhou bronzes and other objects (Bagley, 1995; Underhill, 2002; Li, 2007; Sun, 2008). Related studies have also been conducted on the manufacture of the Qin terracotta soldiers (Ledderose, 2000), on the construction projects and investment of labour (Shelach, 2014), and on the Qin standardised artefacts for administrative and propaganda purposes (Sanft, 2013; 2014), but no similar analysis has been carried out on the production of Qin bronze weapons.

The use of spatial statistics (Bevan and Conolly, 2006) to offer insight into labour organisation in production and the logistics of placement in the pit will be another question that needs to be tackled in this research. The spatial distribution of each type or subtype of bronze weapons will be characterised and related to the possible past technological processes, crafting behaviour and labour organisation. Consideration also needs to be paid to the fact that these bronze weapons were not found in a workshop or arsenal, related directly to a production context, but in a funeral pit where they were arranged to match the battle formation of the terracotta warriors (Institute and Team, 1988; Yuan, 1990). The weapons' spatial distribution therefore has been affected not only by workshop processes, but also by storage organisation, transportation to and placement in the pit, and the battle array of the Army (Bevan et al., 2013).

For this particular project, the study of the Qin bronze weapons will go beyond previous work, and include systematic analysis on a large number of samples and related spatial statistics, to be incorporated within a broader theoretical framework. This book will also bring together both qualitative and quantitative data, as well as some preliminary archaeometric results.

1.2.2 Defining research questions

Against this background, this study sets out to explore three main questions:

- What does the degree of metric variability and standardisation in bronze weapons tell us about technological processes and workshop organisation during the Qin Dynasty?
- What is the spatial distribution of bronze weapons in Pit 1 of the Emperor's tomb complex? How does this relate to actual battle formations, weapons production and craft organisation during the Qin Dynasty?
- What was the role of political influence in the production process?

To tackle these questions, five types of data may be used: inscriptions on the weapons, dimensions and other typological features, spatial patterns, chemical composition, and manufacturing techniques. This book will mainly focus on the archaeological perspective, including inscriptions, dimensions, typology and spatial patterns, contextualised with broader archaeological and historical information. These results are integrated with archaeometric data on chemical composition and manufacturing techniques published as journal papers (Li et al., 2011; Li et al., 2014; Martinón-Torres et al., 2014).

1.3 The Emperor and his tomb complex – defining the temporal and spatial context

1.3.1 The Qin Dynasty and Emperor Qin Shihuang

By the time that King Ying Zheng (嬴政) of the Qin state, later known as the Emperor Qin Shihuangdi, unified China in 221 BC, the Qin state was already more than 600 years old and had experienced a long historical development both as a clan and as a kingdom (Historical Records: Basic Annals of Qin 史记· 秦本纪). The Qin people had been recognised by the Zhou king as a minor subordinate clan at the upper reaches of the Wei river (a branch of the Yellow River), on the northwest borders of the present-day Gansu province, which was responsible for breeding horses for the Zhou (Yates, 2007; Gansu Institute of Archaeology, 2009; Zhao, 2014; Teng, 2014)). Within twenty-five years of becoming king, Ying Zheng was able to eliminate all six other powerful states by military force, and established the first empire of China (Fig. 1.1). Bronze weapons are considered to have been one of the most crucial factors in his military success (Yuan, 1990; Yates, 2007, 2009). Before we turn to discussing the production of these weapons, it is useful to discuss the origins of the Qin Empire and the background to its military supremacy.

There are two modern views regarding the origins of the Qin. One claims that this group emerged from the

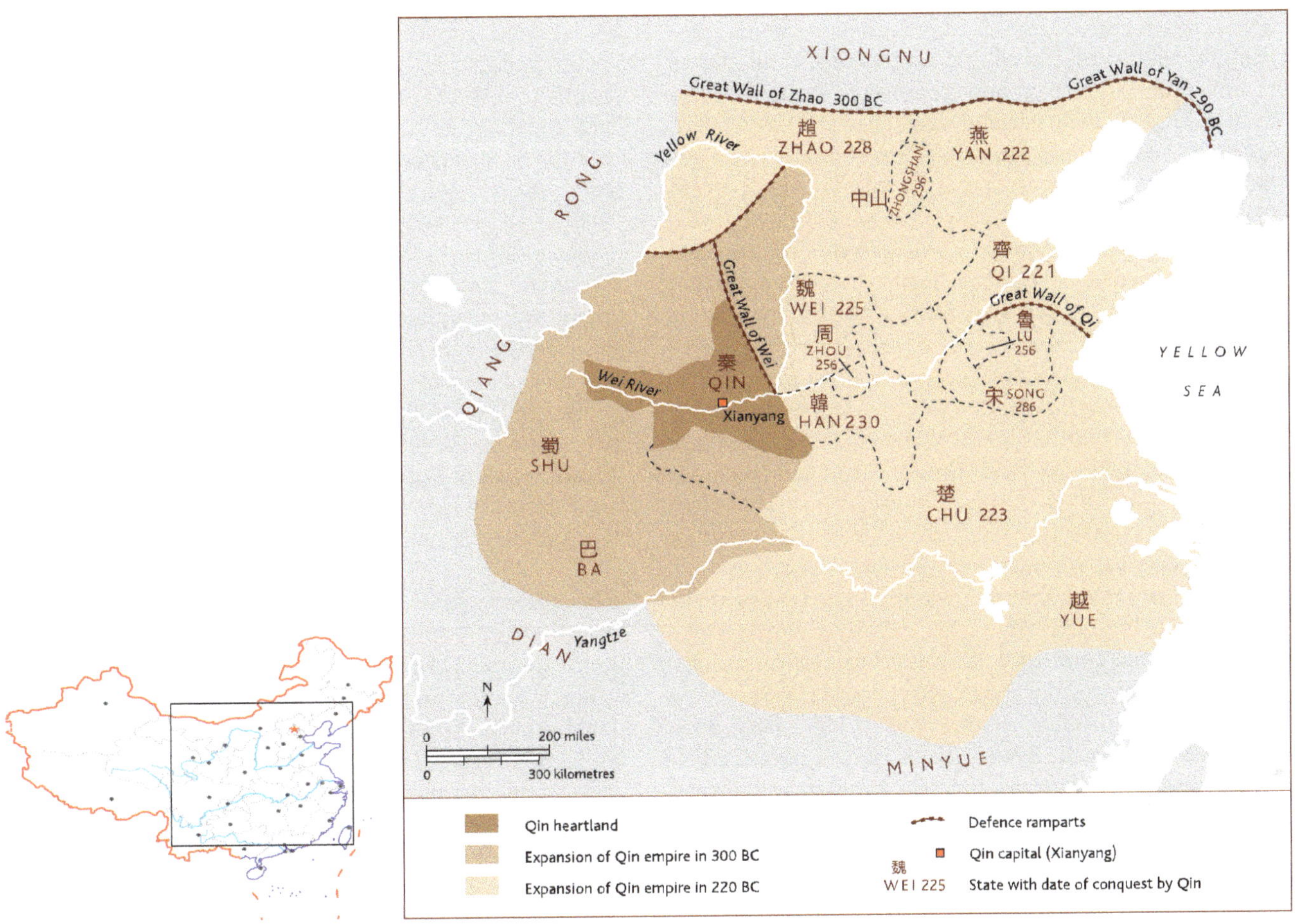

Fig. 1.1. Map of the Qin Empire (right-hand map courtesy of the British Museum).

indigenous peoples of east China, while the other states that they were descendants of the Rong, a so-called western barbarian people (Yates, 2007; Zhao, 1987, 1989, 2014; Li, 2011). According to Shiji (Historical Records: Basic Annals of Qin史记· 秦本记), Qin's ancestor was *Nuxiu* (女脩), who swallowed a bird's egg and gave birth to a son named *Daye* (大业). Daye's son was *Dafei* (大费), who is said to have helped *Yu* (禹), a legendary king of the Xia Dynasty (about 2100 BC), to harness a flood in central China. Recent archaeological excavations in Gansu province have revealed much about early Qin cultural and ritual practice, and show that Qin cultural traditions were different from those of the Rong, and closer to those found in central and eastern China (Collaborative Team, 2008; Zhao, 2008; Gansu Institute of Archaeology et al., 2008). The archaeological discoveries provide crucial evidence that suggests cultural similarities and political proximity of Qin to the Zhou Dynasty (Teng, 2003; Shelach and Pines, 2005; Falkenhausen, 2006: 29–73).

In 770 BC, the king of the Western Zhou was forced to move from Zhouyuan, in the lower Wei River valley (present-day Shaanxi province), to Luoyang, middle reaches of the Yellow River (present-day Henan province) (Yates, 2007). Qin Xianggong (秦襄公), a duke of Qin, was successful in battle and escorted the Zhou king Ping (周平王) in the move to Luoyang. As a duke, he was entitled to establish a Qin territory as part of the Zhou Dynasty, in the western part of China (Lin, 1981). He gradually expanded eastward, occupying the original Zhou domains.

The Qin state experienced fluctuations of fortune during its 500 years of development between 770 and 221 BC. Even though the state became weak after Duke Mu (秦穆公659–621 BC), Shangyang's (商鞅a reformer of the Qin, 385–338 BC) innovations during the rule of Duke Xiao (秦孝公361–338 BC) made it prosperous again (Loewe, 2007; Pines, 2009, 2012). These innovations were considered to lay the foundation for the Qin finally to become a major force among the seven kingdoms that then composed China. Shangyang, also known as Gongsun Yang, acted as the senior official of the Qin government from 359 to 350 BC, and he was permitted to institute many reforms. Groups of five families were bound together, and were jointly accountable for the actions of each other in a legal context. Crimes committed by any member of the whole group were reportable by other members. If this reporting was not done, the whole group was held accountable, i.e. guilty of the same crime that the individual had carried out. As a further extension of the collective concept, one male was taken from each household to form a military squad of five. These five people were also held responsible for each other's safety. If they lost a man, they were required to capture the head of an enemy in exchange. It is conceivable that around this time, the Qin state established an army of conscripts (Yates, 2007). The Qin government also instigated a system of seventeen ranks or grades for all male members of the population. Previously only aristocrats held such ranks. There was recognition that rank based on merit or the capability of the individual was as great or of greater value than inherited rank. Individual talent could be exploited for the benefit of the government, and that in turn bestowed rewards (Loewe, 2007, 2010; Pines, 2009, 2012). Higher rank was awarded for battle successes, manifested by beheading an enemy and reporting it to the army headquarters. Further reform was introduced to the agricultural part of the economy. Farmers were encouraged to cultivate arable land lying fallow, and a new taxation system that promoted agricultural development was created. These policies stimulated grain production to feed both the army and an increasingly complex administrative bureaucracy (Yates, 2009).

Further innovations also changed the organisation of bronze production and a system of craft supervision was established. This information is reflected in the inscriptions carved on some of the bronze objects. One of the earliest examples is a bronze ferrule (the metal cap on the bottom end of the long wooden handle of a spear, lance or halberd) that was excavated from a Qin tomb in Xianyang, capital of the Qin. The fourteen-character Chinese inscription translates as "Shanyang supervised the making of this ferrule in the 19th regnal year (343 BC)" (Yates, 2007: 34). Some of the Qin bronze weapons and bronze ritual vessels made before the Shanyang period were found with cast or carved inscriptions, but most of their contents pertained to who owned it or for what purpose it was produced. No such inscriptions were added to show who had been in charge of the production and which craftspeople had been involved.

In the decades following Shangyang's and Duke Xiao's deaths in 338 BC, a series of rulers built on their legacy and expanded Qin territory by force to the east, south and north, incorporating peoples and territories that had very different cultural and social customs. Qin Huiwen (秦惠文王) proclaimed himself king in 325 BC, and this is generally considered to be the official start of the Qin kingdom. The Bashu (present-day Sichuan province) region was conquered, and subsequently convicts, settlers and officials were sent to the south-west to occupy the land and exploit its extensive natural and mineral resources (Sage, 1992), which may well have later provided the main raw material for Qin bronze weapons. The state of Qin became increasingly keen on annexing other kingdoms with the ultimate aim of establishing a supreme empire (Guo and Wang, 2000).

Ying Zheng was born at the right time after the long development from Qin clan to kingdom. He succeeded to the throne at the age of thirteen in 246 BC. However, ruling authority was initially exercised by the Chancellor Lu Buwei (吕不韦) until 238 BC, when Ying Zheng (Qin Shihuang), then 22, assumed control of state affairs and immediately stripped the minister of his power. With the assistance of a new chancellor, Li Si (李斯), he carried out a series of reforms to develop agriculture and the military. He had the Zhengguo Channel built for irrigation and encouraged famers to have their own land. He adopted Li Si's military strategy, known as "a silkworm devouring a mulberry leaf" and eventually he conquered the other six states and unified China in 221 BC (Historical Records: Basic Annals of the First Emperor of Qin 史记· 秦始皇本纪 ; Lin, 1981; Lindesay and Guo, 1998).

The emperor set about enacting many reforms to consolidate his empire. To strengthen the northern border, he sent slaves and criminals to build the line of defence now known as the Great Wall. Roads radiating from Xianyang, the capital, were built linking the former Yan, Qi, Wei and Chu areas. He also standardised the script used for writing, and introduced a circular copper disc with a square hole in the middle as the standard coin to be used across the empire. Equally important reforms were the standardisation of weights and measures, and the codification of the law. Even the gauges of wheeled vehicles were standardised. These reforms benefited both the economy and cultural exchange during the period (Lindesay and Guo, 1998; Snaft, 2013).

In the long history of the Qin, the production of bronze weapons also went through several stages of development. As mentioned above, after Shangyang's innovation, some bronze weapons were carved with an inscription that would allow the maker's work to be properly scrutinised (物勒工铭). In other words, this was a form of quality control and accountability. According to the inscriptions carved on the bronze lances and halberds discovered from the pits of the terracotta warriors, these were mainly produced from 244 to 228 BC, during the reign of King Ying Zheng but before the unification. After the unification, the ancient documents recorded that "weapons from all over the empire were confiscated, brought to Xianyang, and melted down to be used in casting bells, bell stands and twelve men made of metal. These last weighted 1,000 piculs (1 picul ≈ 60 kg) each and were set up in the palace" (Rawson, 2007: 129). The first emperor of Qin had the bronze weapons melted down in order to put an end to warfare in the newly established empire, and, from these historical sources, it also seems that the Qin bronze weapons were used to recast the bronze statues lined up in front of the Qin palace (Yuan, 1990). However, from an archaeological perspective this appears not to be the case, especially in light of finding such a great quantity of Qin bronze weapons in the Emperor's tomb complex.

1.3.2 The tomb complex, the terracotta warriors and the bronze weapons

According to the *Siji* (Historical Records: Basic Annals of the First Emperor of Qin 史记•秦始皇本纪), Emperor Qin Shihuang was apprehensive about the prospect of death. He tried hard to find an elixir that would make him immortal, but at the same time commissioned the building of his mausoleum. Construction work on this began in the year he became king, 246 BC, and continued for about 40 years, even after his death in 210 BC. The chancellor Lu Buwei took charge of the earlier stages, and the next chancellor, Li Si, was in charge of later stages. Only the fall of the dynasty itself in 206 BC brought work on the elaborate funerary complex to a halt (Yuan, 1990; Lindesay and Guo, 1998; Portal, 2007; Duan, 2011).

The mausoleum is located at Lintong, with Mount Li to the south and the Wei River to the north. Interior and exterior ramparts were built around the edges of the tomb mound. Some traces of the wall are still visible on the surface of the ground. The mound itself was about 115 metres high when it was first built, a truncated pyramid shape, and covered with evergreens. Beneath it, an underground palace in which the remains of Qin Shihuang were buried is thought to lie. As described by Sima Qian in *Historical Records* (《史记》), the First Emperor's tomb chamber reproduced in minute detail the universe over which he expected to rule.

> From the time the First Emperor first took the throne [in 246 BC] work was begun [on his mausoleum] at Mount Li. After he had won the empire, more than 700,000 conscripts from all parts of the country laboured there. The labourers dug through three subterranean streams which they sealed off with bronze in order to make the burial chamber. This they filled with [models of] palaces, towers, and the hundred officials, as well as precious utensils and marvellous rarities. Artisans were ordered to install mechanically triggered crossbows set to shoot any intruder. With mercury the various waterways of the empire, the Yangtze and Yellow Rivers, and even the great ocean itself were created and made to flow and circulate mechanically. The heavenly constellations were depicted above and the geography of the earth was laid out below. Lamps were fuelled with whale oil so that they might burn forever without being extinguished…Finally, trees and grass were planted [on the tomb mount] to make it appear like a mountain. (translated by Hearn, 1980)

The tomb mound has not been excavated so far, but the relatively high concentration of mercury in the soil attested by scientists gives some plausibility to aspects of the description written by Sima Qian (Chang and Li, 1983). The surrounding pits and tombs in the mausoleum complex were not mentioned in the *Historical Records*, and have been discovered by a combination of chance, archaeological survey and excavation (Fig. 1.2).

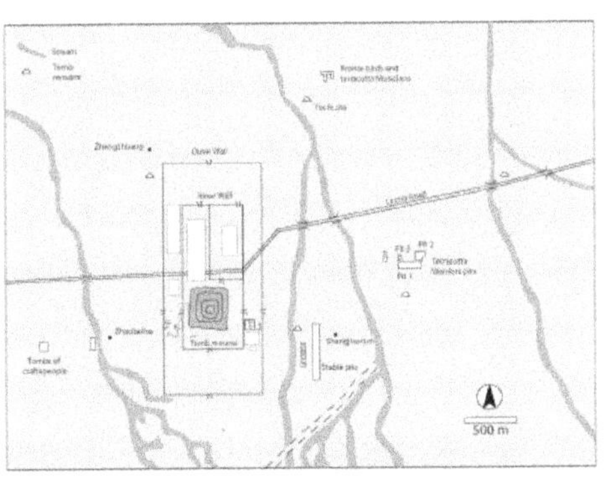

Fig. 1.2. Tomb complex of the Emperor Qin Shihuang (after figure 1 in the Excavation Report, Institute and Team, 1988: 2).

Archaeological excavations carried out during the past four decades have been primarily focused on the surrounding pits and tombs. The three pits with the terracotta warriors and horses of Emperor Qin Shihuang are the most famous archaeological finds in the mausoleum complex. In addition, approximately 600 further pits and tombs have been noted and attested to by archaeological surveys in the area of the mausoleum (Yuan, 1990). To date, only a relatively small number of pits and tombs have been excavated, but they include a pit containing bronze chariots, pits with stables, a pit of terracotta officials, a pit of terracotta acrobats, a pit with a stone armoury, and a pit of bronze birds (Museum and Institute, 1998; Museum of Emperor Qin Shihuang's Terracotta Army, 1998; Shaanxi Institute and Museum, 2000, 2006, 2007; Yuan, 2002; Duan, 2011).

The abundant contents of the surrounding pits and tombs are made in a variety of different materials. The life-sized terracotta warriors, officials, and acrobats; roof tiles; bricks; and some containers are all made of clay. Large amounts of body armour and helmets are made from limestone. Metal objects are also one of the main burial finds in the mausoleum complex, and include the thousands of bronze weapons, two sets of bronze chariots; about forty bronze birds; bronze tripods; bronze bells, coins, mirrors, and weights; iron implements; and gold and silver ornaments (Yuan, 2002; Duan, 2011).

The bronze weapons studied in this particular project originated mainly from the five easternmost trenches of Pit 1, which is one of three pits designed to house the army of terracotta warriors and located about 1.5 km east of the burial mound (Figs. 1.2 and 1.3). Pit 1, the largest of the compound, was partially excavated during the 1970s. To date, almost 1,100 terracotta figures have been restored in the five easternmost trenches (Fig. 1.4), covering 1,000 square metres of the pit. Due to the partial excavation in the past decades, the sampling bronze weapons have been limited to the five easternmost trenches of Pit 1, which was one-fifth of the military formation of this Terracotta Army. However, over 40,000 variety of bronze weapons have been found from such a small section of the pit to provide rich data for quantitative analysis in this book.

According to the density of the figures found to date, it is estimated that the three pits contain about 6,000 terracotta warriors and horses in total. The warriors are of several types: infantry, archers, officials and charioteers. In addition to Pit 1 mentioned above, Pit 2, excavated with trial trenches in the 1970s, contains a battle formation which includes archers, cavalrymen, charioteers and infantrymen. Archaeological work carried out in 1994 mainly concentrated on above the roof layer of this pit. In the archer section the excavation continues, and several well-preserved kneeling archers decorated with bright pigments have been unearthed in recent years (Museum of Emperor Qin Shihuang's Terracotta Army, 2009). Pit 3, the smallest one, is assumed to be the headquarters of the army in the other two pits. It contains only one chariot drawn by four horses and 68 terracotta figures (Yuan, 1990; 2002).

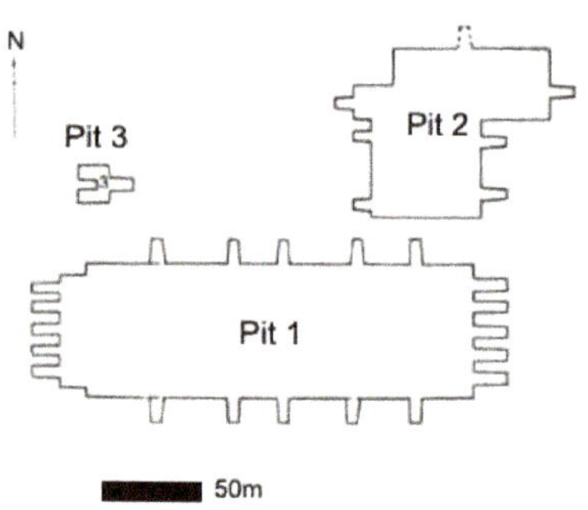

Fig. 1.3. The layout of three pits containing the terracotta warriors (after Figure 2 in the Excavation Report, Institute and Team, 1988: 5).

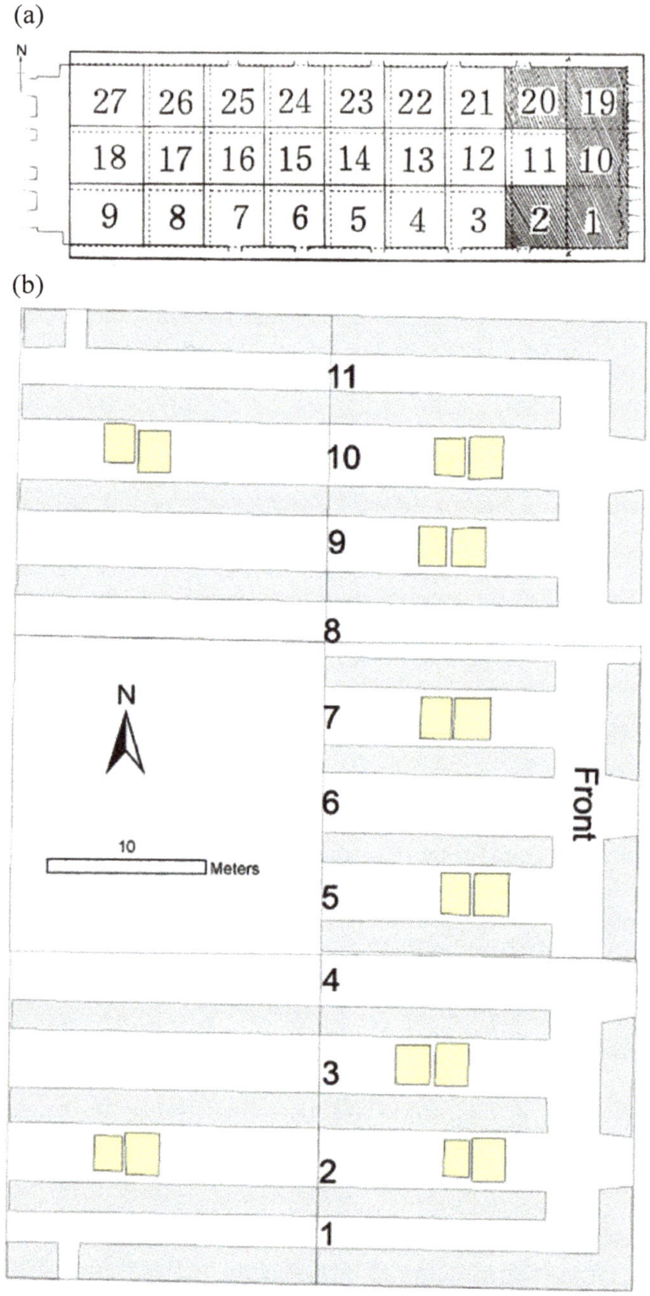

Fig. 1.4. Pit 1: (a) the five easternmost trenches are hatched (1, 2, 10, 19, and 20; after Figure 5 in the Excavation Report, Institute and Team, 1988: 10), and (b) the corridor numbering for the five easternmost trenches (yellow areas are those with chariots).

The weapons (or parts thereof) recovered include over forty thousand arrows (loose or in bundles) and several hundred other weapons such as crossbow triggers, swords, hooks, lances, spears, dagger-axes, halberds, and honour weapons (*Su*), as well as ferrules that were presumably placed on the butt ends of the long weapons (Fig. 1.5). The arrangement of the weapons in the pits is assumed to match a typical Qin battle formation. For example, arrows were mainly unearthed from the front and flank corridors where crossbowmen and/or archers were located, and long weapons, such as lances and spears, were discovered in the middle of the pit.

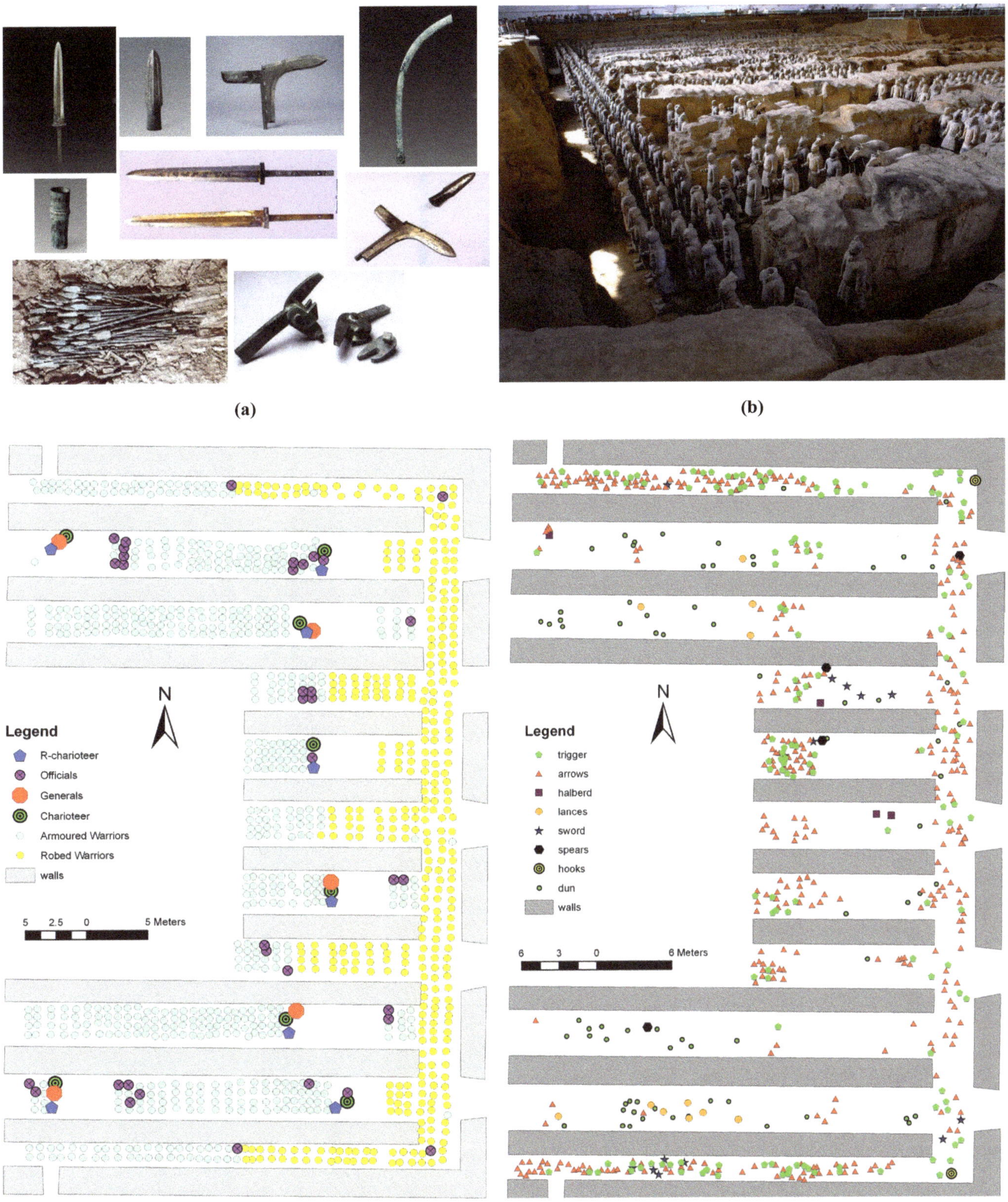

Fig. 1.5. (a) A selection of bronze weapons (image courtesy of Xia Juxian); (b) terracotta warriors discovered in Pit 1 (images courtesy of Xia Juxian); (c) spatial distribution of the terracotta warriors; (d) spatial distribution of the bronze weapons.

Some weapons have long inscriptions, which offer information about supervisors, officials, craftspeople, workers, and the year in which the weapons were produced. Other weapons only have a simple inscription, such as numbers or the name of the main workshop, *Sigong* (Yuan, 1990). The inscriptions bearing chronological information show that these weapons were mostly made before the Qin unification, during the period when seven kingdoms existed in China, engaged in a constant state of war (the period is accordingly known as the Warring States era 475–221 BC). These kingdoms not only competed against each other in terms of the size and organisation of their military forces, but also in the production of bronze weapons of sufficient quality and quantity to overpower the others, and it was the Qin state that eventually unified China with its strong military forces and bronze weaponry. In this sense, the information on the Qin weapons obtained in this project may also be relevant for future studies comparing the weapons technologies of different contemporary kingdoms.

1.4 Chapter summaries

This book is divided into eight chapters: the present introduction (Chapter 1); theoretical framework (Chapter 2); methodology (Chapter 3); a study of the inscriptions (Chapter 4); the main body of results from the metric and spatial analysis of the Qin bronze weapons (Chapters 5 to 7); and a broader discussion and conclusion (Chapter 8).

Chapter 2 covers the theoretical concerns of the research. It first reviews the past literature on the definition of standardisation, the factors affecting it, and the approaches allowing the assessment of the degree of standardisation. I propose that the concept of standardisation is relevant for a whole range of aspects of the bronze weapons production and can be approached by considering technological operational processes, specialised learning of certain skills, the sensory limitations of craftspeople, and labour organisation. I also argue that the spatial pattern of the bronze weapons could be employed to trace back aspects of workshop organisation and coherent activity areas associated with the placement of the weapons into the pit.

Chapter 3 focuses on methodology. It explores methods for data collection, data input and data management, as well as the models for statistical and spatial analysis.

Inscriptions on the bronze weapons offer basic information about the organisation of production, and are considered in Chapter 4. Some long sentence inscriptions on the lances and halberds provide data on the craft organisational structure during the Qin period, and shorter inscriptions can be interpreted as corresponding to simple count, weapons assembly, quality control, and/or the name of the workshop.

Chapters 5 to 7 present the detailed studies carried out on each type of weapon, namely triggers, arrows and long weapons. These chapters aim to evaluate these three main categories of bronze weapons statistically and spatially, and to reconstruct the model of labour organisation involved in such mass bronze production during the Qin period. The relative degree of standardisation is assessed, and a discussion is presented concerning the extents of standardisation and the extent to which they are affected by the human senses and by labour organisation. I argue that, on the basis of typological, metrical and spatial analysis, the triggers were produced by cellular production with a small room of batch mixing. The bronze arrows do suggest cellular production according to the chemical composition tested by portable XRF (Li et al., 2014; Martinón-Torres et al., 2014a).

Chapter 8 returns to overall questions of standardisation and labour organisation, and analyses the motivations behind such priorities. This chapter also aims to establish a model for interpreting the other sites and objects from the Emperor Qin Shihuang's tomb complex. The further potentials and limitations of the methodology employed in this book are appropriately discussed, and questions are raised about the wider spatial and temporal contexts. Given this overall structure of the research, it is to theoretical issues that the next chapter turns.

2

Approaches to Standardisation and Labour Organisation

2.1 Introduction

The Qin bronze weapons industry is an early example of mass production in China (Ledderose, 2000). For example, inscriptions on some of the weapons demonstrate that many of them came from the "*Sigong*" (寺工), a governmental workshop (see Chapter 4 for further details). However, to date there is no archaeological evidence of Qin bronze workshops from which we could gather direct information about weapons production. In this case, the study of the finished products in their deposition context is an important alternative route for understanding the production process, the behaviour of craftspeople and labour organisation during the Qin period. The bronze weapons carried by the terracotta warriors were originally transported to the pit a) from a workshop, b) from a store, or possibly c) after use in battle. The metric, typological and spatial data associated with these objects provide a range of information about their mass production, and this chapter attempts to demonstrate how standardisation and artefact variation can be influenced by technology, the behaviour of individual craftspeople, labour organisation and political events.

2.2 Standardisation

2.2.1 A definition of standardisation

Eerkens and Bettinger (2001: 493) have employed the concept of standardisation as a relative measure of the degree to which artefacts are made to be identical and substitutable. Higher tolerance increases variability, while lower tolerance decreases variability leading to standardisation. Other archaeologists have defined standardisation as homogeneity (Costin and Hagstrum, 1995). This homogeneity is reflected in the consistency of composition, shape measurements, and/or decoration in industries such as metal or pottery production (Costin and Hagstrum, 1995: 622). Rice has emphasised that this is a 'relative' measure of homogeneity, stating that there is 'no single scale, no decontextualised measure, no quantitative index with an absolute zero for comparing variability and standardisation of pottery through time and space' (Rice, 1991: 268).

Standardised artefacts are produced by either intentional or mechanical standardisation (Costin and Hagstrum, 1995: 622). Intentional attributes are controlled by the artisan and include technological, morphological, and stylistic properties. Mechanical attributes are affected by an unconscious motor habit, and are more likely to reflect the number of producers who contribute to the production of a particular assemblage (Roux, 2003). Mechanical standardisation is normally considered as an indicator of emerging social complexity or increasing craft specialisation (Rice 1981; Benco, 1988; Costin, 1991; Longacre, 1999). It may also be an indicator of increasing economic competition (Davis and Lewis, 1985; Torrence, 1986). The degree of standardisation in an assemblage is often used to infer both spatial and social principles of organisation in the production system (Costin, 2001).

2.2.2 Parameters affecting standardisation

Standardisation can be affected by several different processes (Rice, 1984: 111). For example, the number of producers, consumer demand (Hodder, 1983), political constraint, and production costs (Torrence, 1986: 197) can all be contributing elements. Eerkens and Bettinger (2001: 493) have emphasised the importance of human behaviour in artefact production, and proposed that standardisation was related to the life cycle of the artefact type or class in question, reflecting such things as production costs, consumer preferences, replication and learning behaviour, the number of producers, concerns with quality, producer skill, and access to resources. Further parameters affecting standardisation include the scale, efficiency, skill, size and location of the production entity, intensity, and the spatial organisation of production activities (Costin, 1991). Based on this literature, it becomes apparent that documenting artefact variability and standardisation may be a useful starting point for addressing broader issues of socioeconomic structure and labour organisation.

2.2.3 How to define and measure standardisation from archaeological data?

Standardisation is normally studied according to material, technical, and morphological attributes (Rice, 1981, 1991; Costin, 2001: 302). It can be defined according to raw material composition, manufacturing techniques, form and dimensions, and surface decoration (Hagstrum, 1985; Frankel, 1988; Sinopoli, 1988; 1991; Costin and Hagstrum, 1995; Roux, 2003). Metric variation is one of the most common indicators employed (Frankel, 1988; Longacre, 1999). Metric variation for a particular kind of product is assessed by comparison within specific groups through descriptive statistics such as mean, standard deviation, and the coefficient of variation (Frankel, 1988; Costin, 2001; Eerkens and Bettinger, 2001).

Other methods for measuring standardisation have been employed. Rice (1981: 221, 1991: 269) has suggested analysing histograms or distribution curves derived from the observations of a particular technology or stylistic attribute. The shape of the distribution would indicate the degree to which a standard exists and is consistently met. Feinman et al. (1984) investigate standardisation by counting types and relative abundances, with contingency table analysis and the phi coefficient used to assess 'distributional heterogeneity'. Barbara Stark (1991) employs descriptive statistics, including mean, standard deviation, and variance; the coefficient of variation (CV) has been fairly commonly employed in these studies, for example by Stark (1991) and Longacre et al. (1988). Sinopoli (1988) uses histograms and the Kruskal-Wallis analysis of variance in the study of south Indian pottery.

Further innovative approaches to the problem of measuring standardisation have been developed. For example, Hagstrum (1985) uses factor analyses to assess relative standardisation, that is, the number of factors and the percentage of variation explained by the first factor. Costin (1986) uses a combination of principal component analysis of ordinal variables and cluster and canonical discriminant analyses of the components and scores.

Eerkens and Bettinger (2001) have argued that a coefficient of variation is the most robust method for assessing the degree of standardisation, and this will be the basic parameter employed in my initial data analysis. They offer two baselines to assist in the interpretation of standardisation and variation. The upper baseline (the highest degree of standardisation, CV=1.7%) describes the minimum amount of metrical variation humans can generate without such external aids as rulers. The lower baseline (CV=57.7%) describes the amount of variation that will occur when there is no attempt at standardisation. This hypothesis is based on the Weber fraction for line-length estimation which describes the minimum difference that humans can perceive through unaided visual inspection (Eerkens and Bettinger, 2001:494). In the mid-1800s E. H. Weber observed that the ability of individuals to discriminate between objects of different weights depended on the mean weight of the objects involved (Coren et al., 1994: 39–43; Eerkens and Bettinger, 2001: 494). In lifting experiments, Weber discovered that heavy objects needed to be perceived as being significantly greater in weight than light objects. Weber also determined, however, that the relative difference necessary for such distinctions remained relatively constant. Specifically, he found that two objects had to differ by more than about 2% for a difference in weight to be detected, meaning that two large objects had to differ more in absolute size than two small objects. This value (2%) has come to be called the Weber fraction for heaviness (Eerkens and Bettinger, 2001: 495). The Weber fraction for the perception of the length of a line is similar to that for heaviness, about 3% (Teghtsoonian, 1971; Eerkens and Bettinger, 2001).

The Weber fraction and CV both express variation scaled to magnitude, and it is easy to convert the Weber fraction into CV form by using the notion of a uniform distribution, in which the two CV values (1.7% and 57.7%) are derived from this conversion. On this basis, Eerkens and Bettinger (2001) argue that in most circumstances, the CV is a stable and reliable measure of standardisation and variation.

A CV of 1.7% (closely related to the Weber fraction) should represent the minimum amount of variability achievable by humans for length measurements (Eerkens and Bettinger, 2001). Reducing variation below this threshold is not possible without additional aids given the visual perception capabilities of most humans. Sets of artefacts that display CVs of less than 1.7% therefore imply automation of manufacture or the use of an independent standard (such as ruler). In contrast, unstandardised assemblages should display CVs of less than 57.7%. Variation above 57.7% suggests an intentional inflation of variation and may indicate situations where individual manufacturers are actively trying to differentiate their products from those of others, thereby increasing the variation (Eerkens and Bettinger, 2001).

The psychophysical limitation of size discrimination quantified by the Weber fraction can help in recognising different modes of artefact production and degrees of standardisation. Weber fractions also have implications in symbolic archaeology, because humans are limited in their ability to view, interpret, and discriminate artefacts in the same way that they are limited in their ability to produce them in standardised form (Eerkens and Bettinger, 2001). Copy errors reflect human behaviour in the production process and can be identified within a standardised assemblage (Eerkens and Lipo, 2005).

Research demonstrates that people are unable to differentiate subtle differences in the size of objects beyond a certain point. In the transmission of cultural information these limits are just as applicable, affecting how accurately people can copy from and learn from others, and how precisely artefact traits will be transmitted between people (Eerkens and Lipo, 2005, 2007). All of these factors will be considered in the study of the Qin weapons, allowing comparison between different weapon types within the assemblage, as well as between these and others found elsewhere.

This is, however, a significant difference that sets the present study apart from most previous studies of standardisation using CV. As mentioned above, the CV values have been used to assess standardisation in technologies in which human perception is not typically aided by tools such as rulers and moulds, such as much pottery making or stone knapping. Conversely, the production of bronze weapons was based on the use of casting moulds. As noted by Eerkens and Bettinger (2001), the use of moulds can, in principle, help minimise the copy errors and achieve CV values below the established thresholds. Rulers and blank models may also be used to try to produce identical moulds. However, the materials and procedures employed to manufacture the moulds themselves may add variability. Although we do not know the moulding materials employed to cast the bronze weapons, it is likely that these would have been made of stone or clay. If they were carved in stone, then some degree of human copy error between moulds would be inevitable. If made of clay, the variable shrinking of the clay during drying and firing would no doubt add a degree of unpredictability. Furthermore, assuming that several casts took place in each mould, the mould surfaces would progressively be eroded, resulting in slightly larger or otherwise different artefacts. Finally, after casting, the artefacts would be filed and ground to remove casting flashes and seams, and to sharpen the blades (Li et al., 2011). These processes would also add further variation. Altogether, the parameters affecting variability in bronze weapon making are specific to this technology, and CV values cannot be directly compared to those in other crafts. However, CV remains a useful as a dimensionless parameter for quantifying standardisation and facilitating comparisons within the assemblage, as a baseline for cautious interpretations. The Weber fraction can also be used in reverse, to consider which metric differences would be perceptible to workshop supervisors or users of the weapons.

However, 3D scanners have been developed and are currently employed for measuring standardisation in some archaeological research. As this project was carried out ten years ago, the 3D comparison was not used for the bronze weapons in this book, but for assessing the mass production of the terracotta warriors, particularly the ears of the figures. The initial results of the sample of ears strongly indicate that there is considerable variation in contrast to these warriors' highly standardised bronze weapons (Bevan et al., 2014).

2.3 Standardisation and technological organisation

2.3.1 Casting and assembling

Mass-produced bronze weapons or vessels were normally made in ancient China using specific casting technologies that would affect the standardisation of the products (Bagley, 1990; Li, 2007). Models and moulds were used to shape the objects intended for casting. Normally, one 'mother model' was used to make the moulds. As noted above, the material of the models and moulds and their durability would affect their standardisation. For example, a stone mould can last longer than a clay mould, therefore allowing the production of a larger number of nearly identical artefacts, but the difficulties involved in carving the stone, not allowing for corrections as easily as clay, would probably make it harder to produce two identical moulds. In the case of a clay mould, it is difficult to determine whether a single mother model was used to make a large number of moulds or several mother models were used by different craftsmen, although this should be reflected in different degrees of standardisation. The shrinkage of a clay mould also needs to be considered.

Technical requirements such as the need to assemble different parts may have called upon for special efforts to optimise standardisation. In the case of the crossbow triggers from Pit 1, these were made up of three mechanical parts jointed together by two bolts. After the archer released the arrow, the trigger's mechanism would have gone back to its original position and set the bow backward. Thus, it was very important that the five parts were precisely fitted and moved easily, and one might expect a high degree of internal standardisation between these different parts, if the triggers in the Emperor's tomb were really meant to be functional rather than just symbolic representations. These aspects are relevant to the standardisation of weapons, but should be contextualised more broadly in their relevant production or labour organisation systems (for details see Chapter 5; Li et al., 2014).

2.3.2 Technological models

The holistic and prescriptive hypotheses are two classifications proposed by Franklin (1983: 96–97) to explain the technological processes and the organisation of production for Shang bronze making. Depending on the ways in which production procedures were organised, these formed a continuum from simple to complex in technological management (Li, 2006; Sun, 2008). A holistic process is understood as a sequential, linear development, involving a single, stepwise progression towards the final object. In such a process, a relatively small workshop or craftsman is supposed to be in charge of the manufactured objects and all the manufacturing procedures (Franklin, 1983; Li, 2006: 20). A prescriptive process occurs if production is divided into predetermined production units, or groups of workers, and is carried out by individuals with independent skills in each unit. In a prescriptive technology, individual workers do not necessarily possess a complete knowledge of the entire production process; rather, they are confined to a limited range of tasks in one or more steps (Franklin, 1983; Li, 2006; Sun, 2008).

The concepts of holistic and prescriptive technologies have subsequently been employed as models for the evaluation of craft production and production organisation in China's Bronze Age (Ledderose, 2000; Li, 2006; Sun, 2008). Recently, Li (2006) has explored the archaeological application of an alternative tripartite classification using modern terminology: cellular, batch, and flow line production. Standardised mass production is usually associated with modern mechanised factory systems, such as car assembly lines (Wild, 1975; Matthews, 1996). In a flow production line, the total work is divided into small tasks, and workers or machines perform these tasks with great efficiency. Individual parts are mass-produced in separate units and subsequently joined together in an assembly line. The movement of the product is facilitated either by manual labour or by mechanised means, such as conveyor belts (Li, 2006). Much credit for the development and refinement of the production line is due to Henry Ford and his engineering staff at the Ford Motor Company in the early 1900s (Groover, 2010). However, the cellular production is based upon group technology. In this system, machines are grouped together according to the families of parts produced, a family possessing similar design and/or manufacturing features, such as round holes made by drilling and cylindrical shapes by turning. The design of cellular production is related to the spatial layout of the factory, and the products are usually made in batches (Grover, 2010: 884). In modern manufacturing, batches of one product or one part are produced to form stocks; after stocks are formed, facility and production procedures may be revised in order to produce a different product (Groover, 2010). Furthermore, batch production and cellular production are designed to produce more than one type of product in the same production facility, which may or may not require further assembly (Li, 2006). In contrast to the high production of an assembly line, batch production and cellular production are both considered types of medium-output production, as the changeover time needed for equipment to produce different products reduces the overall productivity, and these two factors are closely correlated because the chronological batches may be produced in group or cellular layout.

In modern times, standardisation and mass production have been mostly facilitated by the extensive use of machinery, whereas mass production in ancient China was achieved by the internal division of labour and by the use of standardised parts and procedures (Li, 2006). Ledderose (2000) has argued that through the use of standardised parts that could be prefabricated and assembled, which he called 'modules', and the use of the 'modular system', the Chinese were able to mass-produce artefacts as early as the Bronze Age. According to Ledderose, 'modular systems' can be found in the production of a variety of artefacts from a range of periods, such as Eastern Zhou bronzes and Qin terracotta warriors (Ledderose, 2000). In terms of the 'modular' production proposed by Ledderose (2000) for the terracotta warriors, his emphasis was on the fact that the various warriors were assembled with a limited number of clearly standardised parts. For example, there are three types of plinth, two types of feet, three types of shoes and four types of boots, two types of legs, eight types of torso, etc., with each category having subtypes (Ledderose, 2000: 72). These standardised parts could be produced in batches. Furthermore, the stamps on the back of the warriors suggest that 85 names of master potters have been found, to date, on the 1087 terracotta warriors (Yuan, 1990). This was argued by Yuan (1990) as probably a group production whereby the master potters worked with several apprentices and had responsibility for their group's quality control.

One does need to exercise caution when using modern terminology to discuss traditional mass-production industries. Mass production or large quantity production in ancient China can only be discussed in relative terms (Li, 2006). Discussion of standardisation and production organisation may also be less well defined and less clear-cut, either because we do not have sufficient information about it today or because more than one form, or hybrid forms, of production organisation in the modern sense may have been in operation in ancient workshops. This, however, does not prevent adapting the concepts and parameters of modern manufacturing engineering to the study of ancient craft production, as these concepts may still be useful in formulating assumptions and in defining forms of production organisation.

If the notion that mass production began in China as early as the Bronze Age can be accepted, we need to distinguish further the different forms of mass production that might have existed. Information that might allow for finer distinctions does not always exist, especially since archaeological data rarely sheds light on actual workshop layouts with sufficient physical information, or on a sufficiently large quantity of finished products. These are both very important for determining the types of production organisation in place. For the sake of argument, we may hypothesise, in a very speculative way, that bronze production at Anyang (Shang capital during 1600–1050 BC) was organised differently from the Houma foundry (Warring State era 476–221 BC). According to the workshop layout and the remains of mould fragments and products, the Anyang foundries probably involved both holistic and prescriptive production organisation in the different stages (Li, 2006). In contrast, Houma is considered to reflect a form of batch production system, in which batches of parts for complete vessels were produced separately in different locations of the workshop and then assembled. Bagley (1995; 1996) argues that the workshop might have had some form of cellular layout, in which craftspeople, such as mould-makers and bronze casters, performed separate but related tasks.

Even if we may lack relevant information of workshop remains, the assemblage of weapons from the Emperor's mausoleum should allow us to make inferences regarding the mode of production. In particular, by identifying different artefact types that could have been produced in separate units, studying the way they were assembled together (in the case of composite objects such as triggers), and their spatial layout in Pit 1, it should be possible to obtain information about the way in which production and assembly were organised. The different models outlined above will therefore be useful when discussing the metric and spatial data.

2.4 Standardisation and specialisation

2.4.1 Specialisation

Production entails the transformation of raw materials into usable objects (Rice, 1981). Specialisation is one way to organise this production (Costin, 1991). Specialisation is viewed as 'a differentiated, regularised, permanent, and perhaps institutionalised production system', in which the producers depend at least in part on extra-household exchange relationships for their livelihood, and consumers depend on them for the acquisition of goods they do not produce themselves (Costin, 1991: 4). Specialisation is therefore an important feature of social, political, and/or economic systems, and it is influenced by constraints and opportunities in the environment. It is considered a key factor in the political economy of complex societies and can be defined as a channelling of labour and capital towards the production of a particular good or service (Blackman et al., 1993; Stein, 1996: 25).

Specialisation can be assessed based on two primary parameters: degree and type (Costin, 1991). The degree refers to the ratio of producers to consumers. For any product, there may be few or many producers as compared to the total number of consumers. A product that has a high number of producers relative to consumers will present a low degree of specialisation, while a product made by relatively few specialists in proportion to consumers will show a high degree of specialisation.

The typological classification identifies two main types, which refer to attached and independent production, as originally stated by Earle (1981). This differentiation shows the fundamental distinction between production of special, high-value goods for elite consumption and the production of utilitarian goods for broad distribution (Costin, 1991). More recently, Costin (1991: 11) has developed the theory of attached and independent specialisation. Further, she proposes an eight-part typology for the nature of specialisation, described as follows:

Individual specialisation: autonomous individuals or households producing for unrestricted local consumption

Dispersed workshop: large workshops producing for unrestricted local consumption

Community specialisation: autonomous individual or household-based production units, aggregated within a single community, producing for unrestricted regional consumption

Nucleated workshops: larger workshops aggregated within a single community, producing for unrestricted regional consumption

Dispersed corvée: part-time labour producing for elite or government institutions within a household or local community setting

Individual retainers: individual artisans, usually working full-time, producing for elite patrons or government institutions within an elite or administered setting

Nucleated corvée: part-time labour recruited by a government institution, working in a special-purpose, elite, or administered setting or facility

Retainer workshop: large-scale operations with full-time artisans working for an elite patron or government institution within a segregated, highly specialised setting or facility.

In addition to this typological classification, there are four further parameters used to characterise specialisation in production: the context, concentration, scale, and intensity of production (Costin, 1991). The context of production refers to the relationships among producers and the socio-political demand for their products. The notion of 'attached production' refers to cases when the craftspeople are sponsored and managed by elite patrons or government institutions. As such, attached specialists produce several types of goods of key importance within the political economy in a certain area and time period, and contribute to maintaining status, power, or control structures in a society. Examples of such products include luxury goods, weaponry, and goods that generate further wealth. In contrast, independent specialists produce for a general market of potential customers. The second parameter, concentration, is associated with the geographic organisation of production. In terms of concentration, specialists can be either nucleated or dispersed. The third parameter, of scale, refers to the composition of the production unit. It includes two related variables: size and principles of labour organisation, ranging from a small unit to a factory. The final parameter, intensity, reflects the amount of time producers spend on producing the respective artefacts. In terms of intensity, the production of a certain good could involve casual, part-time specialisation, where a certain commodity or service is complementary to basic domestic production of subsistence products. On the other hand, craftspeople may be engaged in full-time specialisation, producing exclusively one type of product or service, and exchanging these for all the other goods and services used by the household.

The evidence pertaining to specialised organisation in production is of two general types: direct and indirect (Costin, 1991: 18). Direct evidence consists of materials that are directly associated with the specific location where production takes place. In contrast, indirect evidence provides information about the organisation of production without identifying its exact location. Direct evidence thus involves production loci, as well as aspect relevant for context, concentration, scale, and intensity. Indirect evidence is derived from the finished objects themselves. These indirect data include the recognition of standardisation, efficiency, skill, and regional variation (Costin, 1991: 43).

In the absence of archaeological evidence for Qin workshops involved in the manufacture of weapons, all the evidence employed in this research is of an 'indirect' nature. However, it is still possible to assess the various parameters outlined above, in an attempt to characterise further the nature of specialisation in the manufacture of bronze weapons for the Emperor's tomb. As noted in the methodology, this will be achieved by integrating metric, spatial, and technological data, coupled with a review of historical information and, where possible, comparison with relevant archaeological artefacts from other sites.

2.4.2 The hypothesis of standardisation and specialisation

Standardisation is often taken as indirect evidence of specialisation. The linkage between specialisation and standardisation is based on the hypothesis that the more specialised the producers, the more uniform the manufacturing techniques and material used will be (e.g., Longacre et al., 1988; Sinopoli, 1988; Costin,

1991; Arnold and Nieves, 1992; Blackman et al., 1993; Costin and Hagstrum, 1995; Longacre, 1999). In most cases, people assume that standardisation of production is closely related to specialisation, where the goods produced in large amounts by specialists can be recognised in the archaeological records by their high degree of standardisation. Conversely, variation or relative heterogeneity is taken to indicate small-scale or less specialised production (Rice, 1981: 220–221; Feinman et al.,1984: 299; Blackman et al., 1993: 60–61; Roux, 2003). The correlation between specialisation and standardisation has sometimes also been supported by ethnographic and experimental data (Hardin, 1979; Hill, 1979; Miller, 1985; Clark, 1986; Sinopoli, 1988).

One should be cautious about assuming that this hypothesis is always true. Indeed, some of the references cited above bring up examples of cases where either standardisation or specialisation can exist without the other. Whether or not standardisation is an appropriate measure of specialisation depends on the type of object, the technology, its function, and the nature of demand (Costin, 1991). We cannot lose sight of the fact that different types of goods are often aimed at different markets and different functions, some of which demand individuality and others standardisation. The appropriateness of standardisation as an indicator of specialist production is dependent in part on the motivating consumption choices in each particular case. First, specialised or unspecialised craftsmen are equally capable of producing a variety of stylistically or formally different products. Second, variability at this level often reflects sociopolitical processes or functional concerns. Although standardisation can be an appropriate proxy of specialisation, it is then necessary to choose carefully the variable or variables that will best reflect the organisation of production (Costin, 1991).

Standardisation, as a result of indirect evidence gathered through the examination and measurement of the weapons, is an essential means of exploring the production processes in place during the Qin period. However, in view of the above issues, the data will have to be assessed carefully before drawing any conclusions as to the relationship between standardisation and the nature of craft specialisation. The weapons are finished products that were buried as funeral items in the mausoleum complex, not unearthed from a workshop. We have not found any workshop near or far from the mausoleum complex, which means that no direct evidence, such as slag, ingots, crucibles, furnaces, and scrap, are available for analysing the organisation behind the production of these bronze weapons.

2.5 Standardisation versus variation: copying error

In spite of the apparent standardisation of the weapons from Pit 1, we also need to identify and interpret the variation of these weapons within the assemblage. The logistical organisation and individual behaviour of the craftspeople, their skills, and their copying errors also require attention. Many archaeological studies of technology recognise only the role of the physical and social environment in shaping material culture by focusing on how a raw material is modified using various tools, and how different social and physical processes influence the final product (Schiffer and Skibo, 1997). As Bleed (1997: 98) has argued, the human body has seldom been seen as part of this process. The human body, with all of its attendant sensory systems and limitations, is a medium through which technology operates. The human abilities to see, feel, and modify material items are limited, and are affected not only by culture, but also by the physics and psychophysics of the human body (Eerkens and Bettinger, 2001; Eerkens and Lipo, 2005; see also section 2.1.3).

Taking into consideration the physiological limits to the human ability to perceive differences, it is reasonable to assume that a small amount of error, below the limits of detection with the naked eye, will be introduced into human operation processes. Such error is introduced into any situation that involves replication and in which information is transmitted about the characteristics of an artefact and its production process. This source of variation is referred to as 'copy error' (Eerkens and Lipo, 2005). It derives from small errors that are transmitted from one person to another in the dissemination and replication of cultural traits. Even if the error is small and imperceptible at each event, it becomes cumulative and can consequently develop into perceptible differences over time, or whenever many copies are made.

The cumulative result will be a certain drift during the process of transmission, whether horizontal, vertical, or oblique (Eerkens and Lipo, 2005). The simplest simulation involves the vertical transmission of information. This is often referred to as unbiased transmission, entailing several lineages with direct replacement in each generation (Eerkens and Lipo, 2005: 321). With each subsequent generation, error is added to the attribute value and transmitted to the next generation. However, cultural transmission rarely operates in such a directly vertical fashion. Biased transmission, such as conformist transmission or prestige-biased transmission, can potentially influence the range of variation (Eerkens and Lipo, 2005: 321), and normally reduces the amount of variation within the population.

The weapons from the Terracotta Army were most likely produced within a relatively short period (246–210 BC), which prevents a diachronic study of artefact variation as usually performed in cultural transmission studies. However, given the abundance of artefacts and intensity of production, it is probable that a relatively large number of masters and apprentices were involved. Thus, it will be interesting to try to identify and explain artefact variability that may have been introduced through copy errors.

2.6 Spatial signatures of workshop activity and labour organisation

Spatial patterning in artefact assemblages can also be used to interpret the workshop activity spaces and labour organisation in mass production. To date, the most spatially-sensitive investigations of labour organisation have drawn on either workshop debris (Li, 2006) or large resource landscapes (Zhang et al., 2014). In contrast, the Qin bronze weapons are distributed together with the Terracotta Army in the funeral pits of the First Emperor's tomb complex, a single consumption context. In the chapters that follow, a relatively simple spatial relationship is described between the bronze weapons found in Pit 1 and the battle formation of the terracotta warriors. More challengingly, some methods are developed to identify and characterise additional spatial patterns in the micro-attributes of the bronze weapons, in order to identify 'activity spaces' that might be related to workshop practices and/or procedures for placing finished weapons into the pit (see also Bevan et al., 2014b; Li et al., 2014).

2.6.1 Point pattern analysis

A point pattern dataset gives the locations of objects occurring in a study region. Such patterns have been a long-standing concern for archaeologists, who have typically presented their archaeological finds in a distribution map as a first step towards interpreting them. For this particular project, all the locations of the bronze weapons and warriors in Pit 1 are digitised and stored as point data. This is done for purpose of simplification (see Bevan et al., 2013, 2014b).

The points could be represented as two-dimensional detailed shapes or in full 3D, and sometimes with orientation, for example, the direction of the lances in the pit. However, for the purposes of this project, the simple location as a point is sufficient.

There are many questions that arise as part of point pattern analysis, related to the intensity or density of the points across the study region, any evidence for interaction between points, and any role of covariate factors (Baddeley, 2008), as well as to how we relate these observations to practical archaeological significance. Moreover, different approaches are appropriate depending on the specific research context. For the weapons distribution, the most important first step is to achieve a quantitative spatial description of patterns of clustering, regularity, and/or randomness in the artefacts, in order to understand the technological processes, behaviours of craftspeople, and labour organisation behind the patterns.

2.6.2 Spatial randomness, regularity, and clustering

The main baseline for considering a point pattern is what would happen if we scattered points at random across the study region. The baseline model is usually referred to as 'complete spatial randomness' or CSR (Bevan et al., 2013). Such randomness may be the result of a specific behaviour in certain circumstances, but it is also used as a simple null hypothesis. Certain point patterns can depart from CSR by being either more random or more regular. The regularity of artefact distributions can be associated with very deliberate human decisions about placement or with competition between points, for example if they were settlements competing for land. Clustered patterns, in contrast, are often the result of 'attraction' processes. For example, ancient settlements might also cluster close to each other and/or be located close to river valleys for many reasons. For craft activities, we can think of situations in which certain parts of the process of making, storing, transporting, and/or discarding objects lead to spatially-clustered patterns in the archaeological record.

2.6.3 First-order effect and second-order effect

The most obvious factor affecting the layout of the bronze weapons in the pits of the terracotta warriors is the battle formation of the army. However, the Qin bronze weapons analysed in this project were also transported from workshops or stores to the pits, and the labour organisation of the processes of production and transportation was considered to affect their spatial patterns, in addition to the basic layout of the army in battle formation. The different workshops or different units of one workshop might produce distinctive weapons or use different moulds to cast the same type of weapons. The different types of weapons might have been stored in specific places within the arsenal if they were not intended to be mixed. Ideally, the spatial patterns, random, regular or clustered, of these weapons in the pit can offer an indication of original storage or workshop practices, as long as we can distinguish these from the more general pattern associated with the battle formation.

The weapons in the pit are a good example of the problems that arise in spatial analysis when a range of heterogeneous processes are at work. The key is to develop a method that allows us to identify situations when the patterns are driven by factors unrelated to workshop and storage practice and when they are directly related to these behaviours.

Spatial patterns derive from the operation of spatial processes, and can be seen as the result of two sorts of variation – global (large-scale) trends and local (small-scale) effects (Bailey and Gatrell, 1995; Orton, 2004: 299), or as sometimes driven by external factors and sometimes by internal ones. The former is called first-order effect and the latter is usually referred to as second-order effect. For human settlements, we can easily think of cases, where settlements are located both with a preference for being close to rivers (a first-order effect) and not too close to one another (a second-order effect; Bevan et al., 2014b; Li et al., 2014). For these bronze weapons, one first-order effect is the arrangement of certain weapons to match the warriors in battle formation, and one second-order effect might be the way in which labour organisation and production practices affected the pattern of placement into the pit. Several multiple-scale spatial statistic methods, such as Ripley K and L functions (Orton, 2004; Ripley, 1977), can be used to explore the second-order effect based on the patterns of the weapons' subgroups over the whole study area. However, for the purpose of this research I will use a different and arguably more powerful method, known as a pair correlation function (Baddeley, 2008: 213). This function considers the average density of points that fall within each set of concentric circles of increasing size around each known point and uses sets of random points to assess the significance of the pattern (see Chapters 5–7).

2.7 Summary

Patterns of standardisation in the Qin bronze weapons' production should therefore be considered from several theoretical and methodological perspectives, and via appropriate models of labour organisation and human behaviour. The bronze weapons are 'proxy' indicators of changes in past conditions of reproduction, and standardisation in the production of weapons is thus correlated with past technological systems, labour organisation, and human behaviour. Measurements of the bronze weapons, typological features, inscriptions, and spatial patterning will provide indirect information that can help test technological hypotheses regarding the production of Qin bronze weapons and their arrangement in the pit.

Clearly, standardisation and labour organisation were essential in the production of Qin bronze weapons. The issue of whether the weapons were produced in the same workshop or in different workshops needs to be explored, as does the organisation of labour within individual workshops. If the notion that Qin weapons were standardised is accepted, analysis as to whether this was driven by political pressure, efficiency, or functional needs will be a related question of interest. The connections between standardisation and altered patterns of human, social, political, and economic behaviour, rarely discussed before in Chinese archaeology, are also important factors to be considered when discussing the Qin bronze production.

Overall, this literature review has highlighted some of the potentials and some of the problems associated with the study of standardisation and specialisation that are relevant to the present research. By combining systematic data collection with a critical assessment of theoretical models, it is hoped that this study will provide insights into the organisation of production and technological behaviour involved in the creation of the unique weapons assemblage carried by the Terracotta Army. Between the theoretical framework and the archaeological data, a bridge needs to be built with 'stone', 'steel', and 'cement', which are the proper methods discussed in the following chapter.

3

Methodology

3.1 Introduction

The objective of this project is to investigate standardisation in the production and labour organisation behind the large quantity of bronze weapons buried in the pits at the Qin First Emperor's tomb complex. In order to achieve this objective, it is very important to develop a proper methodological protocol that is coherent with the questions asked by the present research and with the archaeological theories outlined in the previous chapter.

The methodological framework discussed below encompasses data collection methods and statistical treatment. Even though a qualitative study of the weapons has been carried out previously, and reported in the Excavation Report (Institute and Team, 1988) and other related publications (Wang, 1980; Liu, 1986; Yuan, 1990), the function, micro-features, and inscriptions of the weapons constitute basic data and, as such, have been considered further for the purposes of the present research. In addition, a suitable sampling strategy had to be devised for the large quantity of bronze arrows, triggers, and other weapons. As regards the extensive metric study, accuracy and efficiency of the measuring processes were achieved by the use of digital callipers and advanced computing software. Furthermore, statistical methods were employed to achieve the quantitative classification of weapon groups, to assess their degree of standardisation, and to identify spatial patterns to facilitate the interpretation of workshop activities and labour organisation. This chapter will present the basic methodological principles that were employed to address an innovative combination of qualitative and quantitative data related to the typology, dimensions, and spatial distribution of the weapons – an approach that has been virtually absent in previous scholarship on the subject.

It should be mentioned that the instrumental analysis of these bronze weapons is another very important research strand, but one that is largely left out of this research due to the time and space limitations. However, as part of a wider cooperation, archaeometric analyses are being carried out by Xia Yin (from the Museum of Emperor Qin Shihuang's Terracotta Army) and Marcos Martinón-Torres (my PhD supervisor from the UCL Institute of Archaeology). This work, involving optical microscopy and portable XRF and SEM-EDS analyses, is already showing some promising results, to be contrasted and integrated with the broader dataset. In the present study, I will be reporting some results kindly made available by the rest of the project team, but the systematic publication of analytical data and its detailed discussion will be undertaken in a separate publication.

3.2 Qualitative classification

3.2.1 Functional classification of the weapons

From the outset of this project, the bronze weapons were generally classified based on their function in the battle formation. They fall into three categories: short weapons, long weapons, and long-range weapons. Short weapons, such as swords and hooks, were used for close combat; long weapons, including spears, dagger-axes, halberds, and lances (and the ferrules fixed to the ends of the long wooden handles), were traditionally issued to infantry and warriors on the chariots; the long-range weapons are preserved in the pit in the form of arrowheads and triggers for crossbows. These three general types of bronze weapons were used to equip the terracotta warriors meant to protect the Emperor Qin Shihuang in his afterlife (Figs. 3.1, 3.2, and 3.3).

The terracotta warriors in the pit were not necessarily equipped with only a single weapon or type of weapon. In fact, sometimes two or three types of weapons were found surrounding an individual figure – an impression that is also corroborated by the posture of some figures, with both hands appearing to hold weapons. Thus, some weapons were probably originally held by the terracotta warriors, such as the long weapon halberds and crossbows corresponding to with the warriors' postures. With the collapse of earthen walls and of roofs after being buried for more than 2,000 years, the terracotta warriors were all broken into pieces and the wooden handles of weapons, as well as the wooden parts of crossbows were largely rotten. The metal parts of the weapons dropped onto the paved floor and were mostly well preserved at or near their original location. Archaeological evidence shows that some warriors had postures that allowed them to hold a long weapon in the right hand and a crossbow in the left hand, as well as being equipped with arrows.

This general classification of the weapons was used to study the battle strategy and formation arrangement during the Qin period in ancient China, in addition to providing a basic structure for organising my work. Furthermore, micro-features of each type of weapons, as presented in detail in the following section, were used to address the variety of models or moulds employed for casting, as well as to infer aspects of the labour organisation behind such processes. Together with the extensive measurement of the weapons, this constitutes the main approach to tackling the research questions this study seeks to answer.

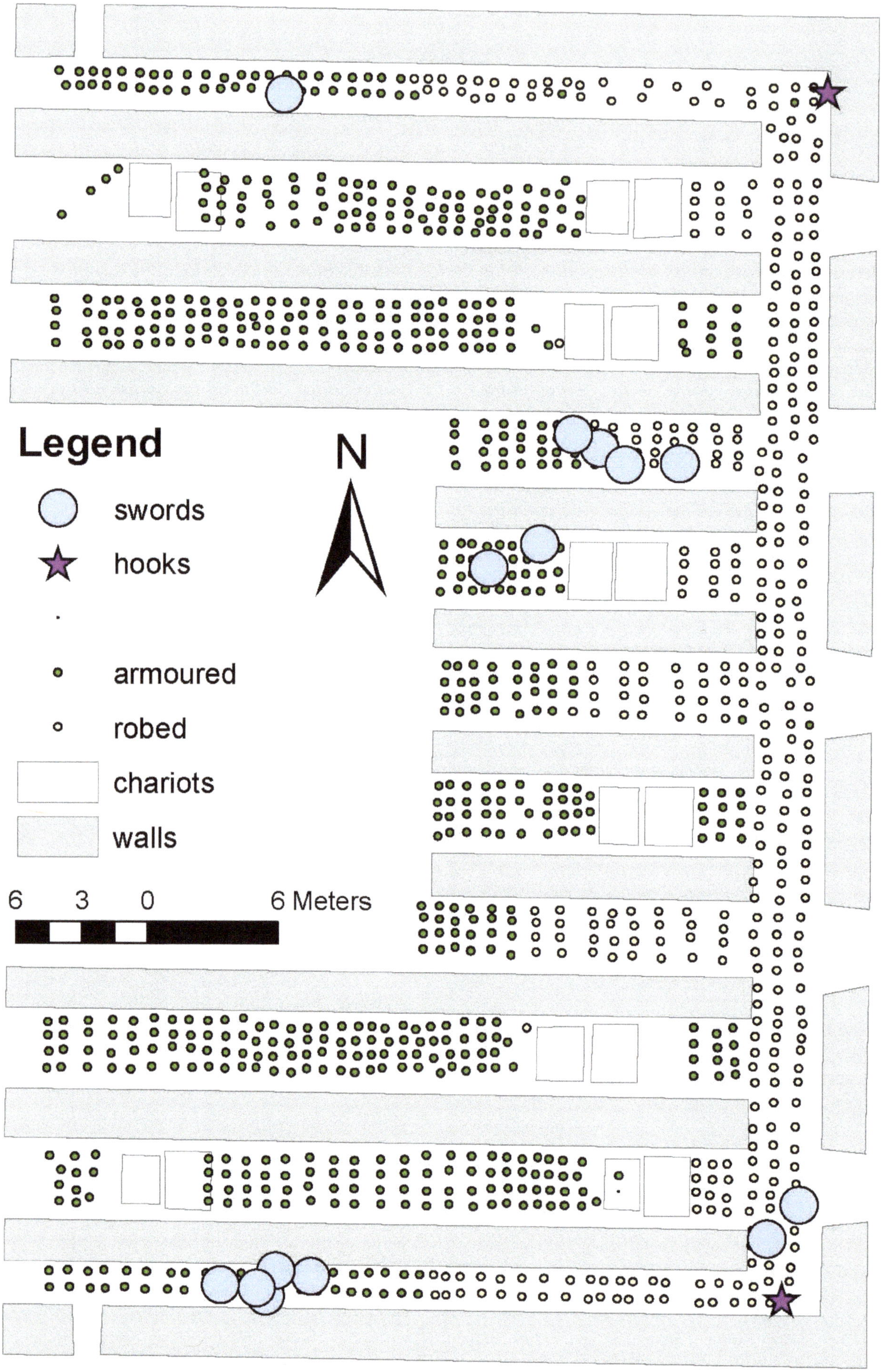

Fig. 3.1. The distribution of swords and hooks in Pit 1, shown on top of the distribution of robed and armoured terracotta warriors.

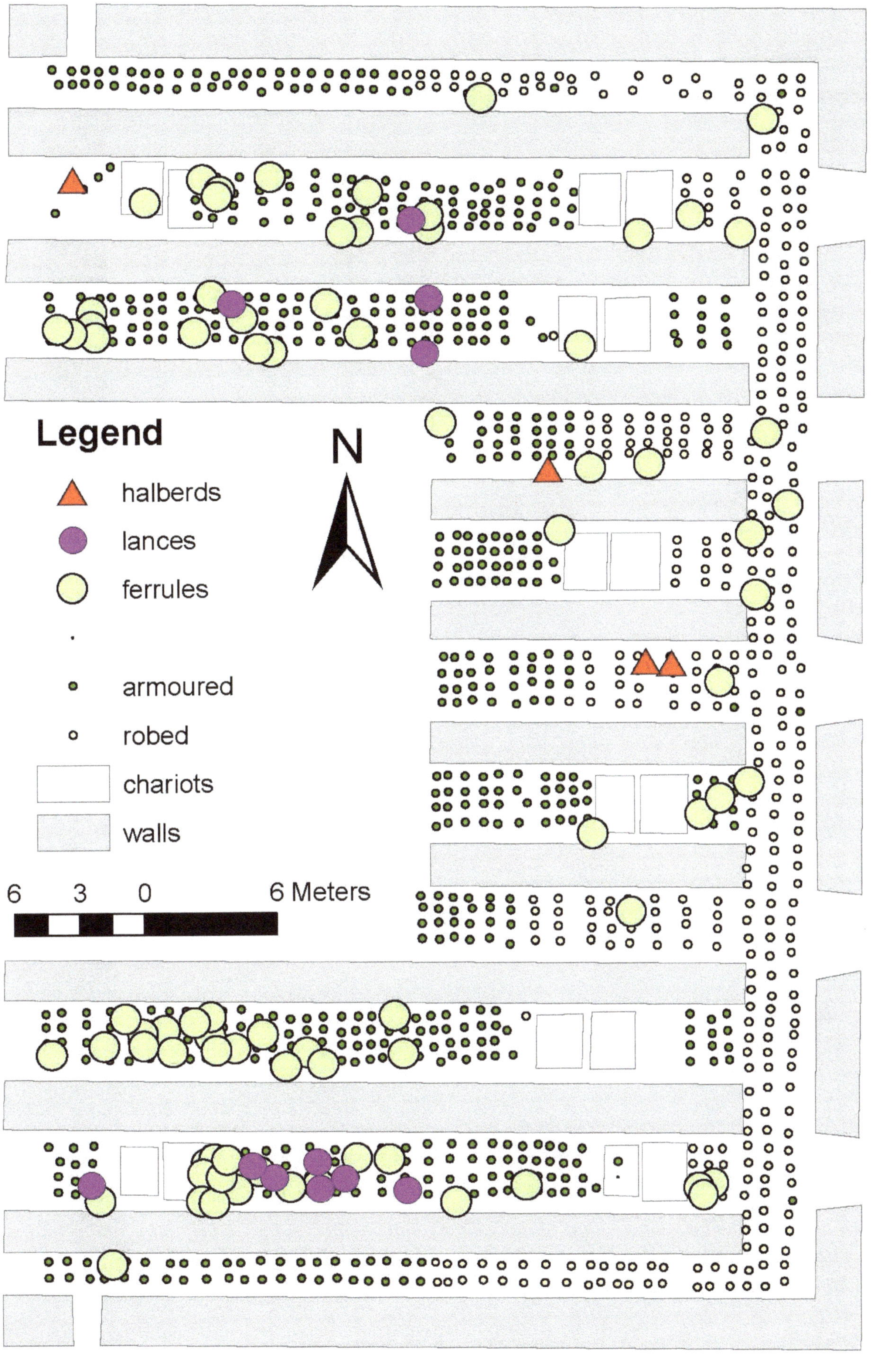

Fig. 3.2. The distribution of ferrules and long weapons in Pit 1, shown on top of the distribution of robed and armoured terracotta warriors.

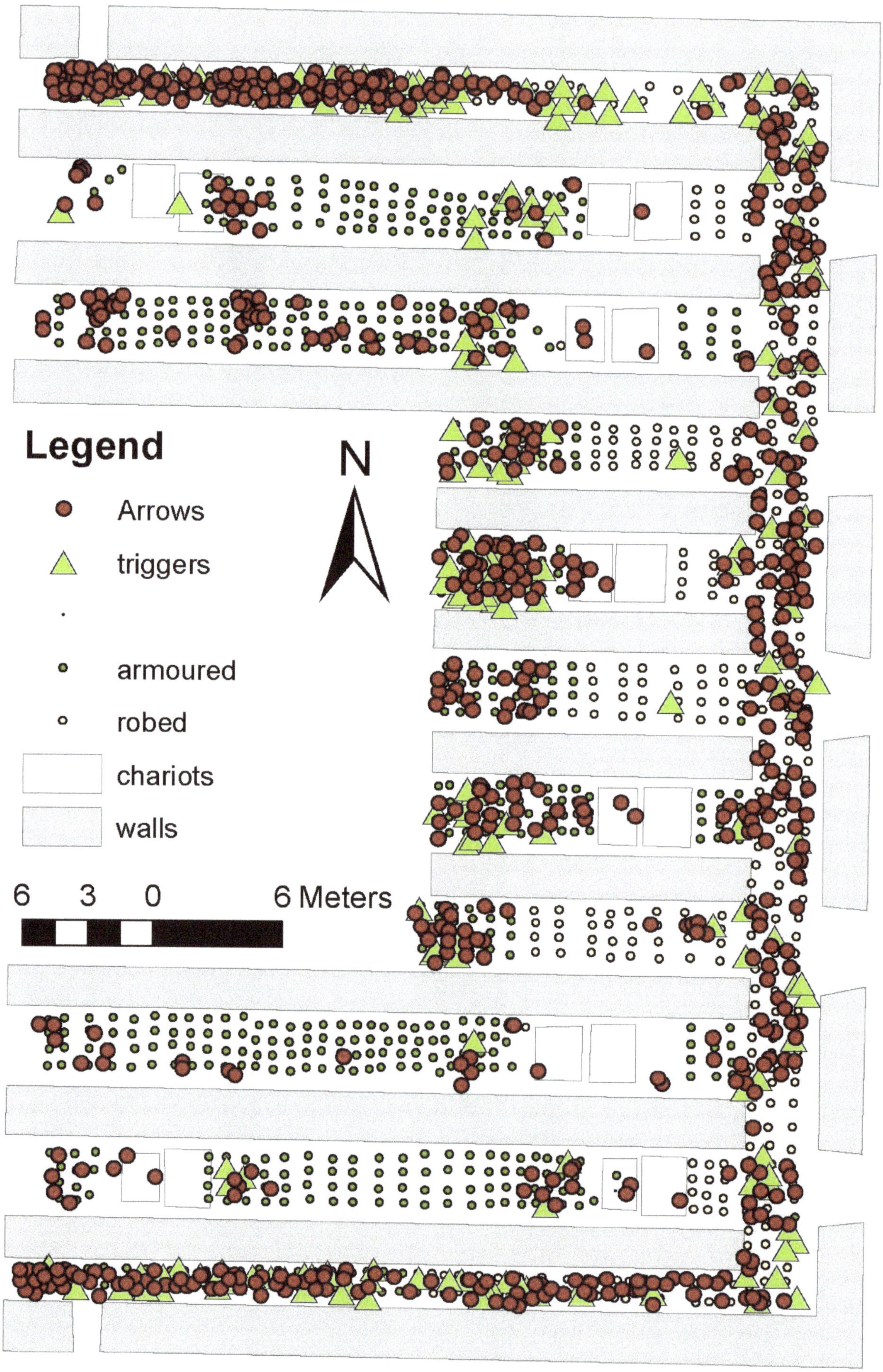

Fig. 3.3. The distribution of triggers and arrows in Pit 1, shown on top of the distribution of robed and armoured terracotta warriors.

3.2.2 Micro-features

Within each of the functional groups outlined above, and notwithstanding the overall impression of formal uniformity, a detailed visual study often reveals small but significant features that distinguish some artefacts from others. For example, when discussing crossbow triggers, I will be differentiating 'bevelled' from 'curved' parts based on a slight difference in the shape of otherwise very similar trigger parts (and a rare difference that is otherwise missed by the choice of quantitative measurements used here). Some ferrules of very similar height and diameter are distinguished according to a horizontal decorative band visible in some of them. These formal aspects will be generally referred to as 'micro-features', to emphasise the fact that they are generally not important in terms of their impact on overall shape and functionality.

The typological analysis based on the micro-features of each type of weapons is important insofar as these features are diagnostic of the use of different models and moulds to cast the weapons or weapon parts.

As such, they allow the identification of weapon subgroups even in cases where the metric data did not readily reveal them (see section 3.4 below). One therefore needs to be cautious in combining both metric and micro-features information because sometimes types appear merged in the quantitative analysis but can be distinguished with regard to their shape, or vice-versa. Both types of information are useful and complementary.

3.2.3 Inscriptions

Some inscriptions cast and/or carved on the surface of the bronze weapons provide important complementary information about labour organisation and standardisation. The long inscriptions on the lances and halberds have received the attention of Chinese scholars ever since the initial excavation was carried out in the 1970s, but they are reconsidered here in combination with the other types of new data generated. These inscriptions provide a general chronology of the production of some of the weapons for the Qin First Emperor, as well as data pertaining to the organisational structure of bronze casting (see Chapter 4). Furthermore, the inscriptions on lances and halberds found to date also recorde a variety of names of craftspeople and workshops.

Comparisons were therefore carried out on the lances and halberds produced in the same year or under the supervision of the same craftspeople for this particular project. The criteria for comparison included types, dimensions, chemical composition and spatial locations in the pit of the terracotta warriors. The comparison subsequently included weapons produced in different years and by different craftspeople, in order to obtain information about the standardisation of production and the extent to which this could be correlated with different workshops, craftspeople, or supervisory arrangements.

In addition to the long inscriptions, single or double characters marked on some of the triggers, swords, and ferrules are also relevant for aspects of standardisation and labour organisation. These characters, chiselled on the weapons' surfaces, may include the name of the workshop, numbers, symbols, traditional stem-branch characters (for details see Chapter 4), or a combination of the above.

In order to obtain a clear image of these characters, both for this project and for future investigation, each one was recorded separately as a vector object (in Adobe Illustrator). The weapons were all photographed in the museum storage and each stroke of each character was drawn to scale before subsequently grouping them as a whole character. Even though this was a time-consuming procedure, obtaining clear images of the characters provided important basic information for further analysis.

Each character was coded and the frequency of its appearance on the weapons was recorded in a spreadsheet. For example, 82 characters or symbols were discovered on the 239 triggers analysed, and these characters were numbered from 1 to 82. Most of these characters were recurrent on the various parts of triggers. The location of the characters on each trigger part was also recorded, as it was possible that the same type of inscription was marked in different locations by different craftspeople.

Statistics were used to calculate the frequency of the appearance of the characters on the weapons, and these data were integrated with information obtained from the other research methods. For instance, following the grouping of weapons with the same inscription, their spatial distribution could be studied and compared to that of weapons with different inscriptions. Such an approach yielded information about the organisation of labour for the transportation of these weapons to the pit and their arrangement there.

Overall, the study and interpretation of inscriptions included three main aspects in this particular project:

- A study of the meaning of these characters or symbols: this aspect required knowledge of relevant ancient manuscripts, and reference to previous research on ancient characters;
- An analysis of the function of the inscriptions, related to the production of weapons: this focused on determining whether the characters indicate the name of craftspeople or a given workshop, and/or whether their presence was related to production and assembling processes, or to quality control procedures;
- Statistical treatment and spatial analysis of the inscribed weapons: this aspect specifically aimed to reveal indirect information regarding the organisation of labour in the production, storing, transportation, and placement of the weapons into the pit.

3.3 Sampling

3.3.1 Principles of sampling

Generally, 'sampling' entails the idea of using information from a part of something to make inferences about the whole (Shennan, 1997: 361). One objective in sampling is to get a representative sample that would allow drawing reliable inferences about a larger population under conditions of limited time and funding resources (Orton, 2000). However, the selection of statistical criteria requires careful consideration, to avoid distortion or misleading results. The aim in this circumstance is to obtain a sample which is an 'honest representation' of the population and which leads to estimates of population characteristics with as great a precision as we can reasonably expect for the cost or effort expended (Barnett, 1974; Shennan, 1997: 362).

3.3.2 Sampling strategy for this project

The collection of weapons from the Terracotta Army is unparalleled in its size and state of preservation. Its size, while offering great potential for research, also poses challenges for sampling. This applies especially to the arrowheads, counting approximately 40,000 pieces, in various size of bundles (from 1 to 200 arrows in each bundle) in Pit 1. Therefore, two key questions were how many weapons needed to be studied in order to properly answer the research questions addressed in this study effectively, and how many weapons would it be feasible to study within the time frame of the research project. The overall sampling strategy is illustrated in Figure 3.4.

For all the weapons except for the arrows, it was decided to study all the preserved items, excluding the few not accessible for this research because they were on display at the museum or elsewhere. The largest group of weapons subject to such a complete study was that of the triggers, comprising just over 200 pieces, each composed of five individual parts. The data collection procedure was rather time consuming, but the wealth of detailed data obtained justified the effort. However, such an approach was not conceivable for the arrowheads.

Concerning the bronze arrows, a pilot study was first conducted on a small number of arbitrarily selected large bundles of arrows (≥90 arrows, for details see Chapter 6). If this showed some consistency within bundles, then a systematic sampling procedure would be applied for the bundles distributed across the pit, with detailed measurements of a number of arrows from each bundle. If the pilot survey showed measurable differences within bundles, larger than those between bundles, then cluster sampling would be used for the arrows. For the small bundle arrows (<90), a random sample would be chosen in any case. The pilot study of the arrow bundles showed that there was comparative consistency within bundles and variability between bundles, both in the metric analyses and in the chemical study by XRF; therefore, a systematic sampling strategy was employed.

3.4 Dimensional measurements and quantitative analysis

3.4.1 Measuring techniques

The detailed measurement of the artefacts represented a starting point for studying standardisation in the weapons' production. The concept of standardisation in mass production refers to the manufacture of nearly identical artefacts for purposes of efficiency in the production process and the manufacture of equivalent, substitutable elements for later use. The degree of standardisation can be assessed based on the dimensions of the bronze weapons using statistical methods. In addition, metric data can be employed to differentiate consistently between different groups of artefacts that could conceivably derive from different moulds or workshops.

The extensive campaign of measurements was extremely time consuming and tedious at times, but the results were

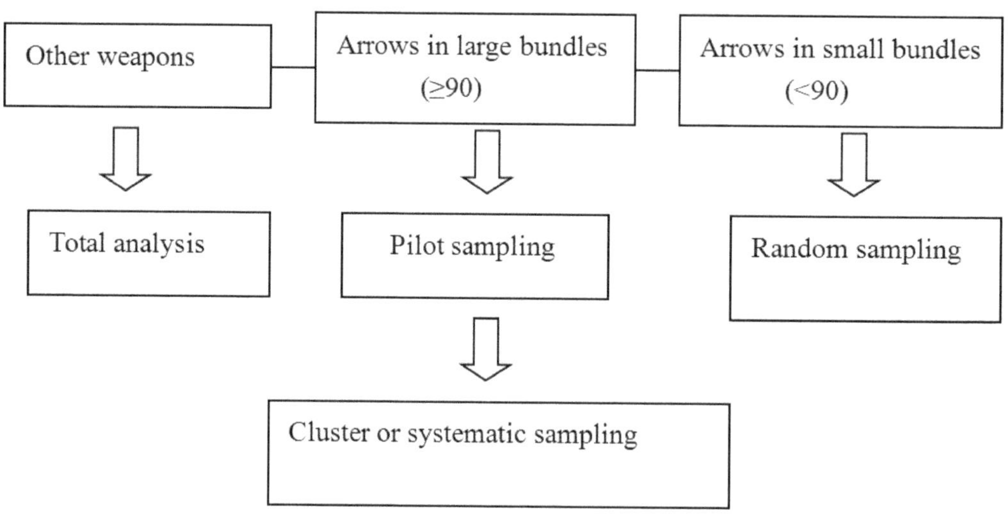

Fig. 3.4. Sampling strategy for the bronze weapons.

often rewarding. Using triggers as an example, over 200 have been measured for this project, and each trigger has three mechanical parts and two bolts. Three-dimensional measurements were carried out for each mechanical part, and two-dimensional ones for each bolt. The process involved taking scaled photos *in situ* in China, followed by rectification with the ArcGIS georeferencing software in London. The digital measurement and the archiving of the data corresponding to over 2,000 measurements were carried out in three months.

Considering the specific characteristics of each type of weapon, such as shape, volume, and size, as well as the practical limitations of their study *in situ*, the measuring technique was adapted to each weapon type. Because the bronze triggers comprise three mechanical parts and two bolts, each with a different shape and size, and as most of the triggers are still assembled together, a normal ruler or digital callipers were difficult to use to obtain proper measurements. As a result, digital measurement with GIS georeferencing software was employed. For the arrows, following the pilot study mentioned above, six arrows were randomly selected from each bundle containing ≥ 90 arrows (for details, see Chapter 6); these arrows were photographed at scale and measured digitally using Adobe Photoshop software. Ferrules were relatively easy to measure for their length and diameters, while they were more difficult to photograph accurately because of their volume; digital callipers were directly used for recording their dimensions, and both digital and hard copies of the data were saved.

Each method mentioned above has its benefits and limitations. ArcGIS georeferencing is a more accurate, convenient, and amendable method for measuring each part of the triggers on a computer. Because most of the triggers are slightly corroded and the mechanical parts are difficult to separate, the measuring process was more challenging. Georeferencing software made it possible to measure them based on the photographs taken in the museum, but at the expense of investing much more time in this procedure. Step 1 involved taking high-quality scaled photographs in the storage of the museum. Step 2 involved rectifying the images in order to integrate them with defining coordinates so that any new data could be positioned and scaled in relation to the existing data; normally, each image took 15–20 minutes to rectify. Step 3 involved measuring each part of the triggers and inputting the data into an Excel spreadsheet.

Measuring arrows with the Adobe Photoshop software is relatively easy and efficient, and amendable on the computer as well, using the scale in the software and then moving or enlarging the picture to measure a specific dimension. However, the two methods mentioned above are both based on the clear images taken of the artefacts. For the ferrules, the measuring process was carried out directly on the actual objects rather than pictures. This meant that much more time was spent with the actual weapons, requiring cooperation with the museum staff, who had to stay in the storage until all the ferrules were measured.

3.4.2 Metric variables

The dimensions selected for measurement obviously varied for each type of weapon, depending mostly on their shape. The metric variables selected for each type of artefact will be detailed and illustrated in the relevant chapters, but they are briefly introduced here. In the arrows, the length of the tang and both the width and length of the arrowheads were selected for measurements, but the tang diameter was not measured because it often varied even within the same tang. The lances were measured on seven dimensions, including overall length, blade length, width and thickness, and handle length, width, and thickness. For the ferrules, the measurements recorded the length, top and bottom diameters, and their wall thickness. For the triggers, three dimensions were measured on each of the three mechanical parts, and two dimensions on the bolts. However, the more complex shape of the triggers dictated that the specific axes for measurements had to be georeferenced to ensure precision and comparability between analyses (for details see section 5.3.3).

3.4.3 Accuracy and precision assessment

Establishing the accuracy and, especially, the precision of the weapons' measurements was an essential step in order to test the validity of this approach before embarking on further statistical analyses. In particular, given the emphasis that this research places on the metric variability (or lack thereof) between artefacts of the same type, it was critical to quantify the extent to which metric variability could be just a result of the limitations inherent to the measuring method and/or the person measuring the objects. Accuracy refers to the degree of 'truthfulness' of a measurement, i.e. the extent to which a measured value corresponds to the actual value, whereas precision is the degree to which repeated measurements under unchanged conditions show the same results. A measurement can be accurate but not precise, precise but not accurate, neither, or both. Ideally, a measurement is both accurate and precise with measurements all close to and tightly clustered around the known value. While the accuracy of the measurements could be relied upon since computer-based and/or digital equipment was employed in all cases, precision had to be tested more carefully through repeated measurements of the very same dimensions on the same weapons.

Reassuringly, precision was found to be very high, as the following example demonstrates. Three triggers were randomly selected, and two dimensions of as many parts were measured repeatedly. The results are presented in Tables 3.1 and 3.2, with summary statistics in Table 3.3. The first five measurements of each dimension were taken after individually rectifying the image five times with ArcGIS, in order to test the reproducibility of the rectification process. The last ten measurements were taken after the fifth rectification, to assess the analytical precision of the

Table 3.1. Test measurements on three random selected triggers: (a) 14 measurements on trigger part B1; (b) 14 measurements on trigger part B2. All values in cm.

(a)

Trigger ID	rect1	rect2	rect3	rect4	rect5.1	rect5.2	rect5.3	rect5.4	rect5.5	rect5.6	rect5.7	rect5.8	rect5.9	rect5.10
2234_B1	8.057869	8.058589	8.053574	8.059678	8.052288	8.052251	8.052288	8.052517	8.052288	8.05219	8.044134	8.052222	8.052251	8.060139
902_B1	8.569663	8.560801	8.549029	8.563039	8.568631	8.559701	8.568645	8.568743	8.568743	8.568701	8.568743	8.559868	8.568743	8.568701
2271_B1	7.967912	7.964624	7.963513	7.968608	7.962646	7.967699	7.961174	7.970436	7.960242	7.961655	7.961174	7.963675	7.962146	7.961174

(b)

Trigger ID	rect1	rect2	rect3	rect4	rect5.1	rect5.2	rect5.3	rect5.4	rect5.5	rect5.6	rect5.7	rect5.8	rect5.9	rect5.10
2234_B2	5.023442	5.017085	5.018681	5.016966	5.016441	5.016487	5.016487	5.016487	5.008291	5.024525	5.016441	5.016487	5.016546	5.024525
902_B2	5.250404	5.245055	5.244324	5.216209	5.230501	5.230432	5.230463	5.230501	5.230463	5.230554	5.239427	5.239548	5.230623	5.221629
2271_B2	5.011712	5.012868	5.005922	5.007758	5.007763	5.010287	5.011159	5.009431	5.011159	5.00241	5.010287	5.011159	5.012047	5.010287

Table 3.2. Test measurements on three random selected triggers: (a) 14 measurements on trigger part C1; (b) 14 measurements on trigger part C2. All values in cm.

(a)

Trigger ID	rect1	rect2	rect3	rect4	rect5.1	rect5.2	rect5.3	rect5.4	rect5.5	rect5.6	rect5.7	rect5.8	rect5.9	rect5.10
2234_C1	5.784274	5.770668	5.769143	5.778719	5.788017	5.786497	5.788017	5.789813	5.785338	5.781975	5.778996	5.778996	5.788314	5.775637
902_C1	5.829257	5.830776	5.825957	5.838628	5.830757	5.82757	5.833928	5.828132	5.822346	5.823436	5.833928	5.829215	5.827057	5.825991
2271_C1	6.356994	6.355188	6.346543	6.349665	6.35683	6.35341	6.356939	6.349929	6.35341	6.355169	6.355534	6.350111	6.350111	6.35341

(b)

Trigger ID	rect1	rect2	rect3	rect4	rect5.1	rect5.2	rect5.3	rect5.4	rect5.5	rect5.6	rect5.7	rect5.8	rect5.9	rect5.10
2234_C2	2.773852	2.788586	2.785036	2.781909	2.782553	2.771096	2.769534	2.775837	2.788667	2.775837	2.770246	2.777304	2.775837	2.782347
902_C2	2.825872	2.831735	2.825004	2.838932	2.848958	2.847124	2.836124	2.834248	2.830202	2.839336	2.844499	2.831559	2.839336	2.827356
2271_C2	3.262287	3.261098	3.279944	3.254636	3.269862	3.259949	3.269282	3.261088	3.268726	3.268194	3.269282	3.261692	3.268726	3.269862

measurements themselves. After each measurement, pre-existing reference points were deleted and then selected again, in order to maximise the chances of error and thus assess precision in the worst-case scenario.

The standard deviations of the measurements for each axis were two orders of magnitude lower than 1 mm (0.003–0.009 cm), resulting in coefficients of variation (CVs) which never exceed 0.3% (generally much lower). While they remained systematically low, the highest relative precision errors were recorded in the measurements of C2 (the shortest axis measured, at ca. 3 cm), whereas the best values were recorded in repeated measurements of B1 (the largest one, at ca. 8 cm). However, absolute errors remained comparable in all cases. This is a reflection of the advantages of carrying out computer-based measurements: as the images can be enlarged or reduced as required to facilitate the measurements, the absolute error is equally small irrespective of the size of the variable measured; the slightly higher relative error is simply a reflection of the smaller size of the variable measured.

In this research, all the arguments founded on comparisons between CVs and other metric parameters are based on metric differences and CVs much higher than those recorded during the precision tests. Hence, the validity of the approach and the statistical significance of any patterns noted can be confidently asserted.

3.4.4 Statistical analysis

3.4.4.1 Cluster analysis

The statistical treatments of the different dimensional data varied depending on the complexity of the data and the number of variables recorded. Cluster analysis, principal

Table 3.3. Standard deviation and coefficient variation of the above 14 measurements for each trigger dimension.

Trigger ID	MEAN (cm)	STDEV	CV (%)
2234_B1	8.05	0.004	0.05
902_B1	8.56	0.005	0.06
2271_B1	7.96	0.003	0.04
2234_B2	5.01	0.004	0.08
902_B2	5.23	0.009	0.17
2271_B2	5.01	0.002	0.05
2234_C1	5.78	0.006	0.11
902_C1	5.82	0.004	0.08
2271_C1	6.35	0.003	0.05
2234_C2	2.77	0.006	0.23
902_C2	2.83	0.007	0.26
2271_C2	3.26	0.006	0.19

component analysis (PCA), frequency histograms, and coefficients of variation were the main statistical treatments, graphic procedures, and variables employed for further analysis and comparison.

In its most basic application, cluster analysis provides an easy way to assess similarities or dissimilarities between objects based on a large number of numerical variables, such as their dimensions. Each object can be thought of as a point in a space, with objects that are more similar appearing closer in the dendrogram, and object that are less similar further apart (Orton, 2004). The aim in clustering classification is generally to discover the pattern of groupings in a set of data, with as few assumptions as possible about the nature of the grouping (Gordon, 1981; Shennan, 1997: 220). It is the archaeologist's job to infer whether the structure in the dataset has archaeological significance. For this project, clustering analysis was used to present the groupings of the bronze weapons based on their measurements.

3.4.4.2 Principal Component Analysis

Principal component analysis was also used to process the multidimensional measurements of the bronze weapons carried out in this project. The potential applications of PCA are similar to those of cluster analysis. However, while cluster analysis facilitates the ascription of each artefact to a given group, PCA makes it possible to determine which variable or set of variables contribute most to the differences between the objects under study. Generally, PCA treatment transforms all the variables in the dataset into a new set of uncorrelated variables. The most important structure-bearing variables are simplified in the first principal component (PC), with further variables compared on the second PC, and so on. There are as many PCs as there are variables, but it is often found that most of the variability is accounted for by only the first few principal components. As this is a mode of analysis which considers objects and variables together, we can see more directly how the two are related (Shennan, 1997: 266).

Some measurements of a few weapon types only include one or two dimensions that could be easily compared using scatterplots, while the triggers, ferrules, and lances were measured in several dimensions. In the case of the bronze crossbow triggers, PCA was very useful in making it possible not only to identify the subgroups within each of the trigger parts, but also to study how different subgroups were assembled together into finished triggers.

3.4.4.3 Frequency histogram

A frequency histogram is one of the common graphical tools used to describe a single population, and to show the frequency of elements that occur within a certain range of values – a graphical representation of a single dataset, tallied into classes. Frequency is defined as the number of values that fall into each class. The histogram consists of a series of rectangles whose widths are defined by the limits of the classes, and whose heights are determined by the frequency of items in each interval. Histograms can depict many attributes of the data, including location, spread, and symmetry.

The bronze arrows considered in this project appeared in bundles that varied in size from a single arrow to over 100 in one bundle. Frequency histograms were used to present the arrows' distribution in bundles, and provided a basic estimate of the original number of arrows in each bundle. This graphical tool was also used to explore for the presence of modes and assess the standardisation of arrowheads and tangs (see Chapter 6).

3.4.4.4 Coefficient of Variation

The coefficient of variation has been widely used for assessing the degree of standardisation (Eerkens and Bettinger, 2001; Eerkens and Lipo, 2005). In order to obtain a CV, a standard deviation is divided by the mean and multiplied by 100 (Eerkens and Bettinger, 2001). As dimensionless variables, CVs make it possible to compare different databases or dimensions irrespective of their size. For example, we can compare the CV of a set of data pertaining to lance thickness to that corresponding to sword lengths in a meaningful way, even if the sword lengths are always significantly larger. Eerkens and Bettinger (2001) have been some of the most prominent scholars arguing that CV is a stable and reliable measure of standardisation and variation. They also provided a baseline of CV values. A CV of 1.7% represents the highest degree of standardisation attainable when artefacts are made by a manual human process, while a CV of 57.7% represents the variation expected when production is random, and accordingly corresponds to the lowest degree of standardisation.

Although CV is a contested method, it appeared relevant in this context for assessing the degree of standardisation of the bronze weapons studied in this project. This method provided some interpretive results for each type of weapon on the standardisation of their production and labour organisation.

The extensive measurements and the statistical treatment discussed above distinguished subtle and undeniable differences within each type of weapon that can be used to define weapon subgroups. Such results suggest that various technological preferences, craftspeople behaviours, and/ or organisational logics may be behind them. However, the spatial patterns of these weapon subgroups can be interpreted with respect to workshop practice, labour organisation for the transport of the weapons, and their placement in the pit. Spatial methods and approaches based on a geographic information system (GIS) will be discussed in the following section.

3.5 Establishing a GIS database and spatial analysis

The database of the bronze weapons including spatial and metric information was designed from the outset to work with a geographic information system. The spatial information for each bronze weapon was derived from the original paper maps published in the primary Excavation Report (Institute and Team, 1988) and from the documents in the museum archives. Digitising the original maps provided the basic information, and further details or corrections were made according to the detailed information in the archaeological archive (accessed in the Conservation Department of the Museum).

3.5.1 Original map

The two original paper maps, drawn in the 1970s, provide the locations of the terracotta warriors and horses in the five easternmost trenches in Pit 1. The distributions of the terracotta warriors and the bronze weapons are correlated, with the warriors lined in each of the corridors, followed by wooden chariots and horses, forming a battle formation. The bronze weapons were originally held by the warriors or laid on the floor to match this battle strategy.

There are three front lines of warriors facing east, as part of the vanguard. At the back of the vanguard, the pit is divided into 11 corridors; two flank corridors, east and west, include two rows of warriors, and each corridor in the middle comprises 4 rows of warriors. The terracotta figures are also distinguished according to their ranks of generals, officials, and common soldiers. Some of the warriors wear armours, while the others are in robes (Figs. 3.1, 3.2, and 3.3). Most of the features are recorded in the maps provided by the original publication, but the posture of each warrior is something that has been reinvestigated as part of this project.

The original map of the weapons was made at 1:200 scale and included all of the types of weapons in the five easternmost trenches of Pit 1.

However, the paper maps only provide the general distribution of warriors and weapons. In order to establish a GIS database and carry out spatial analysis, it was necessary to digitise these maps and cross-check the locations according to the documents preserved in the museum archives.

3.5.2 Digital map

The creation of a digital map was a preliminary task of my research. Many new features not present in the original paper map were added as vector datasets on a local grid. The locations of the bronze weapons and the terracotta warriors are stored as point data, while the corridors, walls, and excavation trenches were digitised as polygons. The centre-points of each arrowhead bundle were then buffered so that within this area, randomly-located points for each individual arrowhead could be created to facilitate object-specific analysis, where necessary. This spatial framework then makes it possible to conduct formal point pattern analysis to consider the spatial structure of the bronze weapons and their relationship to the terracotta warriors. In particular, a pair correlation function was chosen as the standard method of spatial statistical assessment and this will be discussed in the following section.

3.5.3 Using a local grid

The digital map of the warriors and weapons adopted a local grid system with x and y coordinates, both because there are at present very few national coordinate references for the pit and because the collection of further absolute Ground Control Points (GCPs) is not currently allowed, especially for projects that are partly conducted abroad.

3.5.4 Spatial statistics – pair correlation functions

Spatial statistics are a main concern for this project. A range of multi-scalar methods for assessing point patterns is now available but a pair correlation function is comparatively new and arguably a more powerful method among these (Ilian et al., 2008: 218–23; Bevan et al., 2013). This method is a density-based statistical analysis that measures patterns of clustering, regularity, or randomness as they manifest themselves at multiple spatial scales. Essentially, this procedure involves placing concentric circles of increasing size around each known point and then measuring the average density of other points that fall within each circular distance band.

Distribution maps have been common in archaeological research for some time, but most archaeologists have considered the quantitative way with cutting edge facilities and software instead of relying on personal intuition for interpreting spatial patterns and possible environmental processes and the human activities behind them. Spatial statistical methods, such as the nearest neighbour analysis and Ripley's K and L functions, have been employed to characterise spatial distributions in more formal, quantitative ways. However, the nearest neighbour analysis only characterises the point pattern at one scale, and what we want to know is whether there are multiple scales present in our distribution and at what exact distances over the pit these are apparent (Orton, 2004; Bevan and Conolly, 2006).

A pair correlation function can be run in a variety of different software programs, but in the present case it was implemented in R. The standard form of graphical presentation used below combines a pair correlation plot with a distribution map to allow for easy comparison.

The utility of a pair correlation function is partly that the observed results can be summarised as an average point density per distance circle and these can then be compared to sets of random expected values derived from a Monte Carlo simulation. When the observed values (a solid line in the figure) fall above or below a probability envelope created by the Monte Carlo simulation, a significant pattern of clustering or regular spacing can be assumed. Normally, if the values are above the envelope, they suggest more clustering, while if the values are below it, they suggest more even or regular groupings.

The spatial patterns random, clustered, and regular of the bronze weapons in the pit encourage formulating assumptions as to production and distribution processes, as well as further cultural meanings that might be associated with them.

3.6 Integration with archaeometric data

As mentioned at the beginning of this chapter, my research is part of a broader cooperative project, but limits itself at investigating standardisation and the labour organisation of the Qin weapon production from an archaeological perspective. The parameters used for this are therefore mainly metric data, interpreted according to additional statistical and spatial models. At the same time, further scientific research is currently being carried out using portable XRF and SEM analysis *in situ* in China to analyse the chemical composition of the weapons.

However, some of the weapons present clear statistical and spatial patterns based purely on their typological and metric data, such as the triggers, while other weapons show fewer statistical and spatial patterns, such as the arrows. In contrast, the archaeometric analysis of the triggers has produced few clear patterns while the results are far more definitive for the bronze arrows. Thus, in one particular chapter – Chapter 6, dealing with arrows – some archaeometric results will be combined with the metric data (also see Marcos Martinón-Torres et al., 2014a).

3.7 Summary

The methods addressed in this chapter comprise both qualitative and quantitative approaches, including functional typology, measurements, statistical treatment, and spatial analysis, combined with archaeometric data to be introduced in a later chapter. In their ensemble, these approaches aim to draw a comprehensive picture of the Qin bronze weapon production and labour organisation.

In fact, detailed measurements and statistical and spatial approaches are essential pre-requisites for answering the research questions raised here, but have never been a major feature of the research on Qin bronze weapons data. Thus, this methodological framework may also be highly relevant to other research agendas for the Qin First Emperor tomb complex. However, while quantitative methods are a powerful methodology for tackling such issues, it is still worth combining them with qualitative methods, such as a consideration of the inscriptional evidence, and it is to the latter that the following chapter turns first.

4

Inscriptions on the Weapons

4.1 Introduction

The inscriptions found on some of the bronze weapons merit interpretation as one unusual aspect that sheds light on the workshop organisation, the production process, and their storage prior to their intended use in life or for funerary purposes. However, not all of the weapons in the tomb complex have inscriptions. For example, no inscriptions were found on the surfaces of the arrows and hooks. Some of the triggers and ferrules are inscribed with simple characters. The swords and spears all have short inscriptions, while each of the lances, dagger-axes, and halberds have long inscriptions describing the date of production, the workshop in which they were made, the officials in charge of the workshops, and the craftspeople involved. These different types of inscriptions therefore tell us a great deal about the standardisation and labour organisation involved in the production of the bronze weapons.

This chapter will begin with an introduction outlining the types of inscriptions recorded, their technological classification, and the sorts of information they contain. The discussion will focus on data from the longer inscriptions, which include direct information pertaining to the dates of manufacture, as well as to some of the workshops and supervisors involved in production and quality control. Other important aspects related to the inscriptions are their patterns of presence/absence and spatial distribution, but these will be discussed in the relevant chapters for each of the weapon types, so that they can be considered together with the other typological and metric attributes documented.

4.2 Types of inscriptions

The inscriptions on the bronze weapons from the pits of the Terracotta Army fall into three categories from a technological perspective: 1) casting inscriptions; 2) carving and/or chiselling inscriptions; and 3) ink inscriptions. Each type of inscription has its own characteristics, and cast inscriptions are found only on the halberds.

4.2.1 Cast inscriptions

Cast inscriptions were mainly found on the handles of the halberds (Fig. 4.1), and are limited to the characters *Sigong* (寺工), which is assumed to be the name of the Qin governmental workshop (Huang, 1983, 1990; Yuan, 1984; Institute and Team, 1988). In the ancient written documents, there are no specific *Sigong* characters which refer to the name of the workshop. However, the characters appear to be related to the characters *Siren* (寺人), an official who served the king or emperor, and whose name was recorded in *Zhouli* (Ritual of the Zhou Dynasty), and to the characters *Sihu* (寺互), which refer to a governmental workshop responsible for producing weapons and vessels in the following Han Dynasty, and was recorded in the *Hanshu* (History of the Han Dynasty; Yuan, 1984; Huang, 1990). Huang (1983, 1990) has also investigated the similarity in Chinese writing between *Sihu* (寺互) and *Sigong* (寺工), and suggests that the Han Dynasty *Sihu* could have originated from the Qin characters *Sigong* (寺工) and that both refer to a workshop producing weapons and other artefacts. However, if the *Sigong* (寺工) characters are followed by a person's name, it means that the person was an official who was in charge of the production of weapons and other artefacts in the workshop (Huang, 1983; Yuan, 1984). This aspect will be discussed later in this chapter. *Sigong* (寺工) was also carved on other bronze weapons and related weapon fittings, such as lances and ferrules, and has also been found on a Qin chariot fitting and on a bronze vessel, called a *Hu*, which were discovered in the Qin capital Xianyang (Yuan, 1984; Huang, 1990).

Cast inscriptions have a long tradition in ancient Chinese bronze production. The earliest cast inscriptions date from a Shang Dynasty (1600–1050 BC) tomb in Anyang. The tomb was discovered intact and belonged to a woman called Fu Hao (Bagley, 1980). The identification of the tomb's occupant was possible because more than sixty of the bronzes it contained were cast with inscriptions bearing Fu Hao's name. Others showed her posthumous name which would have been used when sacrifices were made on her behalf by her sons or nephews. Cast inscriptions do not seem to occur on pre-Anyang bronzes (Bagley, 1980: 178). Even in recent pre-Anyang archaeological discoveries, earlier evidence of cast inscriptions has still not been found in Erlitou (二里头1900–1600 BC) and Erligang (二里冈 about 1620 BC) culture (Institute of Archaeology, 1999).

After the Shang Dynasty, during the following Western Zhou Dynasty (1050–771 BC) and the Spring and Autumn Period (770–476 BC), cast inscriptions were widely used on bronze ritual vessels, ranging from one or two characters to a long narrative story (Bagley, 1980; Ma, 1986). One Qin bronze bell, *Qingong Bo*, was cast with a text of 135 characters (Rawson, 2007: 118). It is one of a set of three *Bo* bells which, along with a set of five *Zhong* bells, was excavated from the Taigong Temple in Baoji, Shaanxi province. One of the sentences in the text states: *My foremost ancestor has received the heavenly mandate, was rewarded with a residence, and received his state*

(Rawson, 2007: 118). It is thought that Duke Wu (reigned 697–678 BC) had the bells cast during the Spring and Autumn Period (Rawson, 2007), and prophesied that the Qin would take over the Zhou Dynasty.

The technique of cast inscriptions was employed throughout the Bronze Age and continued for a considerable period even after carved inscriptions were introduced on Qin bronze vessels and weapons. Figure 4.1 shows the cast *Sigong* on one side and a long carved inscription on the other side of a Qin halberd.

4.2.2 Carved or chiselled inscriptions

Carving and chiselling were frequently used to make inscriptions on the Qin bronze weapons found in the pits. Carving marks are smoothly straight lines, while chiselled marks are formed with overlapping small triangular impressions. Such inscriptions appear on halberds, lances, triggers, swords, and ferrules. Some characters were made solely by either carving or chiselling, while others were made using a combination of both carving and chiselling to produce different strokes (Li et al., 2011, 2012) (Fig. 4.2).

Carved inscriptions are mainly found among those comprising longer sentences and on some of the halberds and lances. Typically, the lines made to form these characters are smooth and straight. The specific tools used to create these carved inscriptions require further investigation (Fig. 4.3).

The chiselled inscriptions comprise one or more inscribed strokes, which themselves are made up of overlapping marks, each with the overall shape of an elongated triangle (Fig. 4.4). Each line was therefore chiselled step by step, using a finely-pointed tool ending in a wedge shaped like the letter 'V'. The way in which the marks overlap indicates that the chisels were used to hammer into the bronze and slide across the surface at an angle, starting from the bottom of the cone shape. In other words, the base of the triangle was the starting point of the cut, and

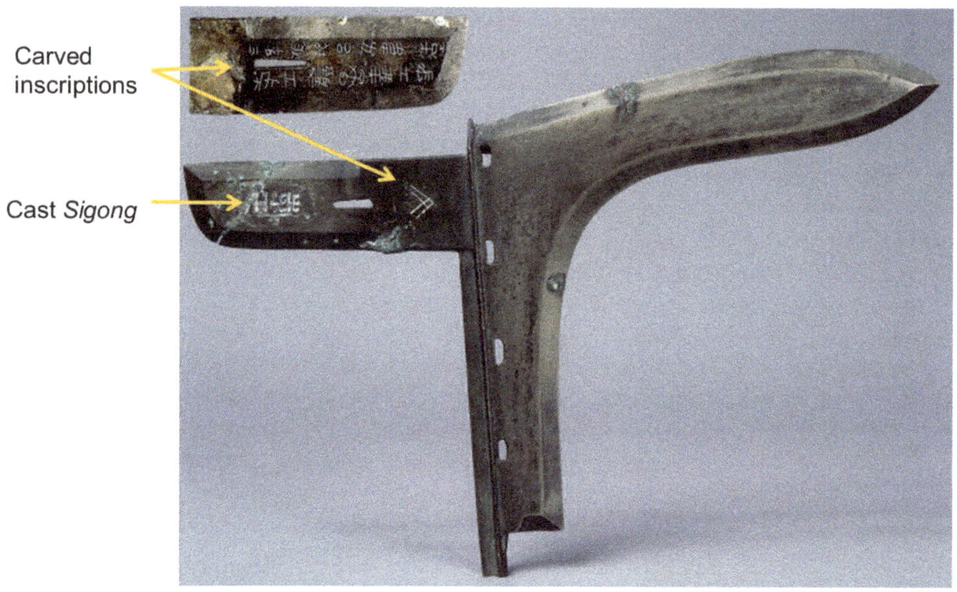

Fig. 4.1. A cast *Sigong* and the carved inscriptions on a dagger-axe.

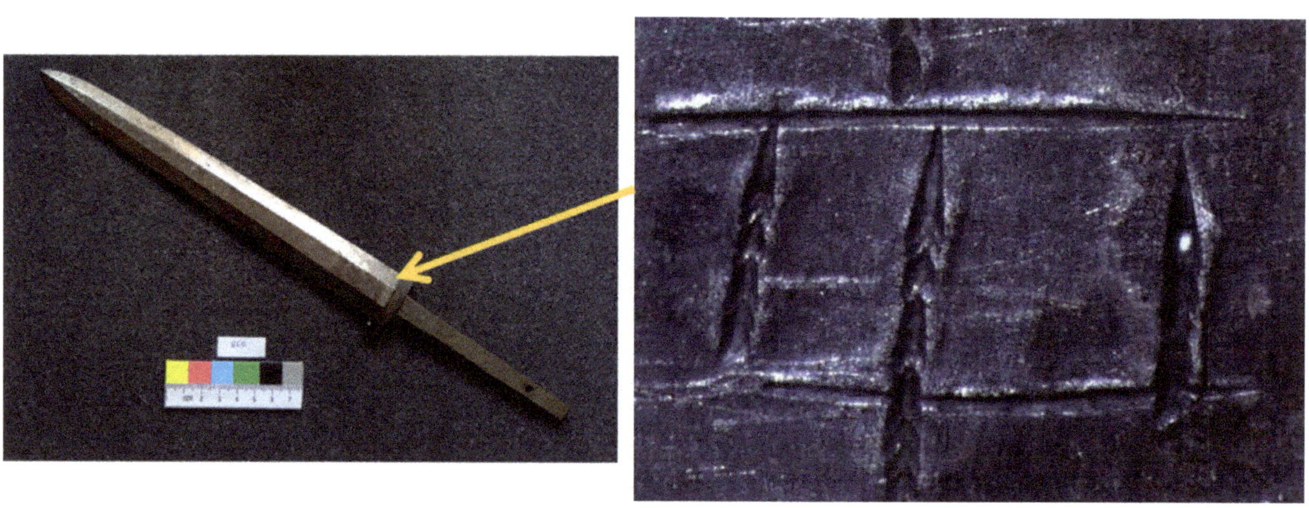

Fig. 4.2. *Sigong* inscription on a lance showing the use of both carving and chiselling techniques.

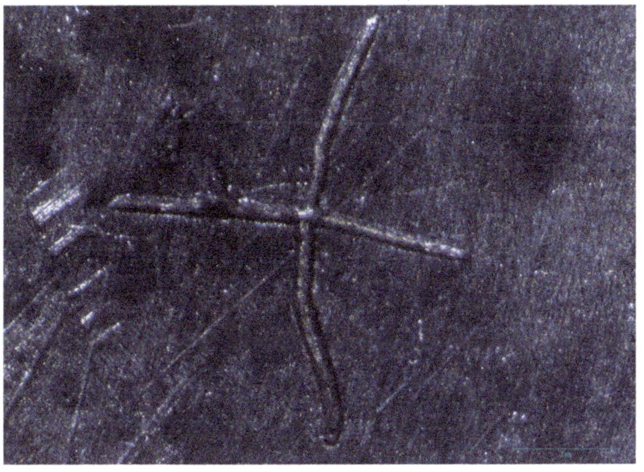

Fig. 4.3. The carved inscription on a lance (ID00860).

the sharp end of the mark was the finishing point (Li et al., 2011, 2012).

Archaeological evidence suggests that carved inscriptions had been used for centuries before the Qin period. A recent excavation at Meiyuanzhuang, Anyang, Henan province, found two Shang ritual vessels bearing single carved marks (Yang, 2000: 100). Further investigations are required to establish whether these marks were intentionally inscribed rather than being unintentional scratches. However, if they were deliberately carved, then these vessels are the earliest inscribed artefacts discovered to date.

Bronze dagger-axes (Qin Zi Ge 秦子戈), made during the Spring and Autumn Period (770–476 BC) and unearthed from Dabushan, Gansu province, also bear carved inscriptions (Zhao, 2008), providing a further precedent for the carved inscriptions on Qin bronze weapons.

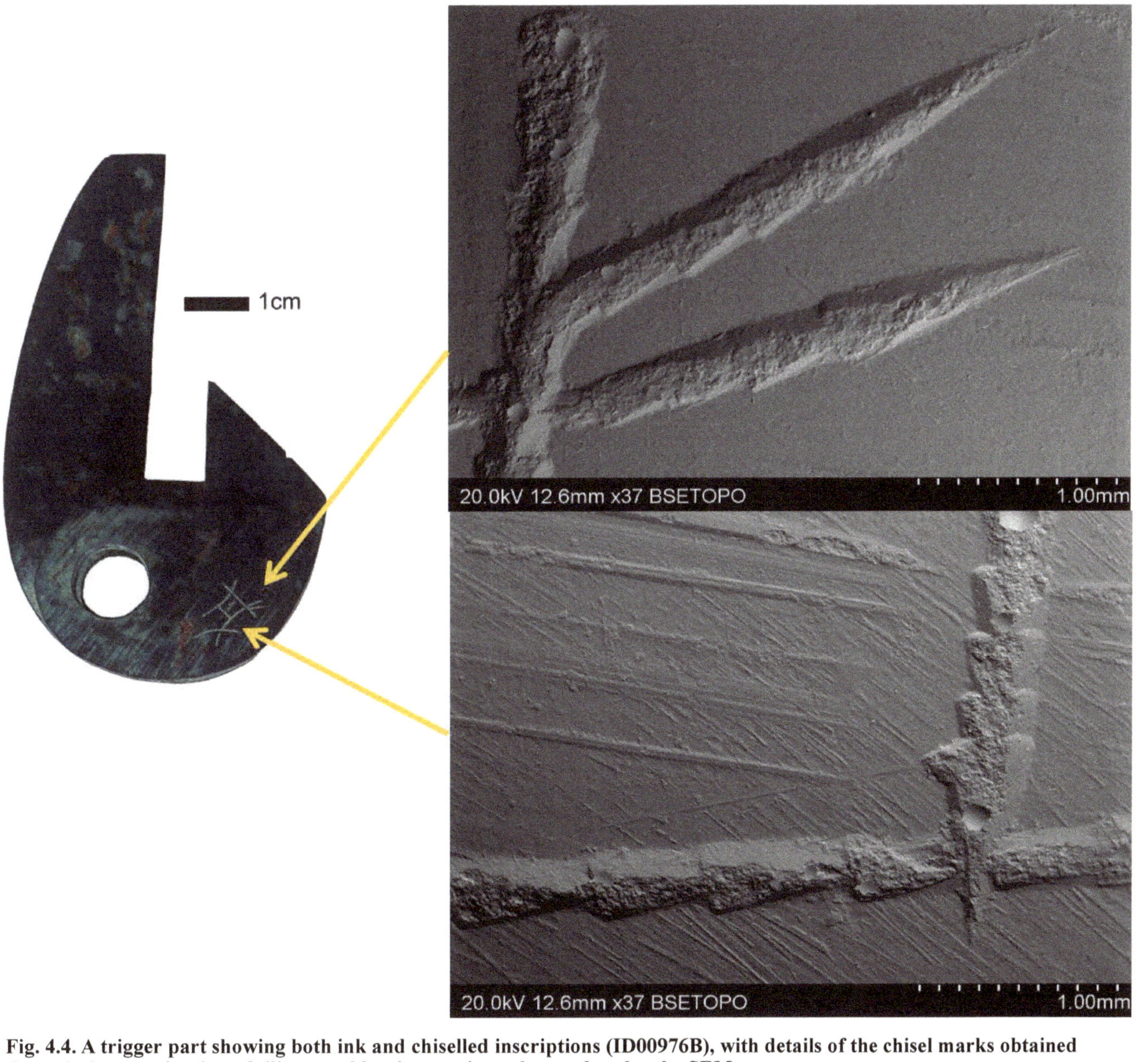

Fig. 4.4. A trigger part showing both ink and chiselled inscriptions (ID00976B), with details of the chisel marks obtained through the examination of silicone rubber impressions observed under the SEM.

Carved or chiselled inscriptions required hard, sharp tools to create the lines on the surface of the bronze. During the Qin period, the fact that carved and chiselled inscriptions became a dominant form on bronzes may well be related to the development of iron steel, which could be used to make tools of the requisite toughness to carve into bronze (Li et al., 2011). However, these were not the only methods of inscribing used by Qin craftspeople, as ink inscriptions have also been found.

4.2.3 Ink inscriptions

Several ink inscriptions have been found on the bronze triggers. One of the ink inscriptions, *Jia* (甲), is to be found on trigger ID2238 (Fig. 4.5). This character is in black, and seems to have been written using a traditional Chinese writing brush. Each stroke looks smooth and flowing (also see Bevan et al., 2018). The ink inscription *Wu* (武) was inscribed on another trigger (Yuan, 1990: 206). Yuan Zhongyi (1984) also noted in his paper that a combination of a stem-branch character and a number, *Wujiu* (戊九), was found on a trigger (ID00976) from Pit 1 (Fig. 4.6). The number *Jiu* (九) is in ink, while the stem-branch (戊) was made by chiselling (see above, Fig. 4.4). The overlapping traces of the filing marks on the surface show that these two characters were both inscribed after filing, but it is difficult to determine with certainty which inscription was made first.

In addition to these, one *Sigong* inscription was written on a lance scabbard in red (Wang, 1994). The scabbard was unearthed from Pit 1, from the context that also included the terracotta warriors.

Inscriptions written in ink on a *Gui* and six *Ge* bronze objects were unearthed from some early Western Zhou tombs at Beiyao, Luoyang, Henan province (Yang, 2000: 101). This discovery shows that casting was not the only method used to produce inscriptions even as early as the Western Zhou Dynasty (1050–771 BC). The presence of ink and carved inscriptions on some weapons raises the possibility that more such combined inscriptions could have been present, but perhaps the ink has not been preserved.

4.3 The content of the inscriptions and the types of weapons

4.3.1 Inscriptions on dagger-axes, spears, and halberds

The dagger-axe and spear are both separate long weapons with wooden handles; however, the halberd is a compound weapon consisting of a spear point and a dagger-axe blade (for details, see Chapter 7). The inscriptions on the five halberds (with dagger-axes and spears together), two independent dagger-axes, and two other individual spears are listed in Table 4.1. Four of the halberds and one dagger-axe were unearthed from the five easternmost trenches during an excavation in the 1970s (Institute and Team, 1988). The other halberd and one dagger-axe were discovered more recently, during the ongoing excavation of the complex (Jiang and Liu, 2006). As mentioned above, a halberd is composed of a dagger-axe and a spear. Seven spears (five assembled as halberds and the other two unearthed independently) bear the carved *Sigong* inscription on each of their handles. The dagger-axes (five of them assembled with spears, and a further two

Fig. 4.5. Ink inscription on a bronze trigger (ID 002388).

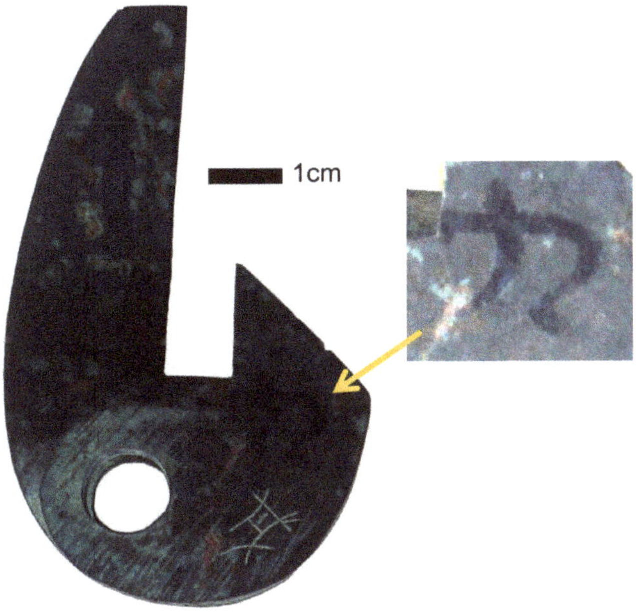

Fig. 4.6. Detail of an ink inscription and a chiselled inscription on a trigger part.

unearthed independently) were carved with long sentence inscriptions on the front-sides and cast *Sigong* inscriptions were found on the rear, with one more carved character alongside (see Table 4.1; the ellipses are used to emphasise the cast characters).

The inscriptions on the halberds provide what is, as discussed above, probably the term for the major central workshop(s): *Sigong*. Unfortunately, there is no indication on the inscriptions as to the location of this workshop or even if it should be thought of as a single workshop space. The rest of the information on the longer inscriptions denotes a form of accountability and hence of quality control (Yuan, 1984). The long sentence inscriptions on these halberds confirm that there was a supervisor called Lu Buwei (吕不韦), officials (workshop *Sigong* followed by a person's name means the official who was in charge of the workshop), craftsmen (*Cheng*丞), and workers (*Gong*, 工), all involved in the bronze production (Institute and Team, 1988; Fig. 4.7). The supervisor Lu Buwei was the chancellor (r. 246–237 BC) of the Qin Kingdom before the unification. An official in the *Sigong* workshop was in charge of the production. The *Cheng* would have been a very experienced worker, responsible for obtaining raw materials, training others in the manufacturing process and

Table 4.1. Inscriptions on the halberds, dagger-axes, and spears.

LocationID	Spear	Dagger-axe		Reference
		back-side	front-side	
T10G6: 00577 Halberd	*Sigong* 寺工	*Sigong* *Zhuo*（左）	Third year (244 BC), Lu Buwei took charge of production, *Sigong Ji*, *Cheng Yi*, and worker *Diao*. 三年相邦吕不韦造寺工詟丞义工窎	(Institute and Team, 1988)
T10G: 00576 Halberd	*Sigong* 寺工	*Sigong* *Wen*（文）	Fourth year (243 BC), Lu Buwei took charge of production, *Sigong Ji*, *Cheng Wo* and worker *Ji*. 四年相邦吕不韦造寺工詟丞我工可戟	(Institute and Team, 1988)
T20G10: 01392 Halberd	*Sigong* 寺工	*Sigong* *Wu* (午)	Fifth year (242 BC), Lu Buwei took charge of production, *Sigong Ji*, *Cheng Yi* and worker *Cheng*. 五年相邦吕不韦造造寺工詟丞义工成	(Institute and Team, 1988)
T19G8: 0710 Halberd	*Sigong* 寺工	Sigong Ren (壬)	Seventh year (240 BC), Lu Buwei took charge of production, *Sigong Zhou*, *Cheng Yi*, and worker *Jing*. 七年相邦吕不韦造寺工周丞义工竞	(Institute and Team, 1988)
T12G5 Halberd	*Sigong* 寺工	Sigong Zi (子)	Seventh year (240 BC), Lu Buwei took charge of production, *Sigong Zhou*, *Cheng Yi*, and worker *Tong*. 七年相邦吕不韦造寺工周丞义工同	(Jiang and Liu, 2006)
Dagger-axe		Sigong	Third year (244 BC), Lu Buwei took charge of production, *Sigong*…(blur) 三年相邦吕不韦造寺工……	(Institute and Team, 1988)
T12G7 Dagger-axe		Sigong Gong (工)	Tenth year (237 BC), *Sigong*, *Cheng Yang*, and worker *Zhao*. 十年寺工丞杨工造	(Jiang and Liu, 2006)
0451 Spear	*Sigong* 寺工			(Yuan, 1990)
0639 Spear	*Sigong* 寺工			(Yuan, 1990)

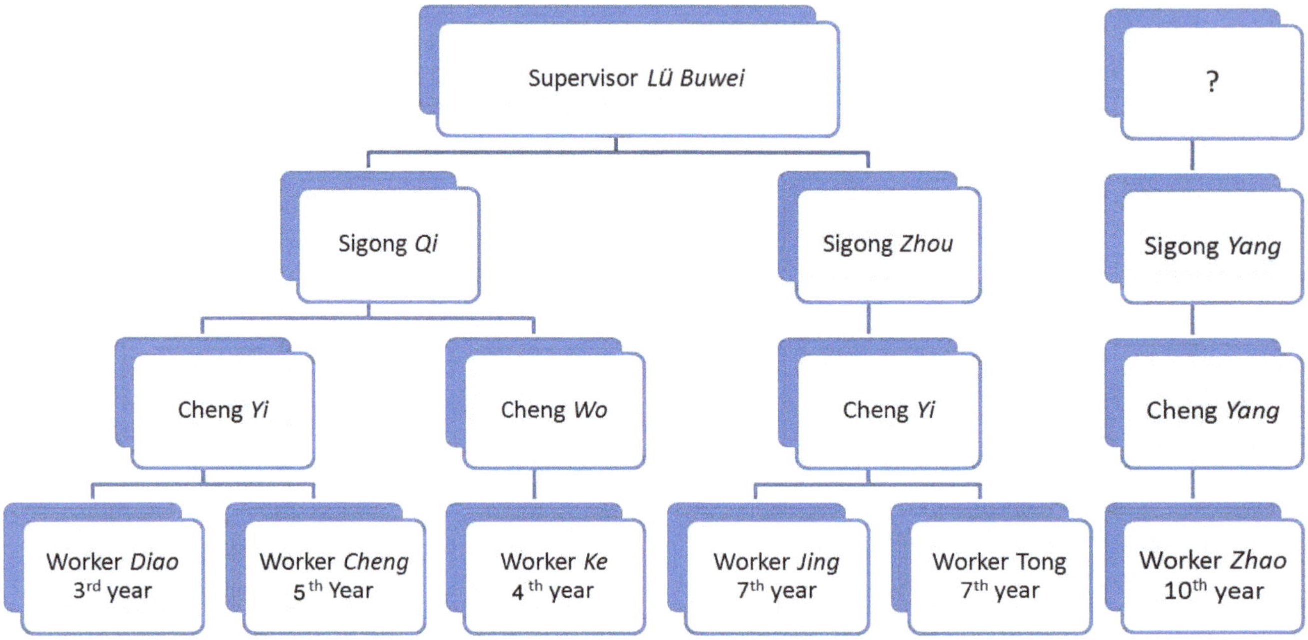

Fig. 4.7. Organisational structure of the production of bronze halberds based on their inscriptions.

technology, monitoring the quality of the bronze weapons, and reporting to the officials. Workers (*Gong*) are the real producers of the bronze weapons (Li et al., 2011).

These workers who actually produced the weapons were usually slaves, convicts, or soldiers based on the bamboo slips discovered in the Qin tombs at Yumeng (Shuihudi Xiaozu, 1978; Yumeng Bianxiezu, 1981; Yuan, 1984). One sentence found on a Shuihudi bamboo slip, *Jungong* (均工), notes that 'a slave with skills can be a *Gong*, a worker in the workshop' (Shuihudi Xiaozu, 1978; Yuan, 1990: 200). One more sentence from a Shuihudi bamboo slip, *Junjulu* (军爵律), records the fact that if a slave or convict won military awards, he could be a Gong in the workshop and avoid any further punishment (Shuihudi Xiaozu, 1978; Yuan, 1990). According to the information from the Qin bamboo slips, some of the slaves, convicts, and soldiers served as skilful worker in the workshop to some extent.

In addition, the long inscriptions on these weapons indicate the regnal year in which they were produced, thus providing basic information regarding the chronology of production. Based on this information, the halberds can be dated from 244 to 237 BC (from third to tenth regnal year), that is, before the unification of the Qin Empire, which took place in 221 BC (in the twenty-sixth year of the reign of Yingzheng, King of Qin Kingdom). The halberd with an inscription from the tenth year is different from the other six, as the supervisor, chancellor Lu Buwei (吕不韦), is not mentioned. He presided over the Qin State when the First Emperor was young, but was condemned for his connection with a rebellion and dismissed in 237 BC, the same year this halberd was made. In contrast, the lances discussed in the following section were clearly made after 237 BC. As such, they do not include mentions of chancellor Lu Buwei (吕不韦) in their inscriptions either. The fact that the chancellor of Qin was named as the supervisor indicates the importance of his political influence over the production of these weapons. The inscriptions were strongly linked with political changes taking place during the Qin era. This high level of influence over the production of the bronze weapons began with Shang Yang (385–338 BC), who was discussed in the first chapter. He carried out a series of military and political reforms in the Qin Kingdom, and some of the bronze weapons were inscribed with his name during the period when he was chancellor.

The function and meaning of the other six individual carved characters (see Table 4.1) on the rear of the halberd handles, alongside the cast *Sigong*, remains unclear. They might be related to the name of the arsenal in which the halberds were stored (Yuan, 1984; Huang, 1990) or to the person who owned them.

4.3.2 Inscriptions on lances

In total, 16 lances bearing inscriptions were found in the five easternmost trenches. Long inscriptions were carved on the spines, providing the regnal year of the production, from 231 to 227 BC (from fifteenth to nineteenth regnal year), the *Sigong* official who was in charge of the production, and a worker's name. The handles of the lances bear one or more Chinese characters whose function and meaning remain unclear. On some of the lances' fittings, called *Ge,* a carved *Sigong* inscription was also found (for details see Table 4.2).

The inscriptions on the spines of the lances are much simpler than those found on the halberds. The names Lu Buwei and *Cheng* are not found on the lances, and only the

Table 4.2. Inscriptions on the lances and their fittings.

LocationID	Spine	Bronze Handle	*Ge* Fittings	Reference
T2G3:0450	Fifteenth (231 BC) year *Sigong Wen* and worker *Hei* 十五年寺工魰造工黑 *Sigong*寺工	丙，左，戊六	*Sigong* 寺工	(Institute and Team, 1988)
T2G3:0463	Fifteenth (231 BC) year *Sigong Wen* and worker *Diao* 十五年寺工魰工窎 *Sigong*寺工	十六子	*Sigong* 寺工	(Institute and Team, 1988)
T20G9:0810	Fifteenth (231 BC) year *Sigong Wen* and worker *Diao* 十五年寺工魰工窎 *Sigong*寺工	五，戊三， 左	No inscription	(Institute and Team, 1988)
T2G3:0448	Sixteenth (230 BC) year *Sigong Wen* and worker *Hei* 十六年寺工魰造工黑 *Sigong*寺工	子　　十三	*Sigong* 寺工	(Institute and Team, 1988)
T2G2:0397	Seventeenth (229 BC) year *Sigong Wen* and worker *Diao* 十七年寺工魰造工窎 *Sigong*寺工	子，五，九	*Sigong* 寺工	(Institute and Team, 1988)
T2G3:0444	Seventeenth (229 BC) year *Sigong Wen* and worker *Diao* 十七年寺工魰造工窎 *Sigong*寺工	左，四工	*Sigong* 寺工	(Institute and Team, 1988)
T2G2:0395	Seventeenth (229 BC) year *Sigong Wen* and worker *Diao* 十七年寺工魰工窎 *Sigong*寺工	子，五，二	*Sigong* 寺工	(Institute and Team, 1988)
T2G2:0400	Seventeenth (229 BC) year *Sigong Wen* and worker *Diao* 十七年寺工魰工窎 *Sigong*寺工	寺工，左	*Sigong* 寺工	(Institute and Team, 1988)
T2G2:0390	Seventeenth (229 BC) year *Sigong Wen* and worker *Diao* 十七年寺工魰工窎 *Sigong*寺工	子壬五	*Sigong* 寺工	(Institute and Team, 1988)
T20G10:0882	Seventeenth (229 BC) year *Sigong Wen* and worker *Diao* 十七年寺工魰造工窎 *Sigong*寺工	子五，丁十	No inscription	(Institute and Team, 1988)
T20G9:0791	Eighteenth (228 BC) year *Sigong Wen* and worker *Diao* 十八年寺工魰工窎	五三	Without *Ge*	(Institute and Team, 1988)
T2G2:0401	Nineteenth (227 BC) year *Sigong Bang* and worker *Mu* 十九年寺工邦工目	左，六	*Sigong* 寺工	(Institute and Team, 1988)
T20G9:0829	Nineteenth (227 BC) year *Sigong Bang* and worker *Mu* 十九年寺工邦工目 *Ten* 十	子乙六	No inscription	(Institute and Team, 1988)
T2G2:0396	Nineteenth (227 BC) year *Sigong Bang* and worker *Mu* 十九年寺工邦工目	八十七	*Sigong* 寺工	(Institute and Team, 1988)
T2G3:0445	Nineteenth (227 BC) year *Sigong Bang* and worker *Mu* 十九年寺工邦工目	左，八	*Sigong* 寺工	(Institute and Team, 1988)
T2G2:0398	Nineteenth (227 BC) year *Sigong Bang* and worker *Mu* 十九年寺工邦工目	子，六二	*Sigong* 寺工	(Institute and Team, 1988)

魰

Sigong official's name and workers' names are inscribed, in agreement with the fact that the lances were made after 237 BC when Lu Buwei was dismissed (Fig. 4.8).

A *Sigong* character was inscribed on most of the *Ge* fittings (Table 4.2). Among the sixteen lances, one does not have a *Ge* fitting, and a further three *Ge* fittings were not inscribed. The remaining twelve *Ge* fittings bear the inscription *Sigong*.

In addition to the long inscription carved on the spine and the *Sigong* inscribed on the *Ge* fittings, inscriptions were also found on the handles of lances. These vary from two to six characters, mainly numerical or relating to dates set out in traditional Chinese time.

Based on the inscriptions on the halberds and lances, two different periods can be discerned, one corresponding to the production of the halberds and another to that of the lances. These two timespans do not overlap, and the production dates make it clear they were made after Ying Zheng (the name of the Emperor Qin Shihuang) came to the throne of the Qin Kingdom, but before the unification, as the following chart demonstrates (Fig. 4.9):

Based on the information derived from the long inscriptions, it is possible to suggest a workshop organisation and supervision system that would have developed over time. According to these data, the workshop would have moved from a structure involving four levels in the organisation of production when the halberds were produced down to an organisation on two levels during the period when the lances were produced. The halberds were made between 244 and 240 BC, with the exception of one made in 237 BC. The halberds produced before 237 BC were all made under the supervision of Chancellor Lu Buwei, a *Sigong*

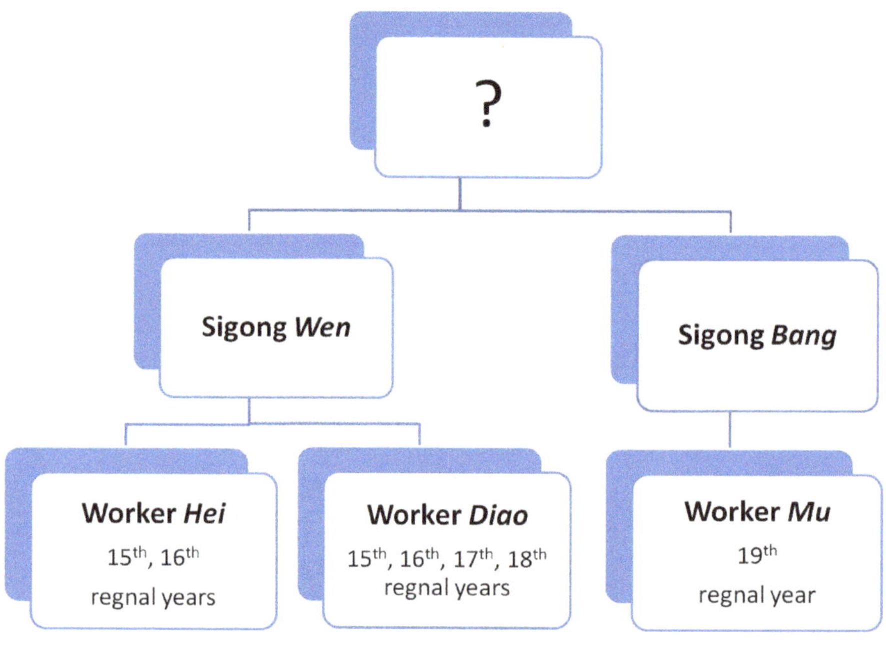

Fig. 4.8. Organisational structure of the production of bronze lances based on their inscriptions mentioned in Table 4.2.

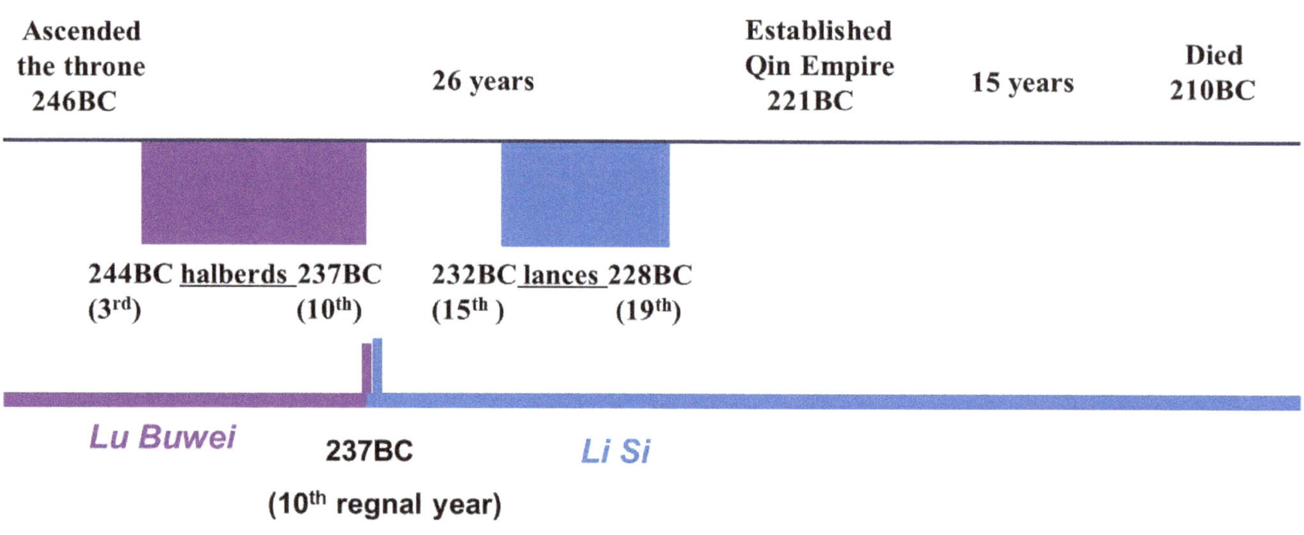

Fig. 4.9. Chronology of the bronze halberds and lances, based on the regnal years recorded in their inscriptions mentioned above.

official, *Cheng*, and a worker. However, the inscription on the halberd made in 237 BC shows only three levels in the organisation of production: *Sigong*, *Cheng*, and worker. This suggests that if a reduction in the chain of authority did occur, it started around this time and was accelerated once Lu Buwei was removed. The inscriptions on the lances, mainly produced in the five years following 237 BC, show only two levels of organisation: *Sigong* and worker. The changes in workshop organisation may therefore reflect political events during the Qin period, indicating the links between the weapons production and the Qin government. The following chancellor's name was not carved on the weapons until Emperor Qin Shihuang died. The name of the new chancellor, Li Si, started to be carved on a halberd produced in the 1st regnal year of the Qin Second Emperor (209 BC), and the workshop was then referred to as *Yueyang* rather than *Sigong* (Jiang and Liu, 2006).

4.3.3 Inscriptions on swords

The bronze swords bear numbers and stem-branch characters mainly on their bronze handles, and include the characters four, six, seventy-seven, eighty-eight, fifty-eight, one, two, five, and a single stem-branch character *Ren* (壬).

Swords and hooks were short weapons for close combat in ancient warfare. Altogether, 17 swords and 2 hooks were discovered in the five easternmost trenches of Pit 1 (Institute and Team, 1988), but they have not been included in the bulk of the analysis presented in this research because they were not available for measurement and first-hand inspection. The two hooks were in the showcase and on loan respectively, and several swords were sealed in conservation boxes that do not allow access for measurement, but their spatial distribution is shown in the previous chapter.

4.3.4 Inscriptions on triggers

Trigger inscriptions are much more complicated in their diversity and location than those found on the other weapons. The distinction partly derives from the fact that triggers are composed of five parts assembled together. Not all of the parts or triggers bear inscriptions. Among the 229 triggers considered in this project, 160 (about 70%) are inscribed and 69 (about 30%) have no inscriptions. Over 70 different inscriptions have been identified, and they were chiselled on the large, flat surfaces or on the edges of the trigger parts. Sometimes the same character is found on each part of one trigger, or on a different trigger. The inscriptions include Qin characters, numbers, symbols, and stem-branch characters.

The stem-branch or Gan-Zi (干支) character refers to the traditional Chinese heaven stems (10 characters) and the earth branches (12 characters), which was used as a time-recording system that originated during the Shang Dynasty (1600–1050 BC) on oracle bones which were used for divination. The names and sequences of heaven stems and earth branches are listed in the following tables (Tables 4.3 and 4.4).

Each heaven stem is paired with an earth branch to form the Gan-Zhi sexagenary cycle that starts with Jia-Zi and ends with Gui-Hai (Table 4.5).

The stem-branch was not only used to record the years, but also to calculate the months, days, and hours. For example, the 12 earthly branch characters were employed both to identify the year using animals and to indicate every two-hour period during each day, starting from midnight (Fig. 4.10).

The numbers, symbols, and stem-branch characters were chiselled on the assembled parts of the triggers. Altogether, 82 different characters were found on the triggers. Some of them appear only once, while others were chiselled on each part of one or more triggers. Inscription codes have been created to facilitate their study (Table 4.6).

These 82 characters can be classified into five types:

Type I: possibly refers to the workshop 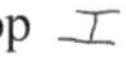

Type II: numbers 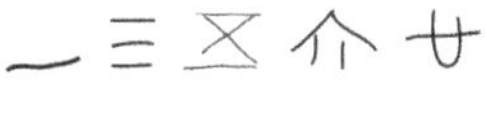 (one, three, five, six, twenty, etc.)

Type III: stem-branch characters 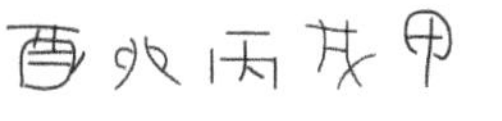

Type IV: Chinese words 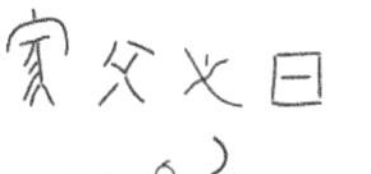

Type V: symbols 

A trigger mechanism comprises three parts and two bolts, respectively identified as parts A, B, C, D, and E (Fig.

Table 4.3. Ten heaven stems.

1甲	2乙	3丙	4丁	5戊	6己	7庚	8辛	9壬	10癸
Jia	Yi	Bing	Ding	Wu	Ji	Geng	Xin	Ren	Gui

Table 4.4. Twelve earth branches.

1 子	2 丑	3 寅	4 卯	5 辰	6 巳	7 午	8 未	9 申	10 酉	11 戌	12 亥
Zi	Chou	Yin	Mao	Chen	Si	Wu	Wei	Shen	You	Xu	Hai

Table 4.5. The sexagenary cycle based on the stem and branch combination.

1 – 10	11 – 20	21 – 30	31 – 40	41 – 50	51 – 60
Jia Zi	Jia Xu	Jia Shen	Jia Wu	Jia Chen	Jia Yin
Yi Chou	Yi Hai	Yi You	Yi Wei	Yi Si	Yi Mao
Bing Yin	Bing Zi	Bing Xu	Bing Shen	Bing Wu	Bing Chen
Ding Mao	Ding Chou	Ding Hai	Ding You	Ding Wei	Ding Si
Wu Chen	Wu Yin	Wu Zi	Wu Xu	Wu Shen	Wu
Ji Si	Ji Mao	Ji Chou	Ji Hai	Ji You	Ji Wei
Geng Wu	Geng Chen	Geng Yin	Geng Zi	Geng Xu	Geng Shen
Xin Wei	Xin Si	Xin Mao	Xin Chou	Xin Hai	Xin You
Ren Shen	Ren Wu	Ren Chen	Ren Yin	Ren Zi	Ren Xu
Gui You	Gui Wei	Gui Si	Gui Mao	Gui Chou	Gui Hai

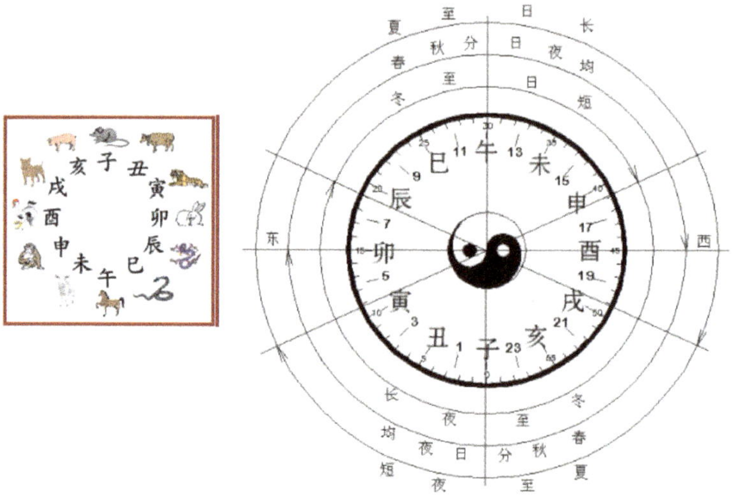

Fig. 4.10. The earth branches indicate the animal year and every two-hour each day (http://en.wikipedia.org/wiki/Earthly_Branches).

4.11; see also Chapter 5 for details). The location of the inscriptions on each part of the triggers varies quite widely, and this may have been affected by the preferences or habits of individual craftspeople or workshops. Hence, it appears important to investigate the locations of the inscriptions. To designate the various locations of individual trigger parts where inscriptions were found, these were labelled with lower case letters following the upper case letter used to designate the trigger part. For example, Cb is related to location b in trigger part C (Fig. 4.11).

For the purposes of individual assessment, the codes for the trigger inscriptions are listed in Table 4.6. Figure 4.11 shows the different location of the codes for where inscriptions occur on the triggers, and Appendix 6 offers information on how certain inscriptions match up with certain locations. Further details and the all inscriptions on the triggers are provided in Appendix 6.

The inscriptions on each specific location of the trigger parts are generally not the same, with the exception of the trigger location Ba (location a on the trigger part B) (Appendix 6), where 28 inscriptions were found, with the dominant character being 工 (*Gong*), accounting for 17 in total. In addition, three 五 (five), four single characters, (酉, 丙, 外, and 午), and four combinations (工十, 工屮, 卜五, 二五) were identified (Fig. 4.12). The character 工 probably refers to the governmental workshop or *Sigong* whose inscription was found on the lances and halberds discussed above (Institute and Team, 1988). The other stem-branch characters and numbers may have been added to aid counting and to register the time of production and/or other parameters related to production or quality control.

As noted above, sometimes the same character was chiselled on the different parts (parts A, B, C, and D and E) of the same particular trigger. Alternatively, one character could be repeated twice (for instance, on parts A and B, or parts B and C, or parts A and C), or three times (parts A, B, and C), or even five times on the trigger parts (Li and Gao, 2010). The following table records those triggers where the same inscription reoccurred three times on one trigger (Table 4.7).

Table 4.6. The variety of characters recorded on the trigger inscriptions

Based on the repetition of inscriptions on different parts, previous researchers (Yuan, 1990; Wang, 1994) have argued that the inscriptions were used to aid in the assembly processes, so that matching parts were carved with the same inscriptions to avoid later confusion or mix-up. If this hypothesis is correct, it could perhaps be interpreted as a limitation to the standardisation, in the sense that parts would not be interchangeable between different triggers despite their similarity (Li and Gao, 2010). As a matter of fact, even though the triggers were cast in relatively standardised moulds (see Chapter 5), during the assembling process it was still necessary to file the metal surfaces to make sure that the trigger parts fitted and could work properly (Li et al., 2011, 2012). Further aspects of these inscriptions relevant to the organisation of production organisation will be discussed in Chapter 5, following the typological and metric analyses.

4.3.5 Inscriptions on ferrules

Typologically, the ferrules were divided into three types: I, II, and III (see Chapter 7). Some type III ferrules were inscribed and, where this is the case, the inscription reads '*Sigong*' (Fig. 4.13), probably referring to the main governmental workshop (see above). Some 41 out of 83 type III ferrules, almost half of the total, bear these inscriptions.

The presence of inscriptions on the ferrules indicates that the ferrules played as important a role as the weapons. They were produced in the same workshop (*Sigong*) as the halberds and lances; this is not surprising, as they were often part of the same long weapons, and were selectively carved with inscriptions. Their spatial distribution and further comparisons will be presented in Chapter 7.

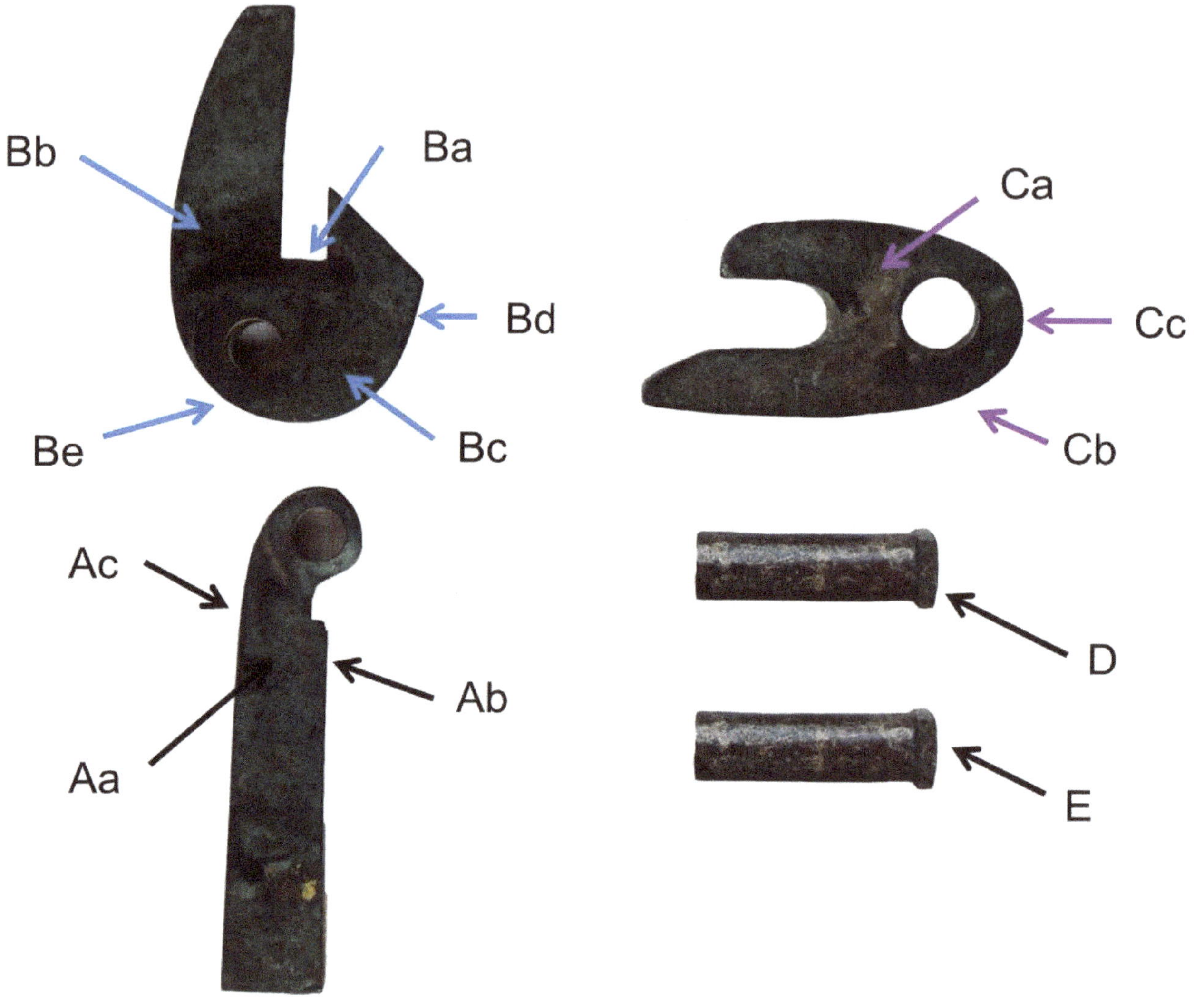

Fig. 4.11. Trigger parts A, B, C, D, E and the locations of the inscriptions.

Fig. 4.12. One of the stem-branch characters (you酉) on trigger location Ba.

Table 4.7. Details of triggers bearing the same inscription in at least three parts.

Count	Trigger ID	Part A	Part B	Part C	Interpretation
1	901	Aa	Bc	Ca	
2	902	Ac	Bc	Ca	
3	946	Aa	Bc	Ca	Three
4	970	Aa	Bc	Ca	Thirty six
5	975	Ac	Bc	Cb	
6	1499	Aa	Bc	Ca	Ninety
7	2231	Aa	Bc	Ca	Eighty two
8	2232	Ab	Bd	Cb	Mountain
9	2239				
10	2243				Twenty three
11	2246				
12	2252				Eleven
13	2261				Twenty five
14	2280				
15	2289				Fifteen
16	2294				Stem-branch
17	2298				Forty
18	2299				Eight, On D and E
19	2301				
20	3790				Forty two
21	5760				Six
22	2237				Three, On D and E
23	2264				Eight, On D and E
24	920				Fifty five
25	2307				Three
26	2323				Twenty seven

Fig. 4.13. A *Sigong* inscription on a ferrule.

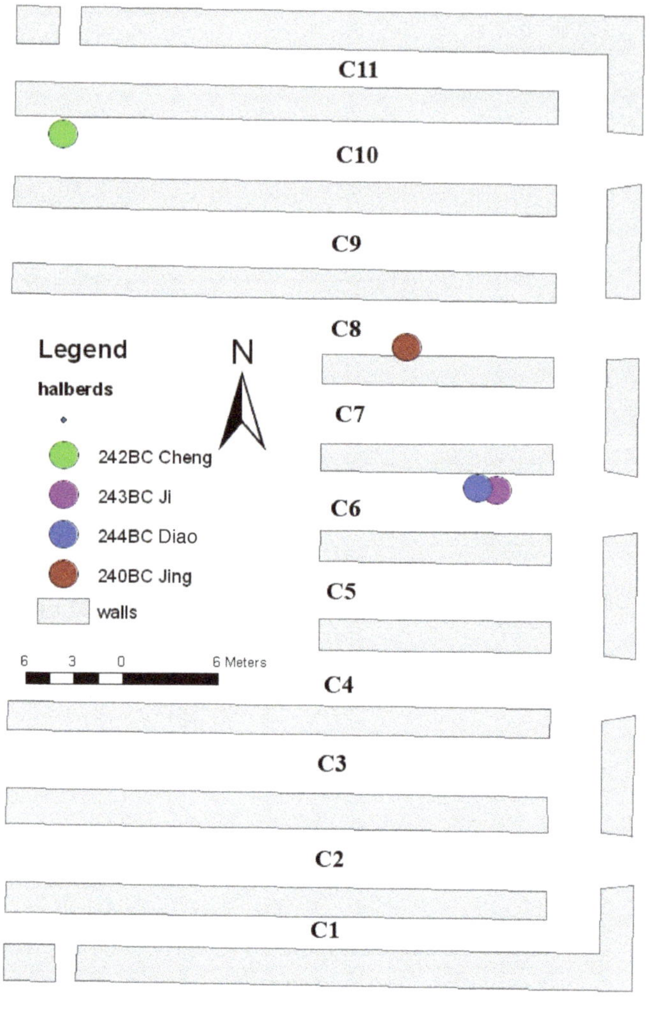

Fig. 4.14. Spatial distribution of the halberds noting their dates of manufacture as recorded in the inscriptions.

4.4 Spatial patterns of the inscribed weapons

4.4.1 Spatial patterns of inscribed halberds

The spatial patterns of the four halberds are difficult to interpret due to the limited size of the sample. In addition, there is no clear correlation between the year of production and individual craftspeople (Fig. 4.14). The bronze halberds were produced by different craftspeople in the third (244 BC), fourth (243 BC), fifth (242 BC), and seventh (240 BC) regnal years, respectively (the halberd produced in the tenth regnal year was found following ongoing excavations). Two of them, discovered in corridor 6 in the middle of the pit, were made in 244 BC and 243 BC, respectively, by workers *Diao* and *Ji*. The halberd inscribed with 240 BC and worker *Jing* was found in corridor 8. In corridor 10, a halberd was unearthed behind the chariot, and bears the inscription of 242 BC and worker *Cheng*. However, the number of halberds is too limited to allow any further statistical analysis. The reason there are so few halberds in Pit 1 is unclear, but may be due to a post-depositional effect, such as looting of the topmost layers of the pit where the tops of the long weapons might have been located.

4.4.2 Spatial patterns of inscribed lances

Based on the measurements of 10 lances, their width shows only differences of a few millimetres (Fig. 4.15). However, the results are not correlated with specific craftspeople or the year of production visible on the inscriptions. The minimal differences in scale may be related to the grinding and polishing processes employed after the lances were cast (Li et al., 2011). The model and mould were probably highly standardised, and the slight changes in the widths of the lances as a result of the finishing processes would not have been detectable with the naked eye.

The spatial distribution according to the lances' width seems to show a pattern (Fig. 4.16), because the relatively narrow ones were mainly found in the two bottom corridors and the wider ones are concentrated in the upper two. However, the sample is still not large enough to provide more than a general impression.

The spatial distribution of the lances according to both the year of production and the craftspeople involved appears random (Fig. 4.17). The reason for this will be discussed in the following chapter, where the weapons will be studied with a view to ascertaining whether they were used in the battlefield before they were buried in the pit or whether they were transported directly from the arsenal to the tomb complex.

4.4.3 Spatial patterns of inscribed triggers

The trigger inscriptions found on part B in the Ba location are very special, as mentioned earlier, and their spatial patterns are interesting as well (Fig. 4.18). The green dots

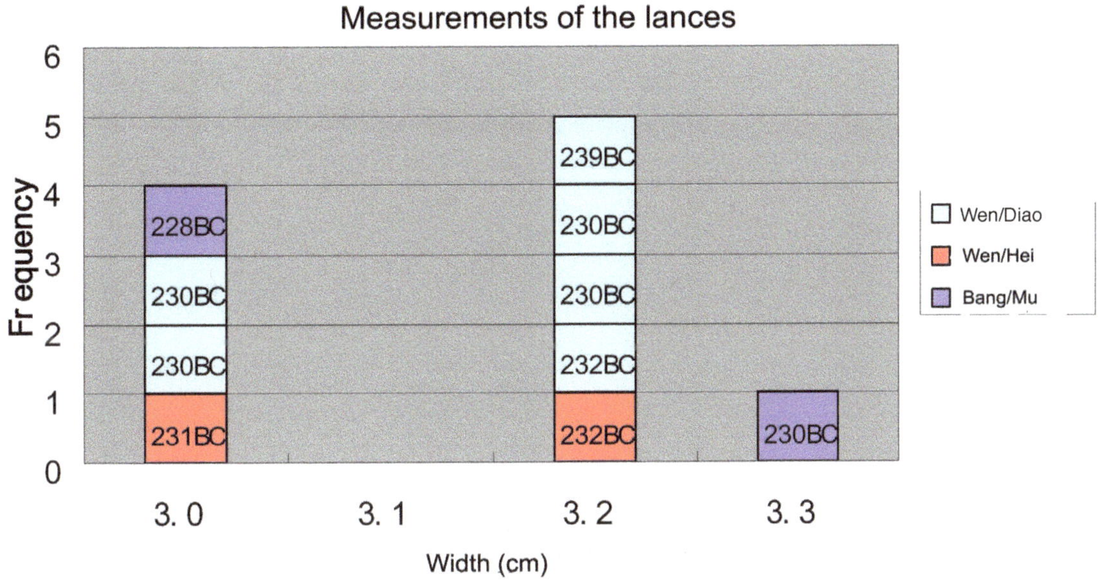

Fig. 4.15. Blade width, craftspeople and years of production for the lances.

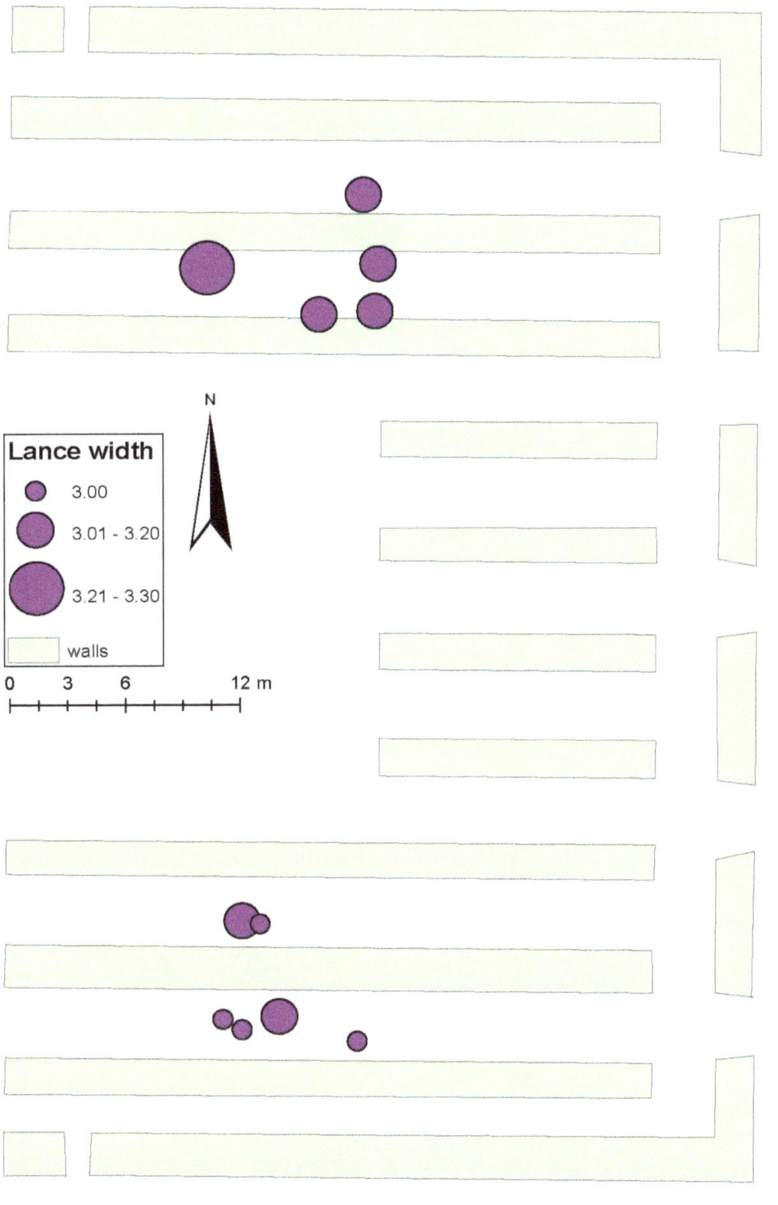

Fig. 4.16. Spatial pattern of the lances, discriminated by their width.

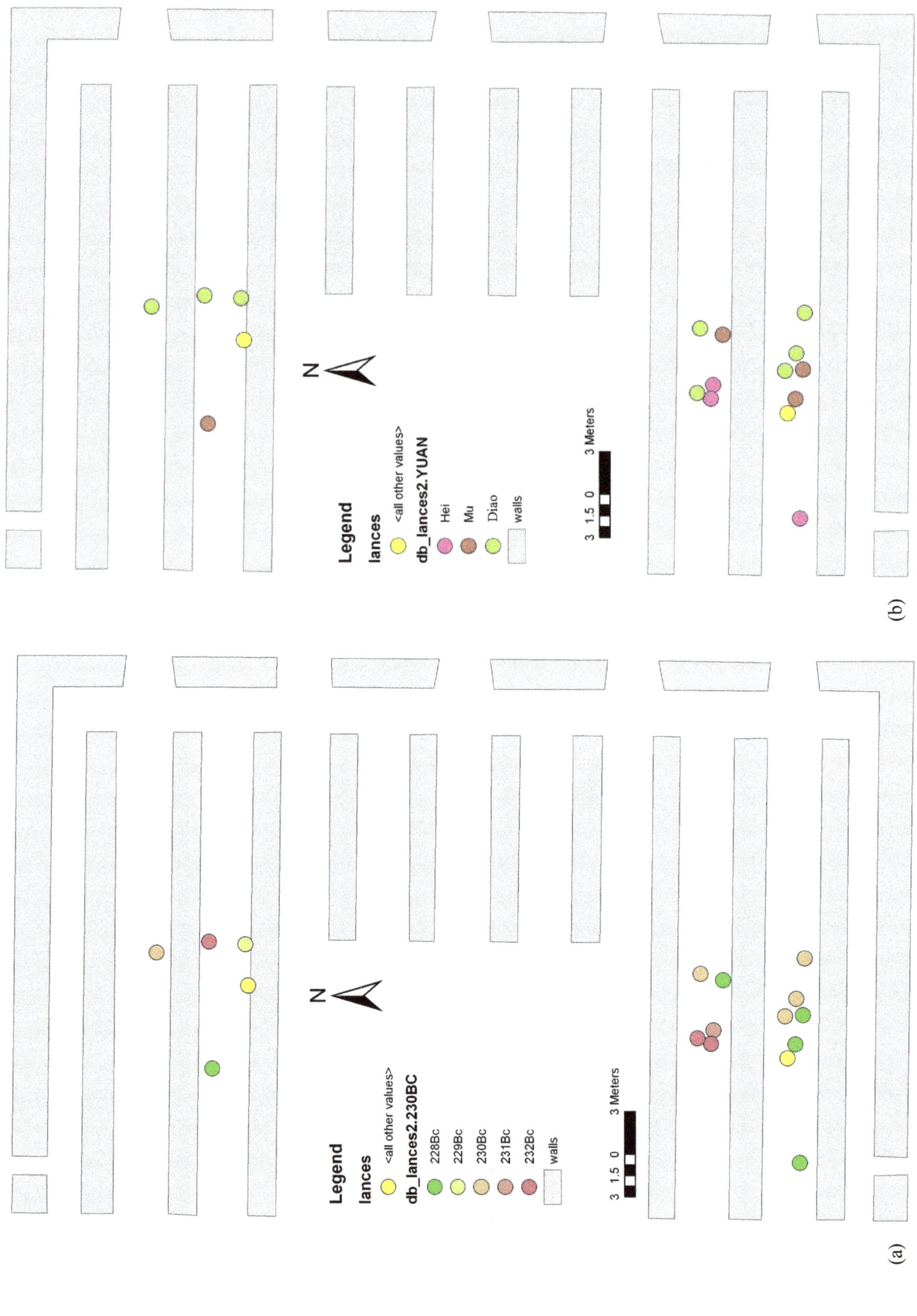

Fig. 4.17. Spatial patterns of the lances according to the (a) year of production and (b) workers involved.

in Figure 18 indicate the triggers with the code 1 inscription (工) (see Table 4.5 and Fig. 4.18), and these triggers were concentrated in two side corridors, 1 and 11. The other two green circle dots, code 1_4 (工十) and 1_50 (工屮), are also located in these two corridors. These 19 triggers all have the 工 inscription which has been interpreted as a workshop (Gong). The codes 14, 15, and 16, stem-branch characters 酉, 丙, and 外, are distributed in a cluster within corridors 9 and 10 (circled in black), while the other three triggers with character 五 inscribed are clustered at the bottom of corridors 1 and 2 and in the eastern crossing path (circled in purple). These spatial patterns are probably indications of workshop practices, storage in an arsenal, and also of the procedure according to which the weapons were placed in the pit.

The repeated inscriptions on each part of a specific trigger support the idea of an assembly process; however, their spatial patterns are relatively random (Fig. 4.19). The two purple dots with a star in the middle indicate the triggers where the same inscription was carved on all five

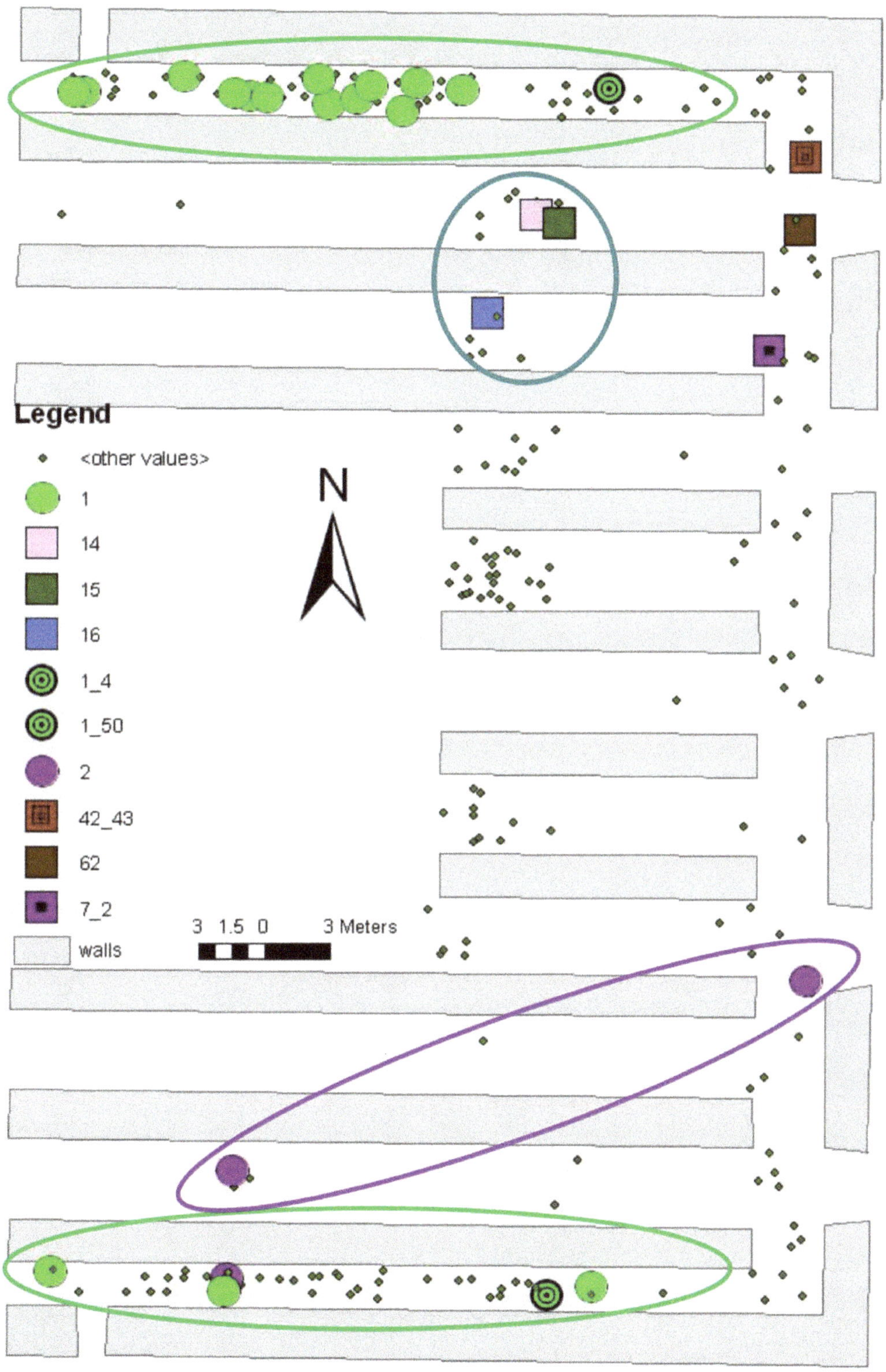

Fig. 4.18. Spatial patterns of the trigger inscriptions on the Ba location.

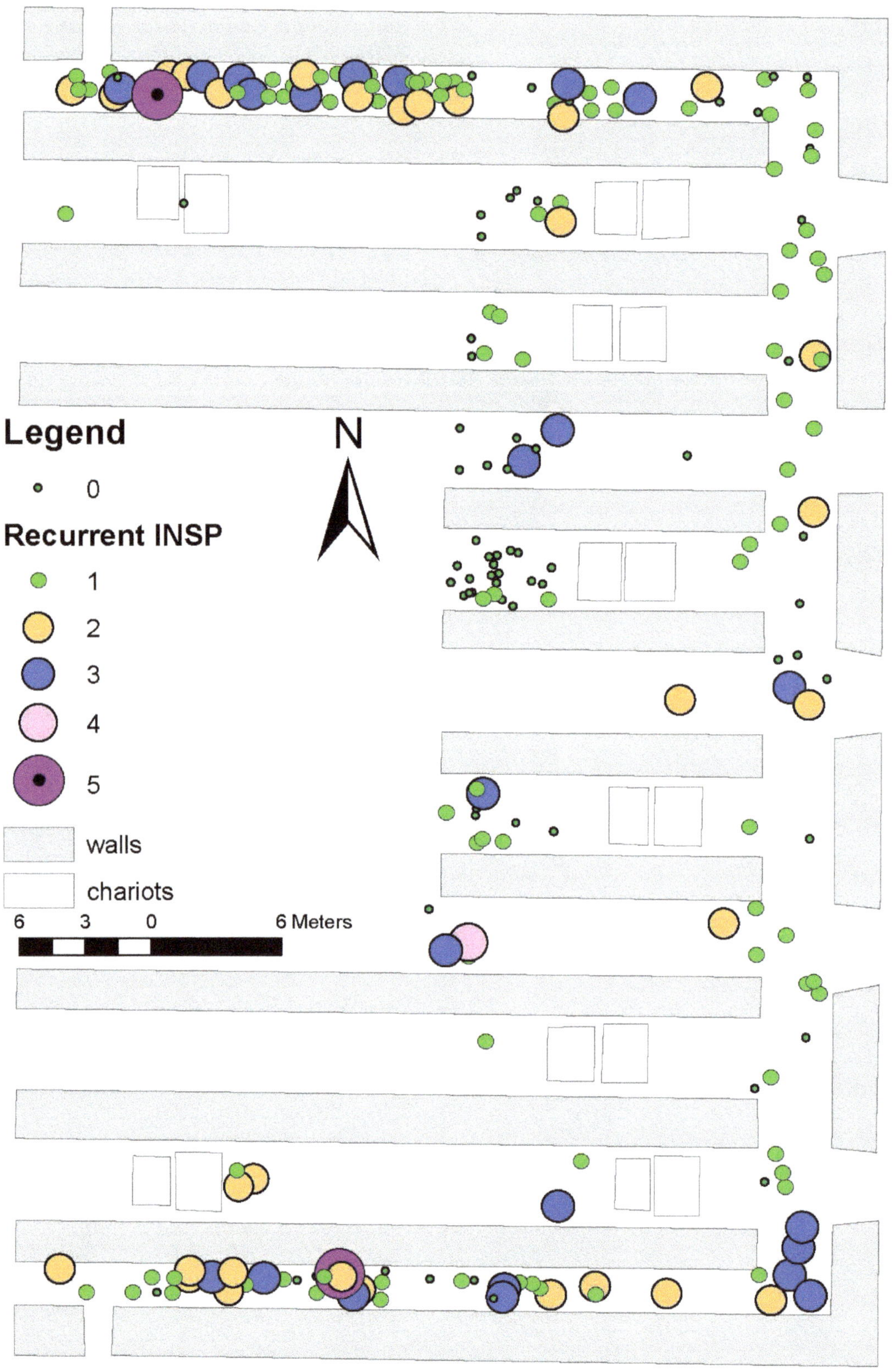

Fig. 4.19. Spatial patterns of the triggers with recurrent inscriptions.

individual parts, and they are located in corridors 1 and 11, respectively. The triggers with two and three repetitive inscriptions appear to be distributed randomly, perhaps as a result of the assembly processes. If the trigger parts were fitted together straight after casting, they might not have to have been inscribed with the same character.

4.4.4 Spatial patterns of inscribed ferrules

The inscribed type III ferrules are concentrated in four corridors (2, 3, 9, and 10), with a fairly balanced number of inscribed versus uninscribed type III ferrules in each corridor (Fig. 4.20 and Table 4.8). These patterns are discussed further in Chapter 7.

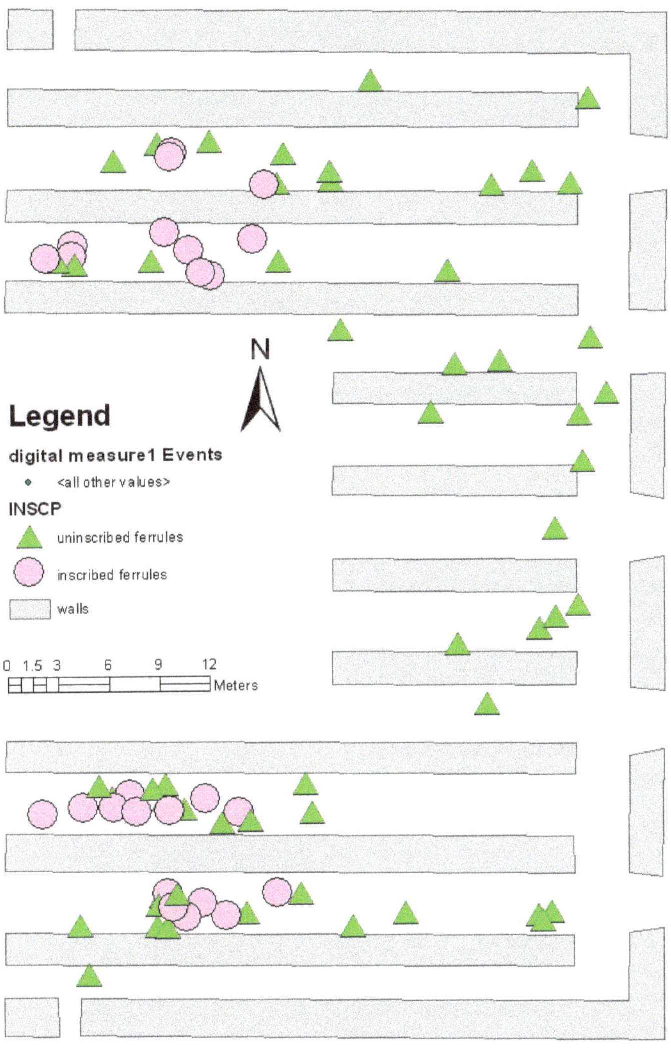

Fig. 4.20. Spatial patterns of the inscribed and uninscribed ferrules.

Table 4.8. Inscribed versus uninscribed ferrules in each corridor.

ferrules	Inscribed	Uninscribed
C2	6	6
C3	8	10
C9	8	3
C10	3	5

4.5 Function of the inscriptions

4.5.1 Quality control

Some of the weapons were inscribed with the names of the workshop and the craftspeople involved in their production. Additionally, the halberds and lances present longer inscriptions, including not only the specific name of the workers involved, but also those of the workshop, the workshop manager, and the chancellor with whom the responsibility for the weapons' production rested. The *Sigong* or *Gong* characters on some of the ferrules and triggers probably denote the workshop of origin. All of these inscription types are likely to have facilitated accountability in case any weapon or batch had to be traced back to its manufacturer. As such, their function is likely to have been related to quality control.

According to the historical documents, this quality control strategy originated with *Shangyang's* (商鞅, 385–338 BC) innovations (see Chapter 1), as noted in the radical changes in the content of the inscriptions cast and/or carved on the bronze objects before and after Shangyang. Before him, the content of inscriptions was mainly concerned with the purpose of the specific vessels and weapons, and/or with their owner. However, after Shangyang, inscriptions relate primarily to the workshop, the time and date of manufacture, the craftspeople involved, and the supervisor who controlled the production. In this sense, the supervisory and organisational structure documented in these inscriptions from the Terracotta Army weapons may merely represent the crystallisation of a practice that had started a few decades before.

The craftspeople who were responsible for the specific weapons might have been awarded or punished by the officials or their supervisors during the Qin period, and this may explain why their names were included on weapons. The fact that not all the weapons are inscribed with these types of characters may reflect the fact that only some items per batch were marked, or, alternatively, that inscribing customs changed over time. Some further discussion on the role of inscriptions, and comparisons between uninscribed and inscribed weapons, will be presented in subsequent chapters.

4.5.2 Assembling triggers

In addition to quality control, some inscriptions may have been added to facilitate the assembly processes, especially on the triggers, as discussed above.

4.5.3 Other functions

Some single characters on several lances and halberds may be an indication of the name of the Qin arsenal, mentioned in the historical documents (Yuan, 1990). However, some numbers carved on the swords, lances, halberds, and triggers might simply have been added as an aid to counting. In addition, some symbols found on the triggers may be the result of workers making temporary notes (although they are carefully carved).

4.6 Summary

The inscriptions on the bronze weapons present relevant information about the labour organisation employed during the production process, and some of the possible measures taken to monitor quality and standardisation. The inscriptions on the bronze lances and halberds provide basic information about workshop organisation, from the involvement of the chancellor down to individual craftspeople. Other characters, numbers, and symbols on the triggers may be related to the requirements of the assembly process.

5

Bronze Triggers

5.1 Introduction

Like the rest of the weapons studied in this book, the bronze triggers considered in this project, 262 altogether, were discovered in the pits containing the terracotta warriors in Emperor Qin Shihuang's tomb complex, with a particular concentration in the five easternmost trenches of Pit 1. In addition to those found in Pit 1, a few triggers were found in the trial trenches of Pit 2 in 1978. The following formal excavation of this pit was carried out in 1994, but only as far down as the roof layer, under which the terracotta figures and bonze weapons are still buried. Pit 3 is thought to be the headquarters for the army in the other two pits, and contains mainly ceremonial weapons without blades. There is no evidence of triggers in this pit.

The bronze triggers in Pit 1 were mainly distributed in the corridors surrounding a group of archers and/or crossbowmen that had been arranged as part of an actual battle formation. Pit 1 is divided into 11 corridors from south to north, with one front corridor at the east end. The warriors in the east corridor are considered to be the vanguard force, and the warriors in the southern corridor 1 and the northern corridor 11 are referred to as the right flank and left flank, respectively. The vanguard and flanks were usually composed of archers and/or crossbowmen in contemporary and later Chinese battle formations (Wang, 1990; Yuan, 1990). They fired arrows at the enemy across the space between the two forces on the battlefield. In these corridors, many arrows as well as triggers were discovered (for details on the arrows see Chapter 6).

As composite items, the bronze triggers are particularly informative about the processes of standardisation and mass production employed in the making of Qin bronze weapons. Each one of the five parts in a trigger had to be precisely made so that, once the parts were assembled together, the trigger could move easily and work correctly. The low tolerance for error presumably required a high degree of standardisation and efficient production and control. As noted above, there are inscriptions carved on some of the triggers (see Chapter 4), including *Gong* inscriptions assumed to refer to the name of one or more centralised governmental workshops, Chinese characters, numbers, and/or unknown symbols. This chapter will focus on the characterisation and interpretation of trigger groups and subgroups based on extensive typological observations and measurements, with a view to inferring aspects of the organisation of production. The information will be collated with that derived from the inscriptions, which will be used as a further variable in the spatial analysis and statistics (also see Li et al., 2014).

5.2 Research review

Within the large assemblage of weapons unearthed from the pits of the terracotta warriors, the triggers have attracted much attention from Chinese and Western archaeologists alike (Wang, 1980a; Yang, 1980; Liu, 1986; Yuan, 1990; Ledderose, 2000). Initially, basic research addressed the typology of the triggers (Institute and Team, 1988), and they were divided into two types (Types I and II) according to formal differences in Part A (the handle of the trigger; see below). Further research by Wang Xueli (1980a) and Yuan Weihua (1981a) laid the foundation for future study of the production of Qin bronze weapons, as these authors suggest that all the parts of the triggers had been standardised and were interchangeable. Although the contributions of Yuan and Wang were based on a relatively small sample and lacked statistical rigour, they are nevertheless important, as an essential initial step guiding further research and discussion.

A good number of publications concerning Qin bronze triggers have focused primarily on their mechanical function and on trigger development during different historical periods. Yang Hong (1980) demonstrates that the earliest bronze triggers appeared in the Chu state during the Spring and Autumn Period (770–476 BC). The archaeological evidence for this was discovered in 1952 from a late Spring and Autumn period tomb in Changsha, Shaobatang (长沙扫把塘) (Gao, 1964). Archaeologists have also discovered bronze triggers (from the Warring States period) in Sichuan, Hebei, and Henan provinces (Sichuan Committee of Cultural Relics, 1956; Ye, 1974; Sichuan Museum, 1974; Hebei Cultural Relics Managing Department, 1975). Evidence from all these excavations show that triggers, containing five mechanical parts, were widely used during the Warring States era (475–221 BC). Yang (1980) proposes that the crossbowmen of this period operated their crossbows using only their hands and arms, in contrast to the later, stronger crossbows that were operated using both the hands and feet. He also devotes attention to gradual innovations in the design of crossbows and triggers from the Warring States period to the Han Dynasty (Yang, 1980). Two main aspects of later triggers are highlighted: whether a bronze case was cast to contain all the other parts of the triggers, and whether a sight was fixed on the top of the trigger. Originally, a bronze trigger was directly fixed into a wooden stock, which was easily broken. In contrast, triggers with bronze cases enhanced the strength of the weapon and increased the distance arrows could travel from the crossbow. To date, only one trigger with a bronze case has been found in the pits of the terracotta warriors, as most triggers were still uncased during the Qin period (Fig. 5.1). Likewise, sights, which

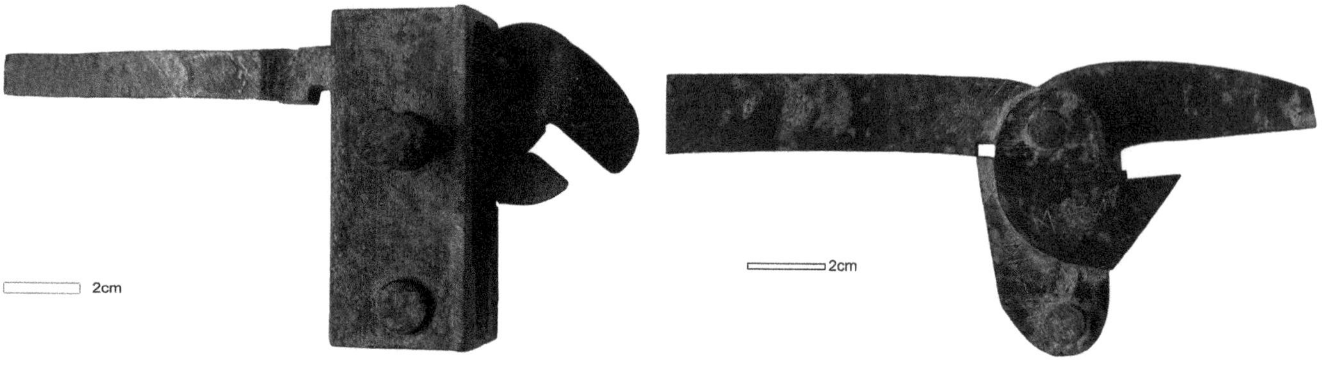

Fig. 5.1. Triggers with (a) and without (b) a case.

were developed later to enable the crossbowmen to fire more precisely into the enemy force (Yang, 1980), have not been identified in the triggers from the First Emperor's tomb complex. Liu Zhancheng (1986) conducted research on Qin crossbows using the traces discovered in the pits of the terracotta warriors. He concentrated on how the crossbow and trigger were assembled together and on the type of material that had been used to make bows, as well as looking at aspects of the production and innovation in the manufacture of triggers. He argued that bamboo, wood, leather or linen, and lacquer were all essential materials for making crossbows, and also commented on the uses of triggers in the Qin and Han Dynasties.

Ledderose (2000) argues that the mass production of weapons started in the Qin Kingdom long before the time of the First Emperor, and emphasises that the lock mechanism (trigger) of the crossbows was an important development. According to him, the mechanical parts were cast with such precision that they fit together perfectly. The margin of error lies within fractions of a millimetre (Ledderose, 2000: 60). Ledderose also argues that it was this precision that helped the Qin state overpower the rival feudal states. Yates (2007) claims that the introduction of the crossbow during the Warring States period revolutionised military warfare, as it required less skill and strength to use than a composite bow. The crossbows would have permitted the archers to fire heavier bolts or quarrels, more accurately, with greater force and penetrating power, and over a longer distance.

Previous research on Qin triggers, as discussed above, mainly focuses on their functioning, development over time, the material used, and crossbow assembly for military benefits. While standardisation and mass production have attracted the attention of some archaeologists, little attention has been dedicated to the way production was standardised or how labour was organised in the making of the triggers. This chapter will therefore put forth an investigation of these triggers from this new perspective. Following a detailed typological study, the analysis will concentrate on both the morphological and spatial data associated with the triggers (also see Li et al., 2014). Statistical methods will be used systematically to identify subgroups, to investigate how the bronze triggers were mass-produced, and to explore how their placement in the pit was organised. This chapter will present the statistical methods also applied to the other weapons discovered from the pits.

5.3 Measurements of the bronze triggers

5.3.1 Sources of data

All the triggers unearthed from the pits have been moved from their original location and are now stored in the Conservation Department in the Museum of the Qin Terracotta Army. They have all been labelled, placed in wooden cases, and documented in the museum database. Relevant archaeological data pertaining to the triggers were obtained from the Excavation Report (Institute and Team, 1988), available in Chinese, and from the museum archive in the Conservation Department, which also contains information on the continuing archaeological excavations of the pits.

Of the 262 triggers, 23 are broken pieces, and another 10 are unavailable since they are exhibited in cases or preserved in plastic bags. Altogether, 229 triggers have been observed, photographed, and measured for the purposes of this study, of which 216 originate from the five easternmost trenches of Pit 1, 12 from the other ongoing excavation trenches in Pit 1, and one from Pit 2.

5.3.2 Triggers' function and manipulation

Before the measurements, it is necessary to introduce the individual parts of the triggers. As mentioned above, each trigger was formed of three mechanical parts (A, B, and C) jointed with two bolts (D and E), and each part had its own Chinese name, character, and function. In order to describe them more easily in English, the parts have been marked A, B, C, D, and E (Fig. 5.2). Part A (the trigger proper) is a handle. Part B (the tumbler) comprises two elements: one side is used to catch the string of the crossbow, while the other extends upward as a sighting pin. Part C (the latch) is cast in one piece and has a hole. The function of part C is to link parts A and B so that they can be manipulated. Parts D and E are bolts used as transverse screws to strengthen the stock and trigger.

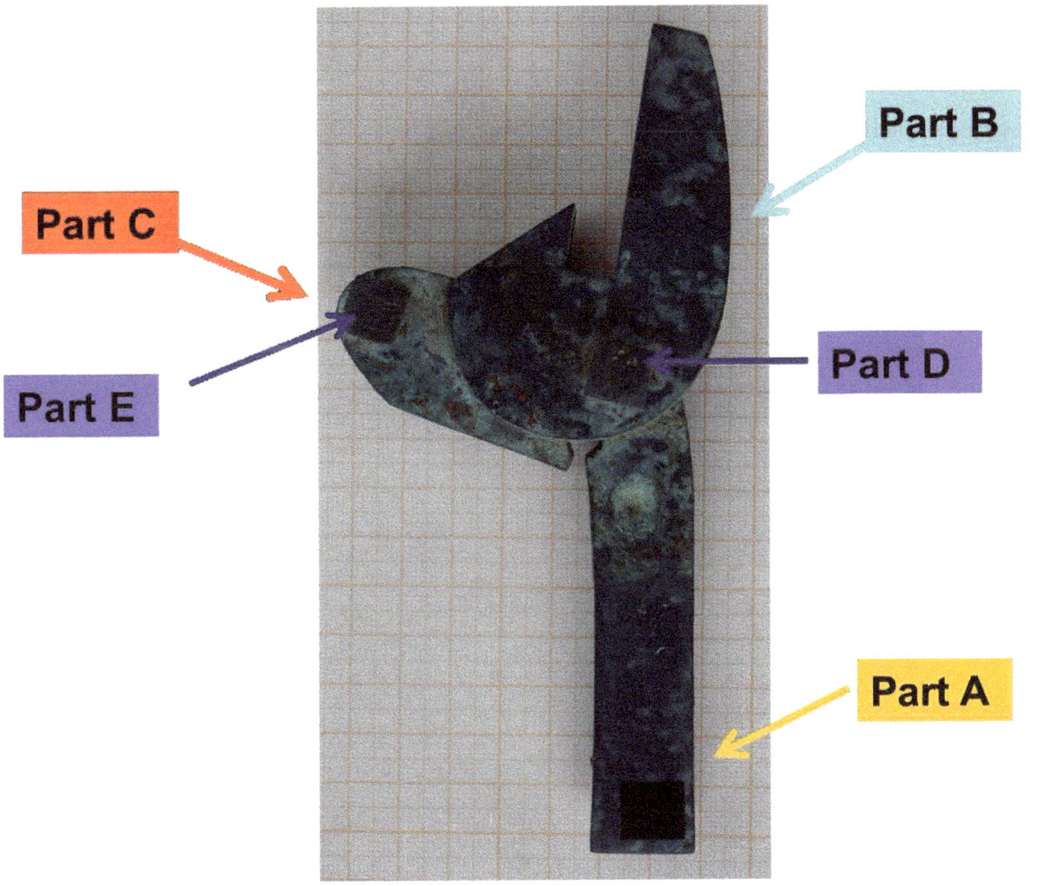

Fig. 5.2. The five parts of a trigger with the labels used to refer to them in this study.

A trigger was fixed into a wooden stock of the crossbow for the archer to manipulate (Fig. 5.3). When the end of the handle (part A) is pressed backwards, part C is then free to revolve clockwise, and part B follows to reverse and instantly release the bow string and arrow. After the archer let off the arrow, he brought the lock back into the original position by quickly jerking the bow backward. The disadvantage of crossbows was that they were much slower to load than traditional bows. However, training the crossbowmen to act in unison could have compensated for this. While the front line was firing together, the line behind would be loading the arrows. Once the front line had fired, the second line could move forward to replace them. In the battle formation of the Qin Terracotta Army, especially in Pit 2, the array of kneeling archers and standing archers seems indicative of this type of manoeuvre.

5.3.3 Method of measuring

In order to identify dimensional differences in trigger parts that could be related to slightly different typologies, casting moulds or workshops, three measurements were taken of each of the mechanical parts and two for each of the bolts (Fig. 5.4).

Considering the large number of samples and the accuracy required when undertaking the measuring, a convenient, efficient, and reliable way to collect this data had to be found. All the bronze triggers were photographed in the museum; the images were then transferred onto the computer, and georeferenced and measured in a geographic information system (GIS). There were many advantages to using this procedure. As some of triggers had some rust, it was difficult to separate them into their five original parts, but it was easy to select a proper point at which to measure them using the GIS. This system was also convenient as it made it possible to review the triggers according to the images taken and to correct any mistakes that might have occurred during the data collection process.

The procedure for measuring triggers and managing data was carried out as follows:

- Taking photos
  a) A frame or tripod was used to hang the camera, which was set to level.
  b) Metric paper was put on the base board, and the trigger was placed in the middle of this. The museum identification number of the trigger was clearly marked on the paper.
  c) Usually two pictures of each trigger were taken from two sides, in order for the five parts to be seen and measured properly.

- Georeferencing and measuring in a GIS
  a) The pictures, as maps or satellite images, needed to be integrated with defining coordinates so that any new data could be positioned and scaled in

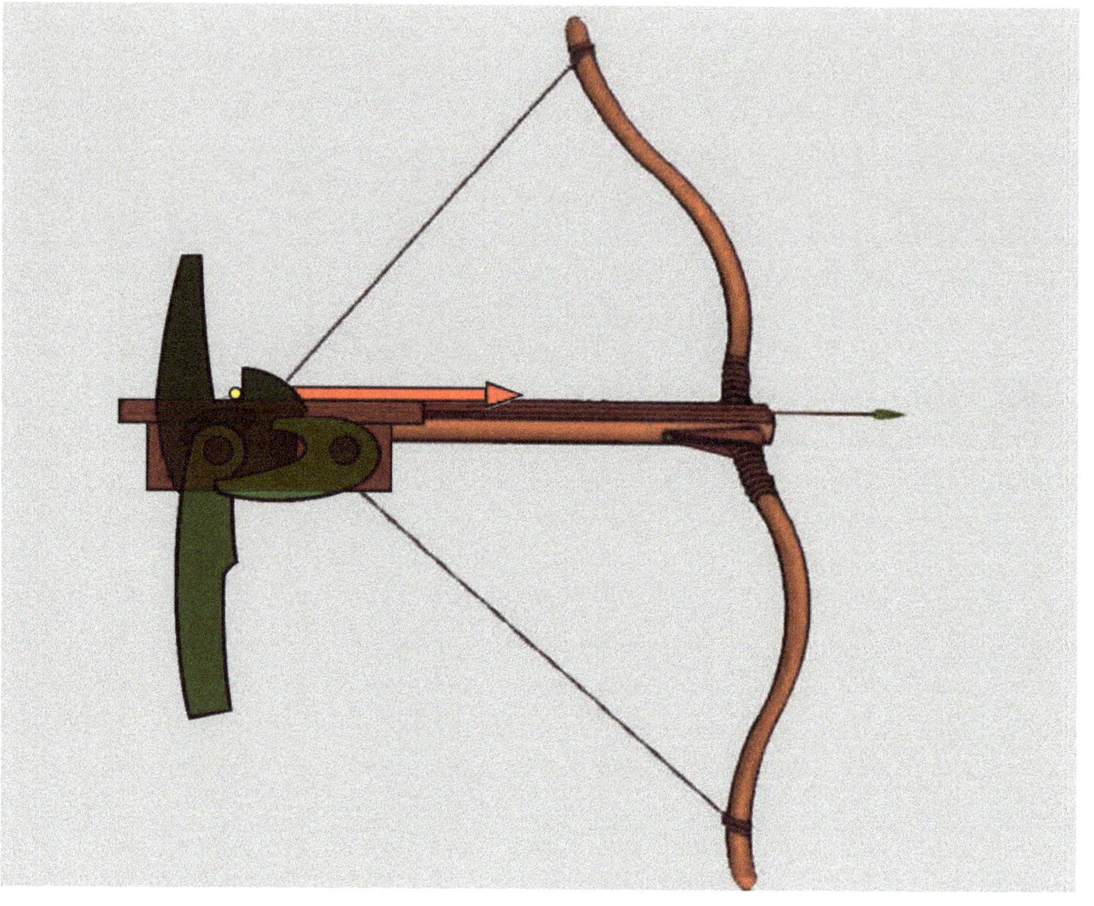

Fig. 5.3. Manipulation of the trigger in the crossbow (image courtesy of Zhao Zhen).

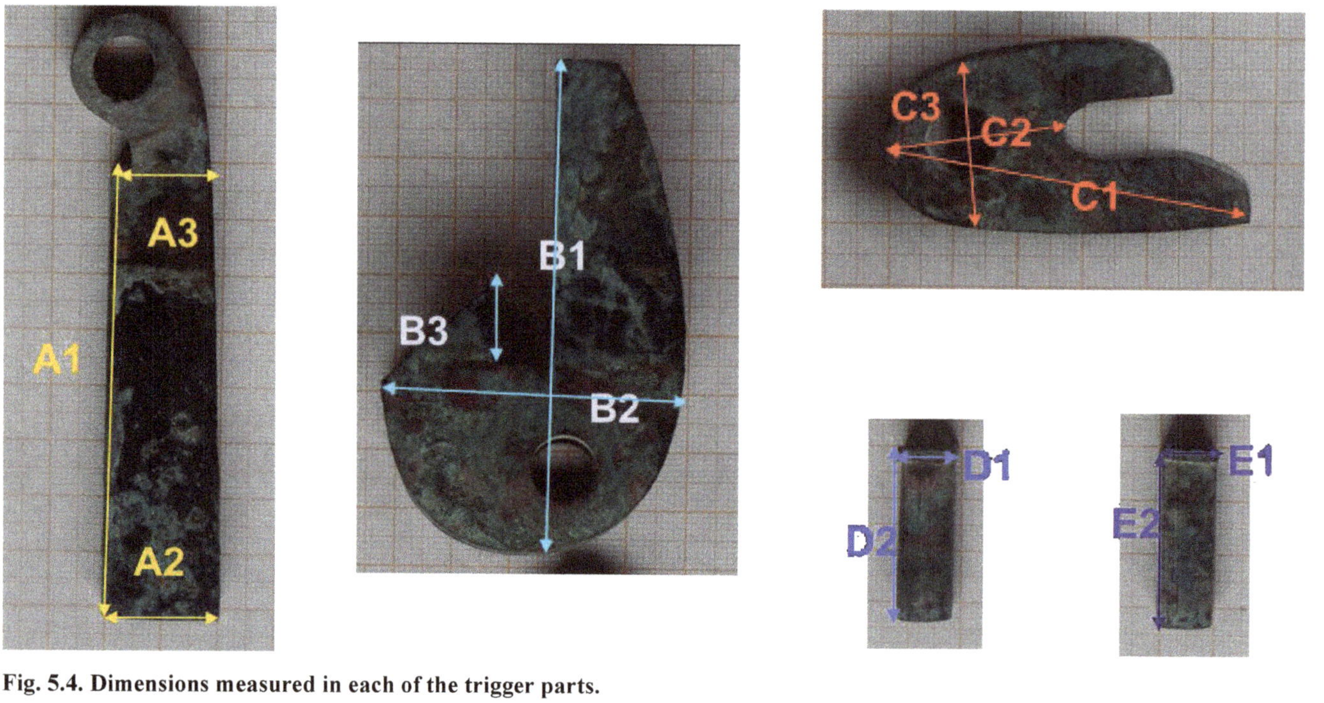

Fig. 5.4. Dimensions measured in each of the trigger parts.

relation to the existing data. This process is called georeferencing. In order to accomplish this, four locations from each picture were used as control points for georeferencing the image in a GIS package (ArcGIS 9.3 was used here). Each picture took between seven and ten minutes to rectify and another seven to ten minutes to scale.

b) Each of the five parts of the triggers was then measured. To ensure the accuracy, the location is very important for such irregularly-shaped trigger parts, especially for parts B and C (Fig. 5.4). B1 was measured along the straight upward line from top to the bottom and B2 was from the right sharp point being perpendicular to B1. C1 was from the bottom left point to the top point. C2 joined the top point to the middle of the curve, and C3 was perpendicular to C2 across the centre of the circle. All these measurements were then recorded on a spreadsheet (Microsoft Excel was used here; Table 5.1).

- Managing data
  a) Bivariate and multivariate principal component analysis (PCA) scatterplots were the main graphical and statistical techniques employed to present and compare the data for each of the trigger parts. The data clusters in these plots were then used, along with further typological information to assign the trigger parts to certain sub-types. These were then compared across the different parts to assign the assembled triggers to different subgroups.
  b) The spreadsheet data were saved as a .csv file, and imported as spatial data into the GIS.

For the purposes of individual assessment, Figure 5.5 maps the triggers with their museum identifiers and the appendices offer a summary version of the measurements taken on each one. Further details and the full set of measurements are provided in Appendices 1, 2, 3 and 4.

5.3.4 Patterns in trigger measurements

Several approaches are possible to explore the trigger measurements and seek to handle the multivariate data. For example, a dendrogram clustering analysis could be used to group the triggers together based on a large number of variables. The first problem with this approach is that the process of creating similarities/distances involves a loss of information. For example, two weapons may have a certain degree of similarity with one another but this is summarised in a single number, which no longer provides information about which variables are the ones that resemble one another and which are those that differ. The second problem arising from this is that we lose sight of the relationship between the objects and the variables that describe them (Wright, 1989; Shennan, 1997: 266). While a standard bivariate scattergram can be used to determine if any trends were present in the distribution of the observed objects, this will not adequately present multivariate data. Hence, PCA is an appropriate method for the data concerned. It is a mode of analysis which considers objects and variables together, and consequently allows us to see more directly how the two are related, and which specific variables are creating the data structure. It also retains the advantages of a technique that can reduce multidimensional data to fewer dimensions.

PCA was carried out using R or SPSS software. The classification of the measurements on the trigger parts A, B, C, D and E respectively shows a variety of patterns, which are discussed in detail below.

Table 5.1. The correlations between the three original variables of trigger part A and the three principal components.

Trigger part A	Dim.1	Dim.2	Dim.3
A1	0.699	0.714	0.008
A2	0.926	-0.259	-0.276
A3	0.919	-0.283	0.272

5.3.4.1 Trigger Part A

Part A measurements A1, A2, and A3 were analysed using principal component analysis. Figure 5.6 clearly displays four main clusters or groups. They are abbreviated as Ag1, Ag2, Ag3, and Ag4, respectively. Ellipses have been added to this and other PCA plots for indicative purposes, to facilitate the discussion. The variables of trigger part A and three principal components are shown in Table 5.1.

The four groups of trigger pieces identified in the PCA plot are also strongly associated with their specific micro-features (Figs. 5.6 and 5.7).

- Ag1 in the middle of the plot (Fig. 5.6) comprises 40 pieces of triggers, and shows a loose distribution. This group is thin and curved at the end of handle (Fig. 5.7), and few inscriptions have been found in this group.
- Ag2 is a relatively big group of 176 pieces. Part A is relatively straight and flat in this group, and thicker than in Ag1.
- Ag3 comprises 12 pieces. Part A shows more flaring at the end of handle, but it is flat and has the same thickness as the pieces in Ag2.
- The far left dot on the PCA plot is Ag4, and represents only one trigger. This part A is much narrower and thinner than any of the others, and the corresponding assembled Parts B and C are both very unusual too. This is a special trigger discovered in corridor 7 of Pit 1. As noted above, triggers were generally composed of five parts, and the three mechanical parts were fixed directly to the stock by means of the two bolts. However, this unique trigger has an extra case, and the three parts were put in the case before being fixed to the stock, thereby enhancing the strength of the crossbow and allowing the arrows to be fired further. This type of trigger was widely used in the following Han Dynasty (206 BC – 220 AD) (Yang, 1980). The appearance of the single-cased trigger, in contrast with those of the large quantity of uncased triggers in the Qin tomb complex, demonstrates that either potential innovation in crossbow technology occurred during that period or that the trigger is in some way a later intrusion in the pit.

Two triggers each selected from Ag1, Ag2, and Ag3, respectively, and together with the single trigger in Ag4, were examined in greater detail, as an initial exploration aimed at identifying further subdivisions within triggers

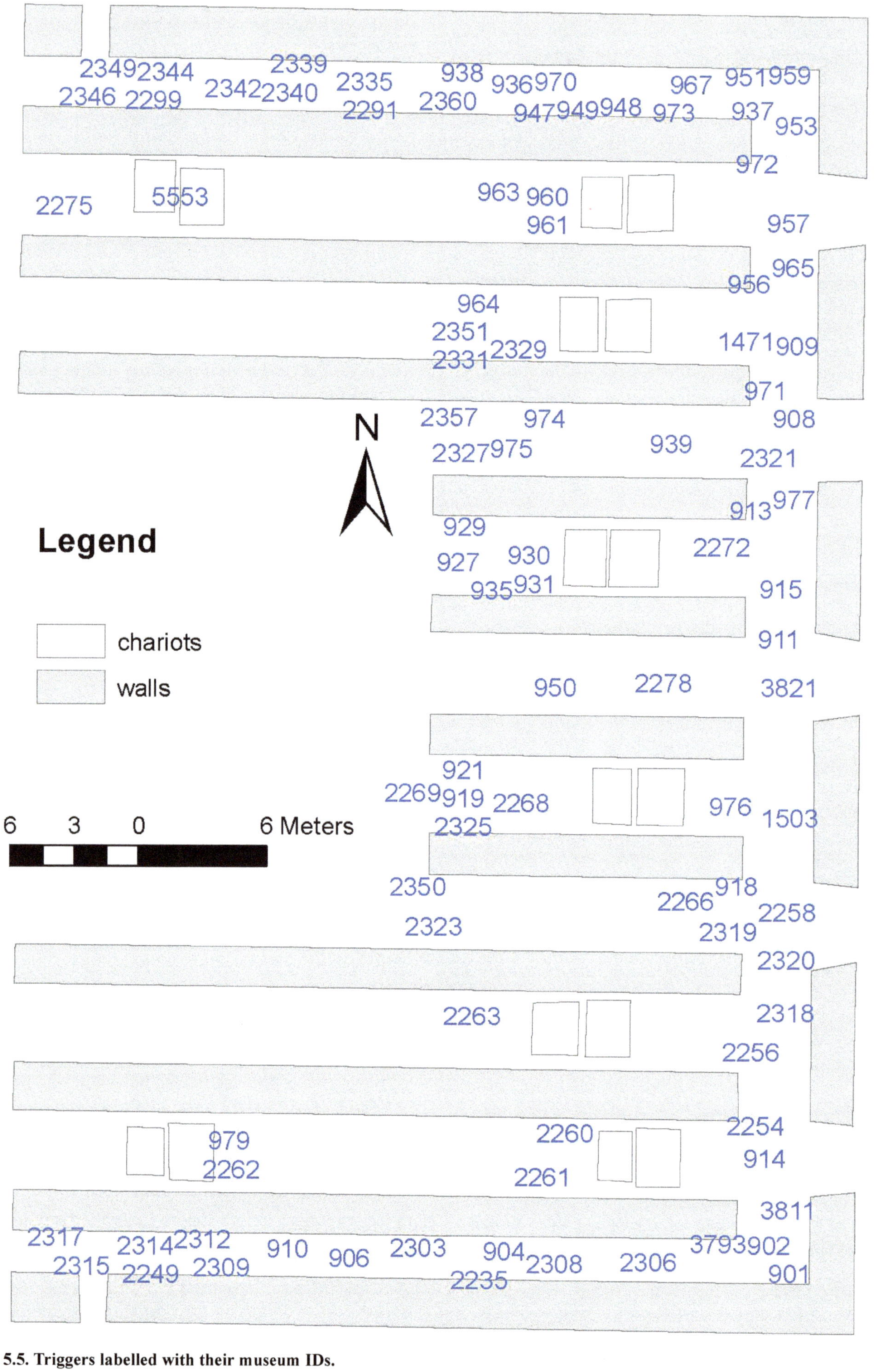

Fig. 5.5. Triggers labelled with their museum IDs.

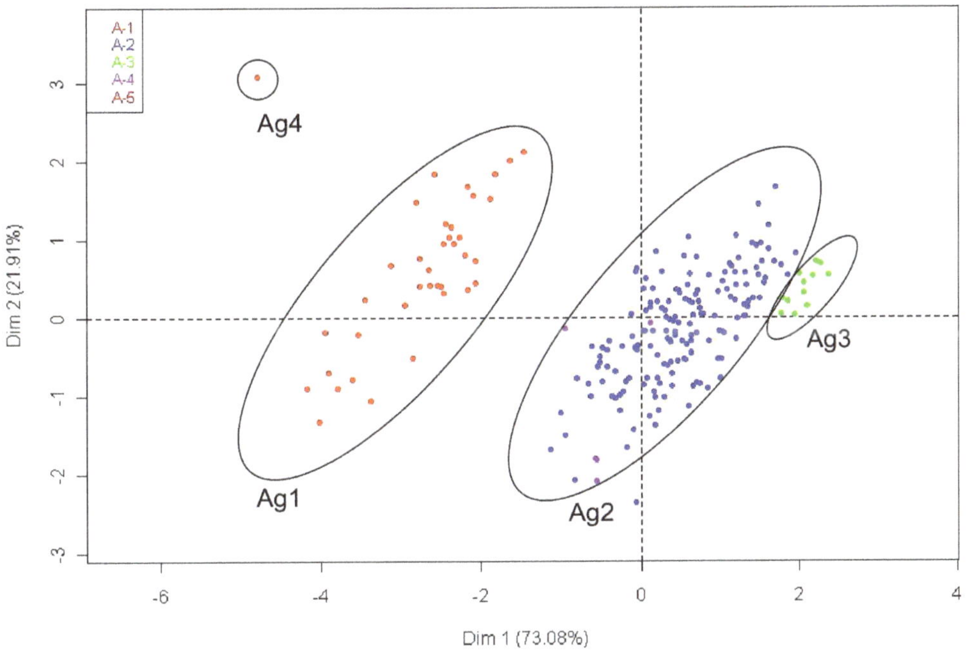

Fig. 5.6. A plot of the first two principal components of trigger part A pieces with the 4 approximate trigger subgroups: the red dots show Ag1, the blue dots show Ag2, the green dots show Ag3, and the orange dot shows Ag4.

Fig. 5.7. Digital reproduction of the four subtypes of trigger part A.

that shared the same type of part A. A more extensive description of the characteristics of these four part A groups is listed in the following table (Table 5.2). Descriptions of parts B, C, D, and E assembled to Ag 1–4 are also listed in this table.

5.3.4.2 Trigger Part B

Part B components of the triggers can be macroscopically divided into two types depending on whether they have a clearly defined notch (see Fig. 5.8). The group of triggers displaying a notch on part B comprises 35 pieces of all the triggers, while the remaining 193 are unnotched. The notch protrudes on the side that holds the string, and its function remains unexplained. As with the parts A above, from a technological perspective, the difference could be interpreted as related either to workshops or different craftspeople working with different models or, at least, to the use of different moulds. Following this initial subdivision, we can search for the presence of further subgroups based on the metric data.

Figure 5.9 plots the measurements reduced to the first two principal components, and Table 5.3 shows the correlation between the variables of trigger part B and the three principal components. The plot shows a rough division into three clusters or groups, Bg1, Bg2, and Bg3; however, these groups do not overlap with the groups macroscopically identified above as 'notched' and 'unnotched'. Notched and unnotched part Bs are dispersed in both groups. Even though dimensions within a given group may be very similar, the presence/absence of notches clearly indicates the use of different moulds in each case. In fact, if we return to the triggers to check for further differences among the notched parts B in the two groups, Bg1notched and Bg2notched (Fig. 5.10), we can observe not only differences in size, but also differences in shape. To retain both the measurement groupings and the typological differences, the following groups are used hereafter: Bg1notched (Bg1n), Bg1unnotched (Bg1u), Bg2notched (Bg2n), Bg2unnotched (Bg2u), and Bg3 (corresponding to the special cased trigger). From the information above, one gets the impression that different groups of artisans could have worked under unified supervision to ensure an acceptable degree of standardisation. The differences may also be related to different phases of production, as discussed below.

Table 5.2. Variation in the different trigger part A groups and assembled parts.

	Ag1		Ag2		Ag3		Ag4
ID Number	**915**	**926**	**902**	**2284**	**2256**	**2277**	**5525**
Part A	Thin and curved end	Thin and curved end	Straight, flat	Straight, flat	Flaring, flat	Flaring, flat	Narrow and thin
Part B	Notch	Without notch	Inscription on middle of short side, notch	Inscription on middle of short side, notch	No notch, narrow, inscription on high side	No notch, narrow, inscription on high side	Short and round
Part C	Flatter end, no inscription	Flatter end, no inscription	Bevelled	Bevelled	Elliptical end	Elliptical end, inscription on high side	Thin, flat
Parts D and E	Different thickness of bolt head, bolt put in from different direction	Different thickness of bolt head, bolt put in from different direction	Square head, thin	Square head, thin	Bolt head small square	Bolt head small square	Strong, no bolt head

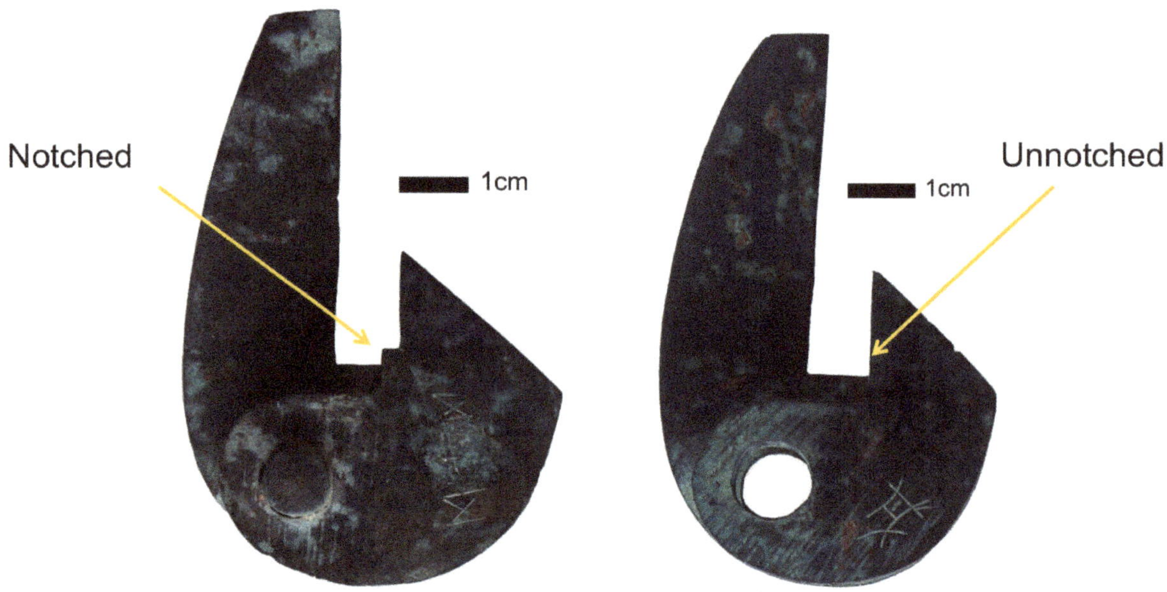

Fig. 5.8. Typological variations in trigger part B.

5.3.4.3 Trigger Part C

Simple typological analysis indicates that part C can be divided into more curved (203 examples) and more bevelled versions (25 examples; Fig. 5.11). The plot of the first two PCA components displays the groups deriving from their measurements: Cg1, Cg2, Cg3, and Cg4 (Fig. 5.12). Table 5.4 shows the correlation between three variables of trigger part C and the three principal components. It shows a strong association between the typological observations and the measured dimensions within trigger part C, but also allows us to differentiate two subgroups within Cg3. The fact that they are slightly segregated from each other indicates that they were produced in different moulds, since all the bevelled parts C belong to group Cg3; this group can be further subdivided into Cg3curved (Cg3c) and Cg3bevelled (Cg3b). As with the previous parts, the analysis demonstrates that several types of models and moulds were involved in the production of the part C pieces: Cg1, Cg2, Cg3c, Cg3b, and Cg4. It further suggests the existence of different working units or groups of artisans in the workshop, or possible different phases of production, which will be discussed in the following section.

5.3.5 Analysing the patterns of the triggers

Generally, typological variation in archaeological artefacts can be interpreted as being caused by either an evolutional sequence or regional differences. However, in the case under consideration here, the timespan for the manufacture of the triggers is known to be short. Moreover, from the

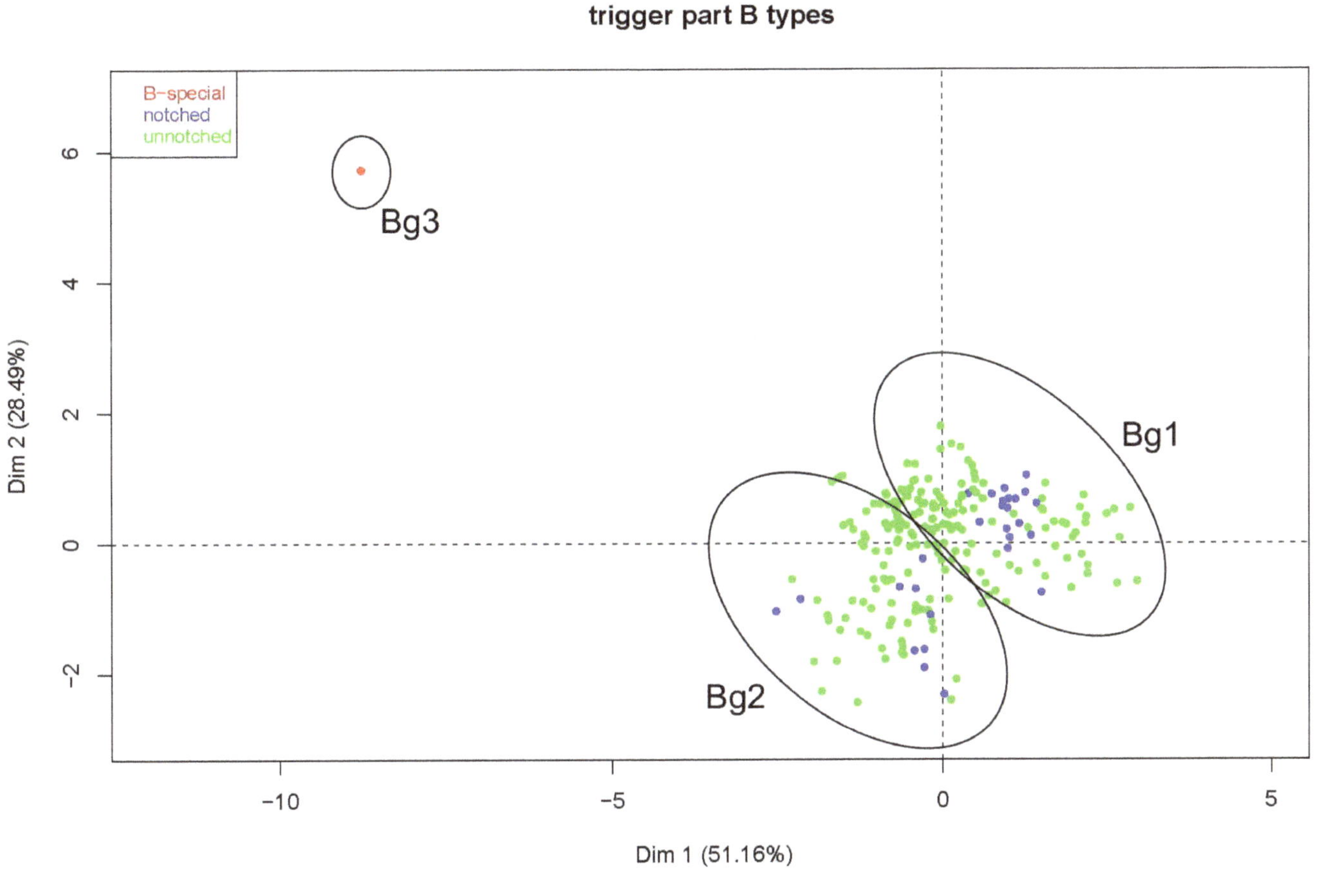

Fig. 5.9. A plot of the first two principal components of the trigger part B pieces; the green dots show the notched and the blue dots the unnotched triggers. The diagram also shows the approximate three groups, Bg1 (n and u), B2 (n and u), and Bg3.

Table 5.3. The correlations between the three original variables of trigger part B and the three principal components.

Trigger part B	Dim.1	Dim.2	Dim.3
B1	0.692	-0.592	0.411
B2	0.803	-0.052	-0.593
B3	0.639	0.708	0.299

inscription '*Gong*' on some of the triggers, seemingly the equivalent of '*Sigong*' on the lances and halberds, at least some of these weapons can be assumed to have been made in a governmental or royal workshop. Based on the information available at present, it is difficult to ascertain whether some variation was due to different phases of production or to the existence of further regional workshops. However, the single special trigger (ID5525) can be linked to a later dynasty, as mentioned above, and one can assume that it might have been produced later than the other types.

Another explanation for this phenomenon is that different artisans were involved in the production of the triggers, even if working with the same master plan and near each other. As such, the Ag1 group was probably produced by one group of craftspeople, while Ag2 and Ag3 might have been produced by another group of craftspeople. Most trigger parts in Ag2 and Ag3 were carved with inscriptions, but there is no evidence of inscriptions on those of Ag1 (see Chapter 4 for detailed information about the trigger inscriptions).

A total of 35 part B pieces are notched, while 193 are unnotched. Craftspeople therefore produced more often unnotched parts B, and some of those have '*Gong*' inscriptions on them, implying that they might have been made in the '*Sigong*' workshop. It is possible that some craftspeople from a non-governmental workshop or from other states joined the '*Sigong*' workshop, and that they either employed special technical procedures or had been taught to make triggers in a slightly different way. Considering the three part B groups according to their measurements, Bg3 is associated with Ag4 and corresponds to the special later stage trigger, but the other two groups Bg1 and Bg2 seem dimensionally very similar to each other, and the slight overlapping boundary probably represents the transition of two phases, or some other form of segregation between production units. Moreover, these two groups contain both notched and unnotched parts B.

The curved and bevelled parts C are, of course, due to a difference in the mould used, but can also be regarded as being the result of different preferences of craftspeople or the result of different working units within the workshop.

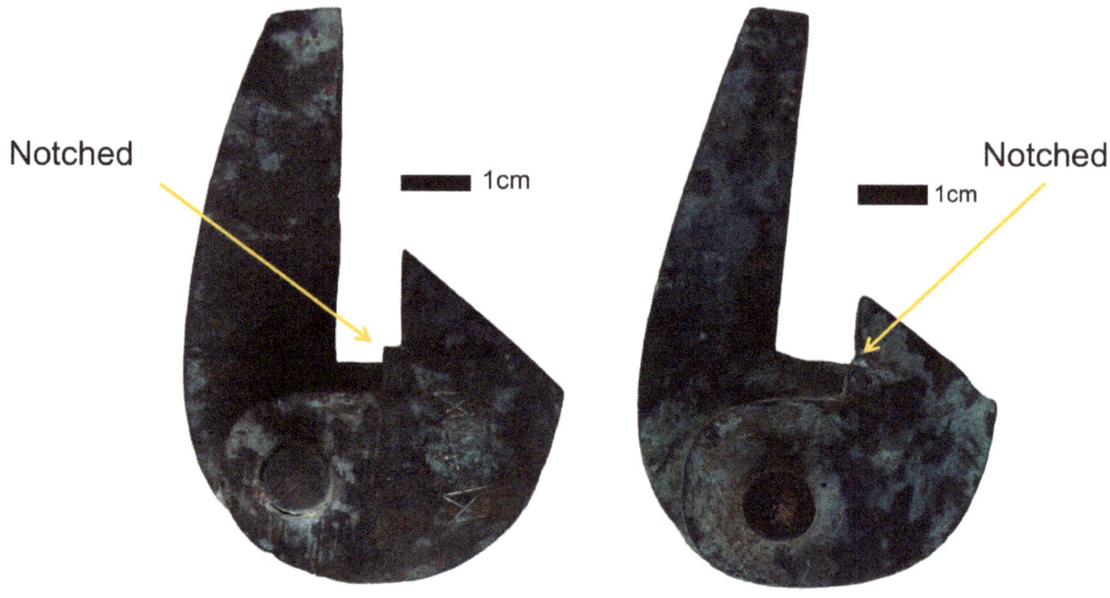

Fig. 5.10. Notches on otherwise different groups, Bg1 and Bg2.

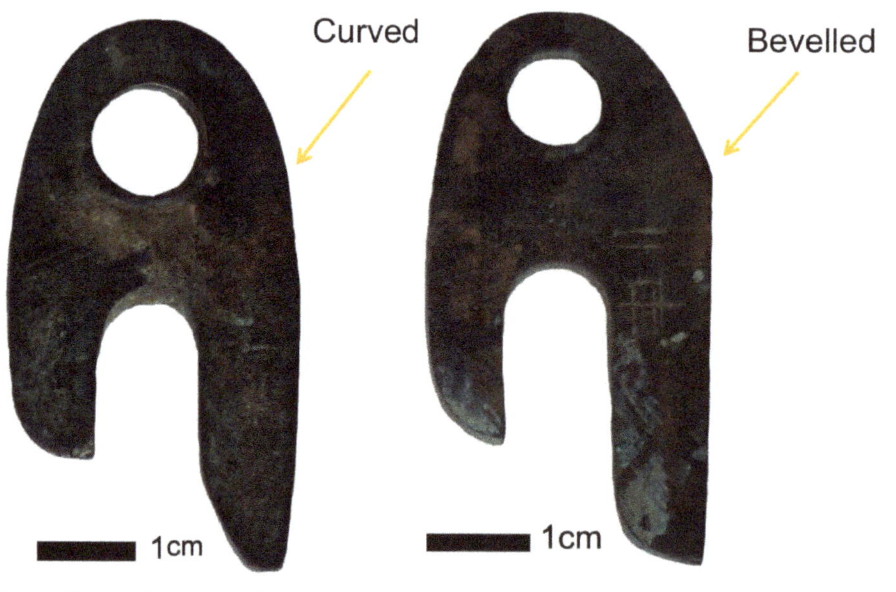

Fig. 5.11. Typological variations in trigger part C.

Further classifications, Cg1, Cg2, Cg3 and Cg4, based on the dimensions, provide further information about possible production units or groups. Cg4 is associated with Ag4 and Bg3 as part of the single cased trigger. Cg3 includes both curved and bevelled examples, and Cg1 and Cg2 are all curved. How are these typological and measured differences related to the organisation of production and craftspeople during the Qin period?

The differences in parts A, B, and C may be due to the working units and operations in the workshop, as well as to ongoing innovations over time in the manufacture of the triggers. While these differences in each of the parts mentioned above were probably often too small to affect the functioning of the triggers, it is interesting that the various part A, B, and C groups are almost consistently matched. How was standardisation ensured within the workshop and also in ongoing production processes? What was the overall degree of standardisation on each part of the triggers? Furthermore, how did the Qin craftspeople assemble the parts A, B, and C (and the bolts D and E) together? These issues will be discussed in the following two sections.

5.4 Coefficients of variation and the degree of standardisation

The trigger part groups defined above were further investigated for their relative levels of standardisation. Assessing the degree of standardisation allows, on the one hand, us to understand wider technological issues, such as changing models or moulds for casting, and various production processes. On the other hand, the degree of standardisation also makes it possible to investigate the crafting behaviour involved in the trigger production and labour organisation in the workshop (also see Li et al., 2014).

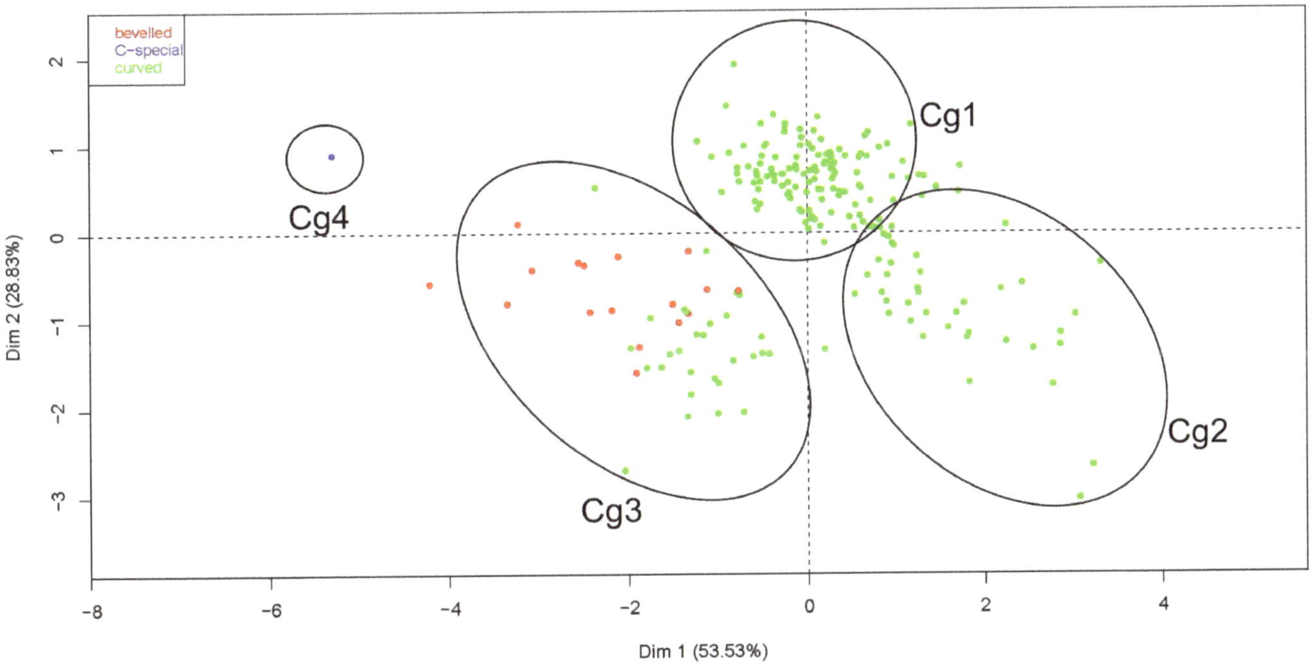

Fig. 5.12. The four groups resulting from the measurement of trigger part C: Cg1, Cg2, Cg3 (curved and bevelled), and Cg4. The dots marked in red show the bevelled examples and the green dots refer to the curved ones, with the blue dot for the special, cased trigger.

Table 5.4. The correlations between the three original variables of trigger part C and the three principal components.

Trigger part C	Dim.1	Dim.2	Dim.3
C1	0.720	-0.563	0.405
C2	0.836	-0.064	-0.544
C3	0.623	0.737	0.262

Coefficients of variation (CV) have been calculated for each of the trigger part groups based on the measured values. Starting with part A, Ag4 contains only one trigger, which is therefore of no statistical significance, and has thus been excluded from this section. The CVs for the other three groups are shown in Figure 5.13.

CV values in these three groups of bronze trigger parts range from 0.5 to 4.5%, which indicates a relatively high degree of standardisation as defined by the theory of the Weber fraction and CV conversion (Eerkens and Bettinger, 2001).

Figure 5.14 shows the CV results for trigger part B. CVs are lower than 5% for the measurements of B1 and B2, but the highest point is roughly 7.3% in the B3. This may indicate that there was more than one mould involved in the production of this group.

CV values were also employed to assess the standardisation of trigger parts B with '*Gong*' inscriptions (19 examples) versus those without (Fig. 5.15). As noted above, '*Gong*' is supposed to refer to the name of the governmental workshop during the Qin period. The CV values do not show any differences on B1 measurements, but on B2 and B3, those with '*Gong*' inscriptions show a slightly higher degree of standardisation (3.5–6.3% CV values). This finding is suggestive, as it might indicate a tighter control of standardisation in the weapons produced under the direct supervision of governmental workshops.

The CV values employed to assess each group of trigger part C vary from 1.4 to 6.5% (Fig. 5.16). The bevelled part C (Cg3b) is slightly less standardised than the curved part C (Cg3c).

One needs to be cautious when interpreting these results. On one hand, the trigger parts were cast using moulds, which minimised the craftspeople copy errors; nonetheless, shrinkage of the moulds would still slightly affect the degree of standardisation on the final products. Theoretically, their CVs should be close to 1.7% (Eerkens and Bettinger, 2001); however, the results show that only groups Ag3, Bg2n (at B1 point), Cg3b, and Cg3c (at C1 point) are below 1.7%, and all other types present relative higher values, even though a mould was used in their production. This situation may be the results of several factors. Firstly, the number of models used to shape the parts of the triggers is a main factor. If a single mother model was employed to make all pieces of one type of part A, there should be a high degree of standardisation within the group. If, however, several mother models were used by different craftspeople in the workshop, one should expect a lower degree of standardisation. Secondly, as mentioned above, the material used in the making of the moulds and its durability would also affect the CV results. A stone mould is more stable and lasts longer than a clay mould and could therefore be employed to cast a larger number of virtually identical objects. A clay mould might shrink a

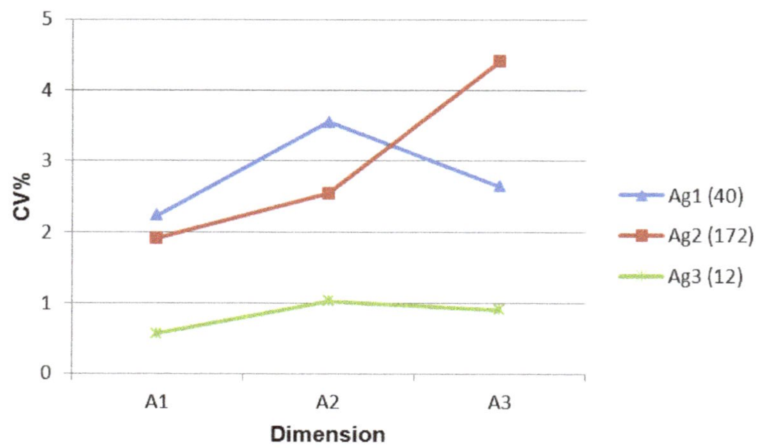

Fig. 5.13. CV results for trigger part A.

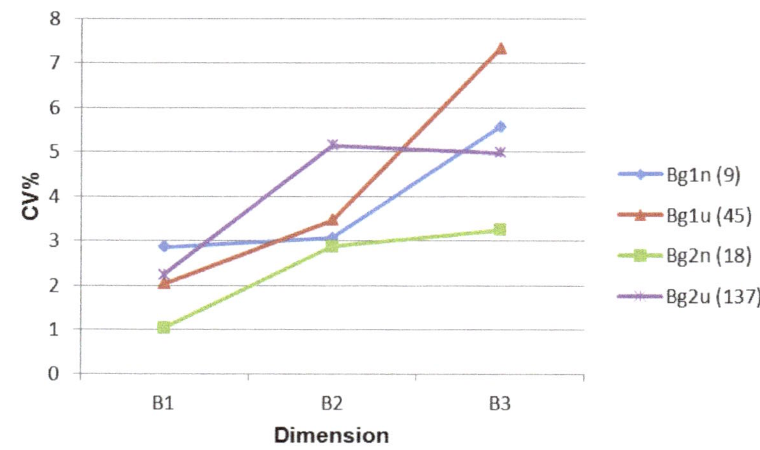

Fig. 5.14. CV results for trigger part B.

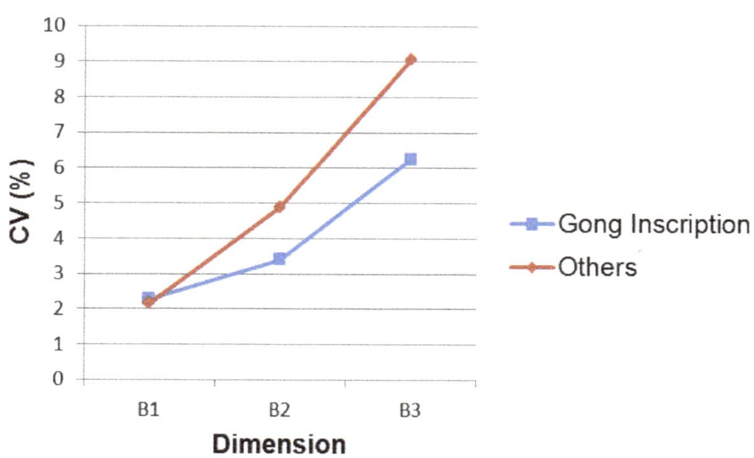

Fig. 5.15. CV results for trigger part B pieces with '*Gong*' inscription versus those without.

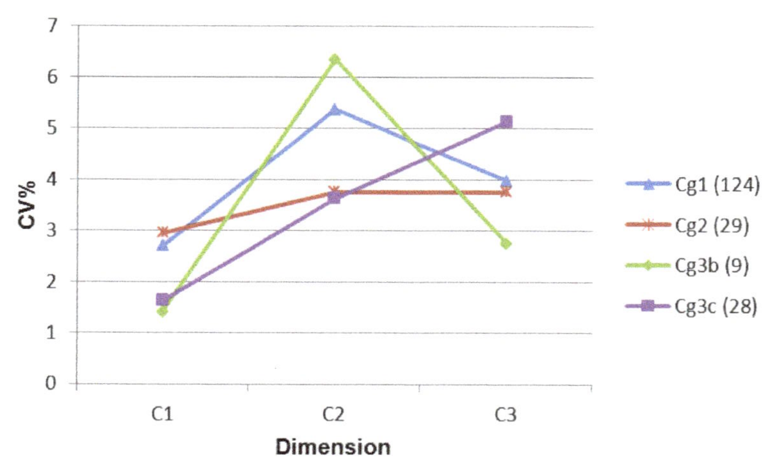

Fig. 5.16. CV results for trigger part C.

little bit during the drying process. Thirdly, the behaviour of individual craftspeople affects the uniformity of the production process. Their skill and sensory limitations result in a small margin of error during the mould making and casting process. Although the error may be small and imperceptible in each cast, it becomes a cumulative perceptible variation when passed through many hands over time. The resultant variation is called 'copy error' by Eerkens and Lipo (2005) and is caused by small errors that could be transmitted from one person to another in the casting process. For example, if a master taught several apprentices, then the work of these apprentices would be slightly different from that of their master in making the same type of objects. Fourthly, the sample size of the types of triggers used in the CV application affects the final outcome. For example, Ag3 is a much smaller group than the other two groups of part A pieces, Ag1 and Ag2. The CV value of Ag3 includes only 12 examples and varies from 0.5 to 1.1%, while Ag1 contains 40 examples and shows CV values that range from 2.3 to 3.6%, and Ag2 has 172 examples and CV values ranging from 2 to 4.5%. Generally, it is suggested that the bigger the sample size, the more craftspeople were involved in the production, and the more copy errors result with their sense limitation, even the tighter control with mould production. Conversely, a small group implies that a smaller number of craftspeople were involved, and consequently shows a relatively high degree of standardisation.

A final aspect worth noting is that the triggers reflect variation at various levels: the making of the mould, the casting itself, and the filling of excess metal on the finished objects. It is quite likely that variation/errors were introduced when making the mould, even if using rulers and models. This would explain from another perspective why larger groups would show higher CVs: they were probably made using a larger number of moulds.

In any case, overall, the small intra-group variation indicates high standardisation. If we bear in mind that a CV of 3% is the lower threshold of human perception of dimensional differences without measuring devices (Teghtsoonian, 1971; Eerkens and Bettinger, 2001), then we can assume that observers or supervisors would have seen most trigger parts in one group, with CVs typically lower than 3%, as identical. The significant differences between groups, which are more clearly noticed, and especially the presence of the distinctive Ag4, are perhaps more difficult to explain. This would seem to suggest that, in spite of the overall standardisation, there was some room for workshop variability, and perhaps for special needs or special technological or cultural preferences.

Once all parts were produced, assembling parts A, B, C, D, and E together as a functional trigger was the final objective of the production. As mentioned above, each part presented typological and dimensional differences, and their patterns of combination can provide the model for the organisation of production, even though these triggers were found far beyond the workshop context. Which part A goes to which parts B and C? This question is explored in the following section.

5.5 Trigger assembly and technological organisation

Before assembly, most trigger parts were ground and polished with hand files after mould-casting, and traces of these finishing techniques can be observed clearly under the optical (OM) or scanning electron microscope (SEM) (Li et al., 2011; 2012). This process was aimed at removing the excess metal and casting seams, and at the same time, making the parts fit together properly. On some triggers, inscriptions were carved, which include symbols, characters, and numbers. For example, the '*Gong*' inscription has been discussed above. Other relevant information and discussions regarding the inscriptions have already been presented in the previous chapter (Chapter 4), while the techniques and tools employed have been fully discussed in the separate publications (Li et al., 2011, 2012).

This section examines the ways in which triggers were assembled. In particular, I explore the presence of patterns and correlations between the assembly of the different groups of parts A, B, and C identified above. From this information, inferences can be made regarding production procedures, labour, and technological organisation, making reference to the various production models outlined in Chapter 2. In particular, I will examine the extent to which the holistic and prescriptive hypotheses, proposed by Franklin (1983) and termed as batch (cellular) versus flow line models in modern terminology by Wild (1975) and by Li Yunti (2006), are associated with bronze weapon production. These models have been identified from workshop remains or mould debris for the Shang (1600–1050 BC) bronze production (Li, 2006). However, this project is concerned with the production models for finished products that were found far beyond the workshop context, in the funeral pit of the Qin First Emperor's tomb complex. I argue that these models could also be verified in the production of triggers for the Terracotta Army, based on the data from the finished objects and their spatial distributions (also see Li et al., 2014).

The organisation of labour would have had an impact on the trigger's assembly. As a starting hypothesis, if the production of these triggers was organised in a flow line model, we could expect large numbers of each part to be cast by separate units of craftspeople, and then these parts would be assembled together later, at a different stage of the flow line, perhaps by another set of craftspeople. In this case, we could expect a higher degree of mixing between the different part groups identified above. On the other hand, in a batch or cellular production model, a single set of craftspeople would cast and assemble all the required parts. Using the triggers as the case in point, one single unit in the workshop could cast a relatively small number of parts A, B, and C, and assemble them together before producing the next batch. Such a production model would result in more homogenous and concentrated patterns in trigger part groups.

To illustrate an initial approach to this question, a visual chart was created to show, in Figure 5.17, the assembly of different parts into a trigger. All bevelled parts C have a notched part B, and are assembled with a part Ag2. However, there does not seem to be a perfect correspondence between subgroups, as some notched parts B go with Ag1 and are assembled with curved parts C.

To recapitulate, the recurrent association between part C bevelled and part B notched, and between these and Ag2 could indicate some form of holistic production (batch production or cellular production), where a single work unit would make and assemble all the parts of a given trigger. However, Ag1 pieces also appear assembled with notched part B and curved part C pieces. This suggests that these connections are not systematic, and perhaps it is more likely that the production operated as a flow-line: each unit working on a separate part, and a different unit assembling them together – which would explain the mix-ups. In any case, it is possible that the parts were not completely interchangeable for the triggers to function, which would limit the degree of freedom for combination even if different units worked on them, requiring them to operate quite closely and according to a similar plan. On some triggers, the same character was carved on each part of an assembled trigger, which is indicative of such limits on trigger assembly (Li and Gao, 2010).

The analysis of metric data made it possible to subdivide the above groups, based on micro-features, into further subgroups that could be related to different work units or casting moulds (also see Li et al., 2014). Part A pieces are divided into 4 groups, while parts B and C include 5 groups each – in addition to Ag4, Bg3, and Cg4, assembled as a special cased trigger. Further investigation into the assembly patterns incorporating the dimensional differences was therefore carried out using the PCA plots, in order to better define the correlations between the subgroups of these three different trigger parts and approach the organisation of production accordingly.

Figure 5.18 presents three PCA plots of the metric data for trigger part B: (a) part B subgroups marked according to whether the parts were notched/unnotched, as also seen in section 5.2.4; (b) part B subgroups coloured depending on the part A to which they are attached; and (c) part B data plot coloured depending on the associated part C. In plot (b), it is clear that Ag1 corresponds to a dimensionally defined Bg1, the left cluster with the smaller types of part B. Most of Ag2 assemble with Bg2, both notched and unnotched. The parts Bg2 attached to Ag3 tend to be clustered towards the left corner of Bg2. The plot (c) shows that all the pieces in group Cg3b correspond to Bg2n, as shown in the bar chart above, and Cg2 mainly corresponds to Bg1n and Bg1u.

Figure 5.19 shows the PCA of dimensions and assembly corresponding to part C: (a) part C groups marked as curved/bevelled, as shown in section 5.2.4.3; (b) part C data plot coloured depending on the associated part A; and (c) part C data plot coloured depending on the corresponding part B. Again, we can see a reasonably good correspondence between part A, part B, and part C subgroups.

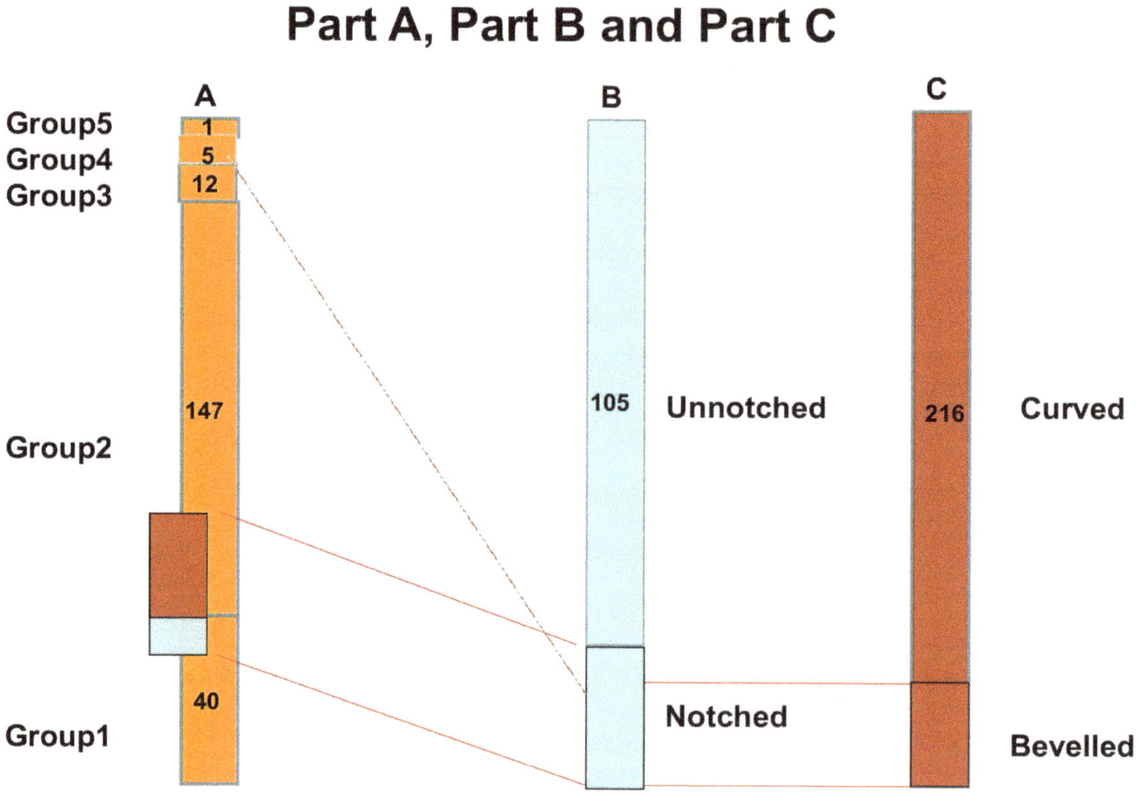

Fig. 5.17. Assembly patterns of trigger parts based on micro-features.

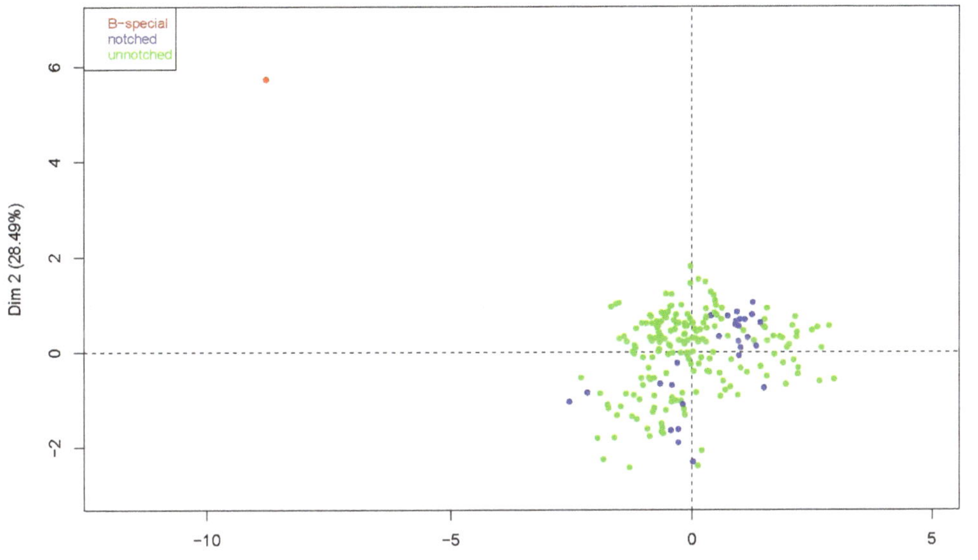

(a) part B

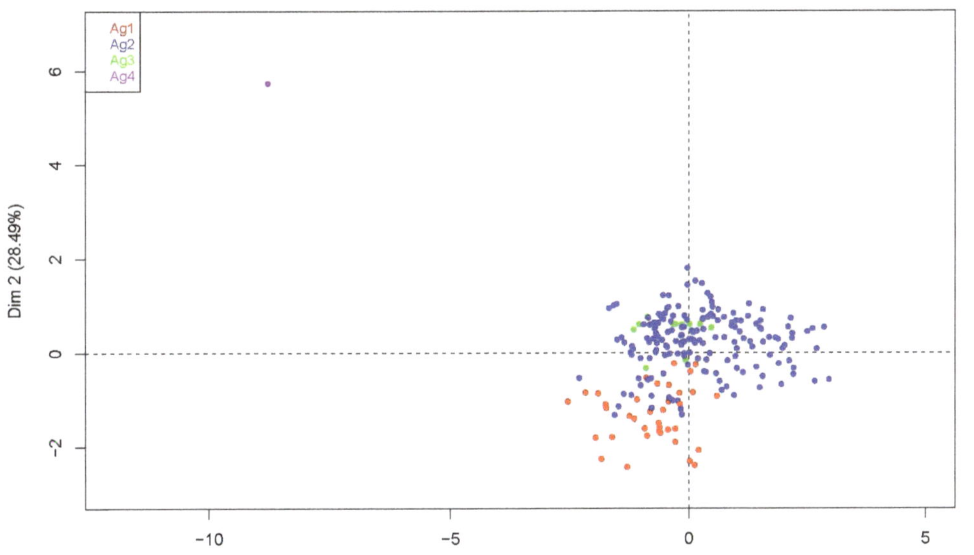

(b) A to B

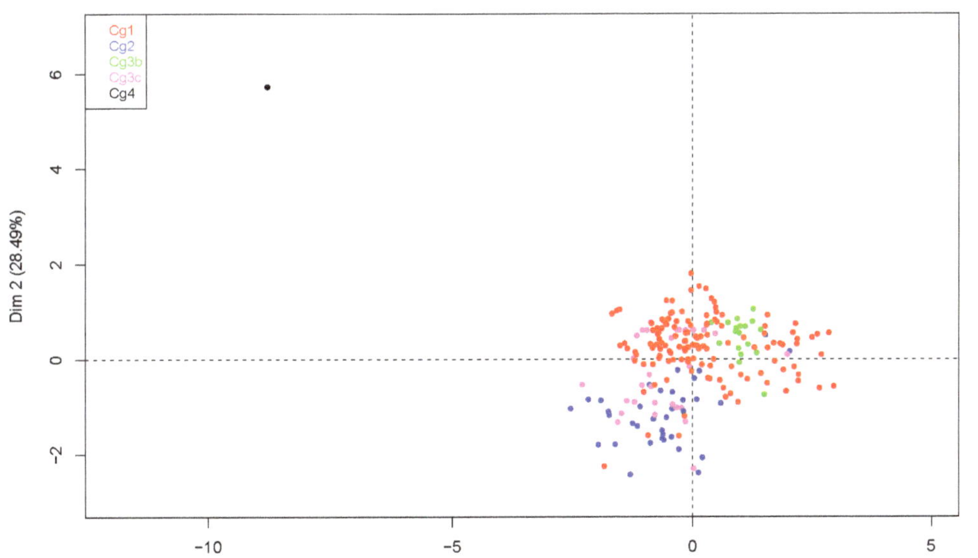

(c) C to B

Fig. 5.18. Trigger assembly, part A to B and C to B.

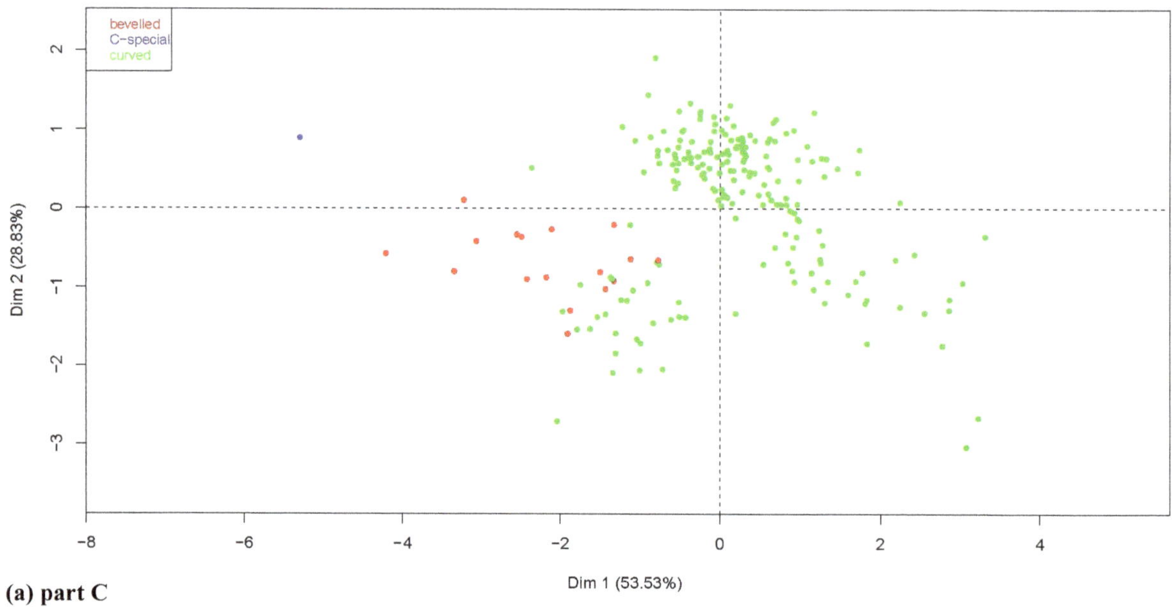

(a) part C

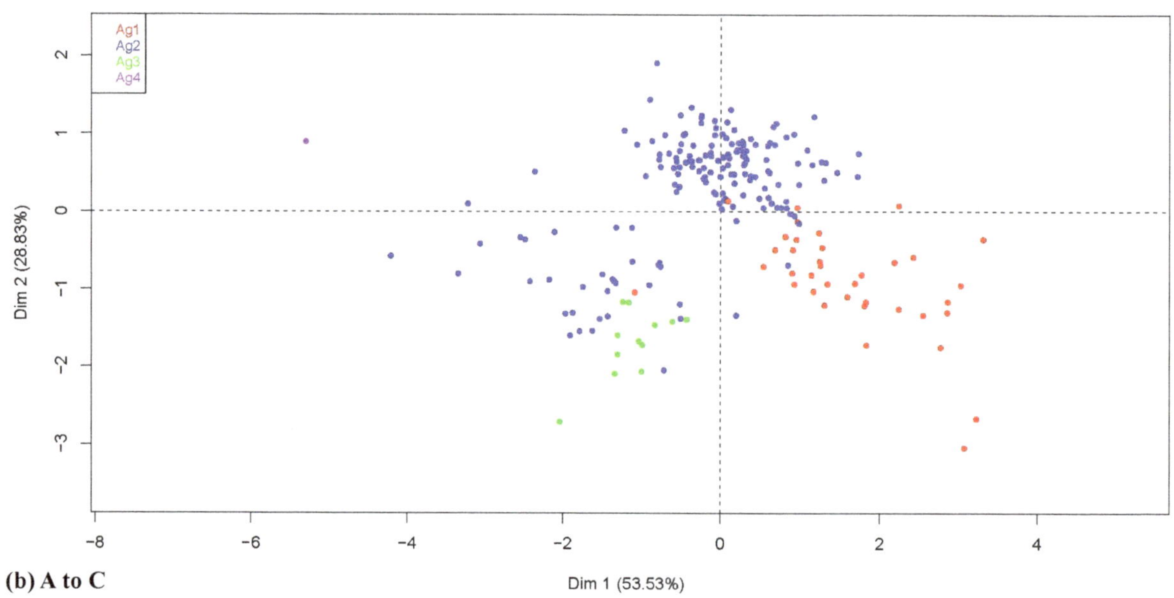

(b) A to C

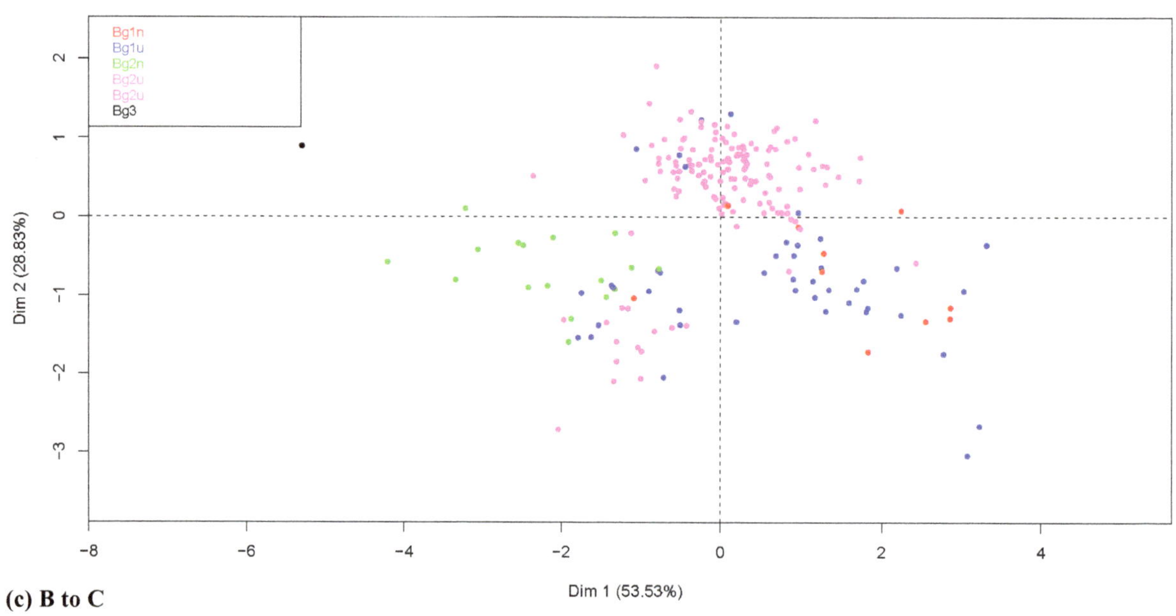

(c) B to C

Fig. 5.19. Trigger assembly, part A to C and B to C.

Figure 5.20 shows the PCA plots of trigger part A: (a) part A subgroups as shown in section 5.2.4.1; (b) part A data plot coloured depending on the assembled part B; and (c) part A data plot coloured depending on the attached part C. Bg1n and Bg2n appear attached to Ag2 and Ag1, respectively, which suggests two different events of notched parts B being assembled separately with two groups of parts A. The correspondence between dimensional groups of parts B and C seems less obvious, with significant overlaps between subgroups, but they do not appear to be random.

Based on the above PCA plots, an illustration of the complicated assembly of the trigger part groups is presented below (Fig. 5.21):

Altogether, 14 groups of complete triggers with their assembled parts (henceforth named AssemG) are listed in the following table. The trigger count emphasises that there are 8 main assembled groups (AssemG), in addition to 5 possible marginal or batch mixing groups with a single or two triggers each, as well as the exceptional cased trigger, AssemG14 (Table 5.5).

We can reconfirm the identified AssemG groups of triggers in the following PCA plots (Fig. 5.22), where the main AssemGs 2, 5, 6, 7, 8, 9, 11, and 12 are all relatively well clustered.

Overall, the typological, metric, and statistical analyses show a good correspondence of the different subgroups of parts A, B, and C. Even though some groups are too similar dimensionally to distinguish them statistically based on measurements alone, the overall impression is quite convincing: each assembled group is relatively homogeneous internally, and marginally different from the next (Fig. 5.22). This pattern strongly suggests that a single unit would have produced parts A, B, and C and assembled them together, rather than obtaining these from different production lines. The finished triggers were thus mainly made according to a cellular production model, with little room for batch mixing in the organisation of production.

Interestingly, some inscriptions on the triggers discussed in the previous chapter (for details, see Chapter 4.3.4; and also Li and Gao, 2010) correspond with assembled trigger groups (Fig. 5.23; also Appendix 6). For example, four groups of triggers have inscriptions made at Ba locations. Of these, triggers with 工 (*Gong*) inscriptions (n=17) were mainly from AssemG9 (Fig. 5.22), but with two exceptions from AssemG6 (which is closely related to AssemG9 in terms of its component parts). Three stem-branch characters (酉, 丙, 外 and) found at the Ba location are all in AssemG7 (Fig. 5.22). The specific linkages for this example are shown in figure12 and demonstrate a possible association with cells in the workshop.

This workshop cellular production applies at least to the entire trigger as discussed above, but possibly also to the entire crossbow. The latter possibility is difficult to ascertain because the wood or bamboo making up the crossbow are all rotted and it is thus impossible to verify their details now. The assembled triggers ended up in the pit with the terracotta warriors after being stored in the arsenal and being transported to the tomb complex (with no conclusive evidence for them ever having been used on the battlefield). These triggers were then placed into the pit to equip the terracotta warriors in battle formation. The interest of this research project is not only in the trigger assembling procedures for interpreting the organisation of production, but also in the spatial patterns concerning where they are found in the pit. These spatial patterns of the triggers in the pit can be employed for further investigation regarding the organisation of the storage, delivery, and placement, as well as for establishing links with previous workshop practices.

5.6 Spatial analysis

The point pattern analysis discussed in Chapter 2 is a method that has become popular in fields such as ecology and astronomy over the past 20–30 years, and has also been employed in archaeology (Orton, 2004; Gelfand et al., 2010; Bevan et al., 2013). As discussed earlier, this method appears highly relevant to this project, both for clarifying aspects concerning the spatial patterns of artefacts in Pit 1 of the Qin First Emperor's tomb complex in general, and specifically for the bronze triggers discussed in this chapter (also see Li et al., 2014).

The bronze weapons were digitised as vector points in the funeral pit of the tomb complex of Qin's First Emperor, based on the distribution map of the bronze weapons published in the original excavation report, and verified against the archived excavation records (see Chapter 1, section 1.3.2).

5.6.1 Parameters affecting the spatial patterns

The spatial patterns of the bronze triggers in the pit were affected by two main factors: i) the relation to the battle formation of the terracotta warriors as laid out within an underground earth-wooden tunnel structure; and ii) the organisation of the labour force during the production of the triggers and during the placement of weapons in the pit. During the Qin period, armies consisted of chariots, cavalry, infantry, and crossbowmen; the crossbowmen were normally arranged in front or on the flanks of the force to enable them to fire the arrows over a long distance before the enemy came close (Yuan, 1990; Yates, 2007). The posture of the terracotta warriors can also provide us with information about the weapons they originally held, as the wooden and bamboo parts of the weapons have all rotted away, with only metal pieces left on the floor. The first factor in the spatial analysis relates to the battle formation (Fig. 5.24) with some unarmoured crossbowmen as part of the vanguard in the front, some very selectively positioned crossbowmen at the back of the chariots in the middle corridors, and a predominance of crossbowmen along the army's flanks. The overall clustered pattern of

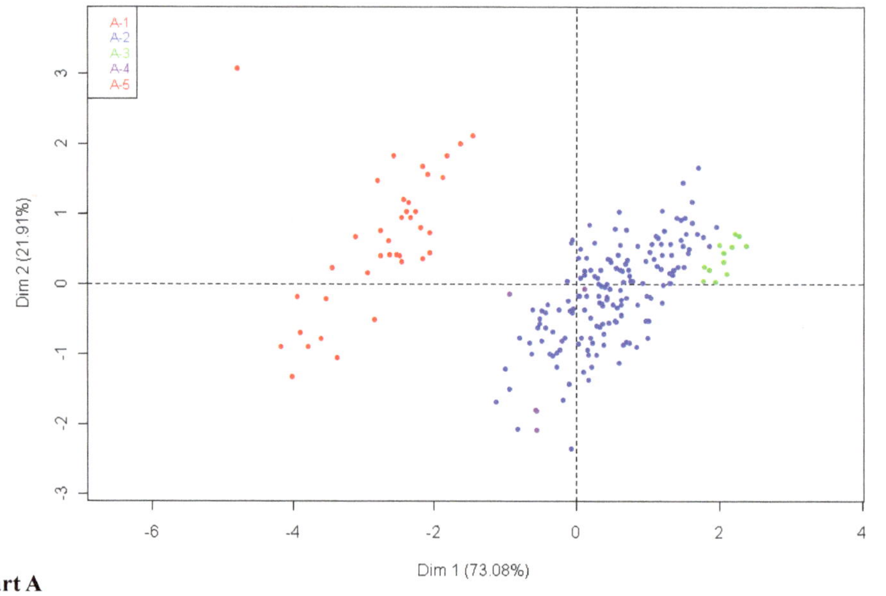

(a) Part A

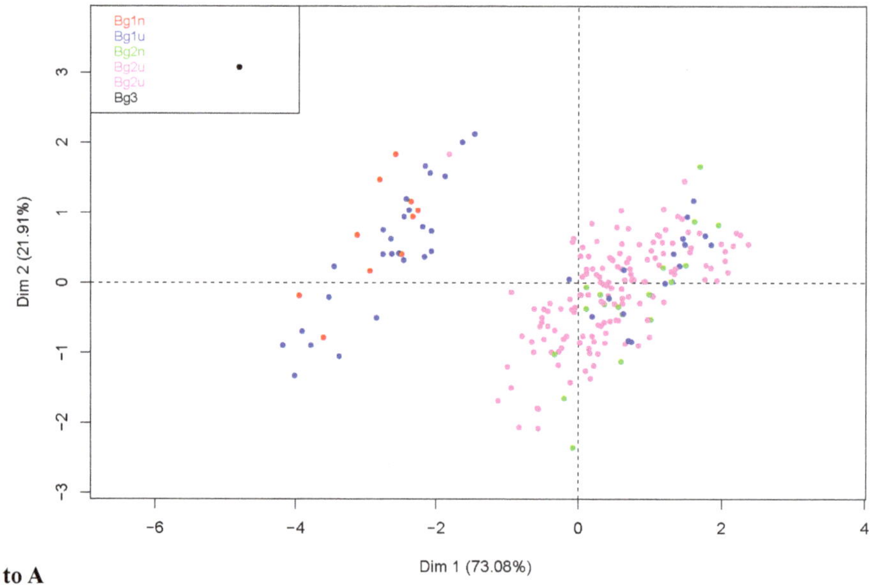

(b) B to A

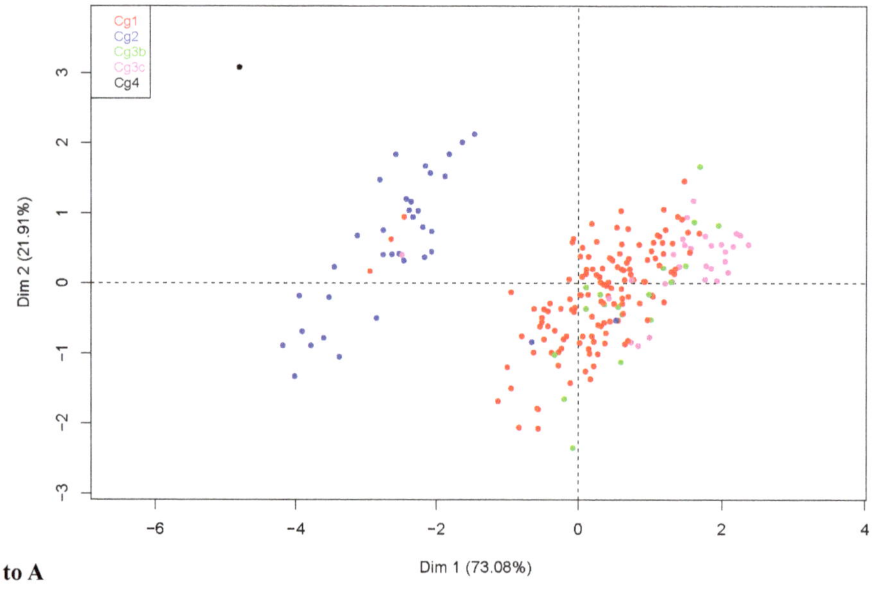

(c) C to A

Fig. 5.20. Trigger assembly, part B to A and C to A.

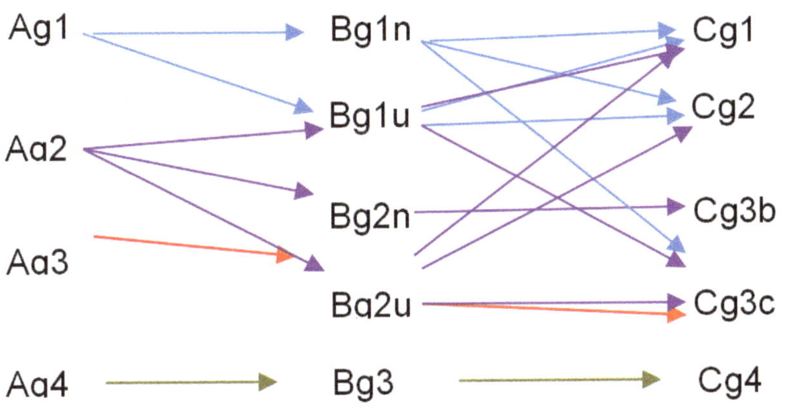

Fig. 5.21. Assembly of trigger parts A, B, and C.

Table 5.5. Assembled triggers and their count.

No.	Assemble ABC	Count of triggers	comments
AssemG1	Ag1-Bg1n-Cg1	1	
AssemG2	Ag1-Bg1n-Cg2	8	main
AssemG3	Ag1-Bg1n-Cg3c	1	
AssemG4	Ag1-Bg1u-Cg1	2	
AssemG5	Ag1-Bg1u-Cg2	24	main
AssemG6	Ag2-Bg1u-Cg1	5	main
AssemG7	Ag2-Bg1u-Cg3c	13	main
AssemG8	Ag2-Bg2n-Cg3b	18	main
AssemG9	Ag2-Bg2u-Cg1	125	main
AssemG10	Ag2-Bg2u-Cg2	2	
AssemG11	Ag2-Bg2u-Cg3c	4	main
AssemG12	Ag3-Bg2u-Cg3c	12	main
AssemG13	Ag1-Bg2u-Cg2	1	
AssemG14	Ag4-Bg3-Cg4	1	special

the triggers at a large and coarse scale is therefore driven, first by the tunnel structure of Pit 1 and then by the battle formation itself, and this general trigger distribution also matches the spatial arrangement of arrows in the pit (Fig. 5.25).

The second factor relates to the different types of triggers, and it is important to distinguish it from the first factor in spatial analysis. In this chapter, typological, dimensional, and assembled patterns have been used to distinguish groups of triggers with distinctive characteristics that suggest production procedures, labour, and technological organisation – most likely, batch production in semi-autonomous cells (also see Li et al., 2014). In addition to the relationship between the triggers, the arrows, and the battle formation, it is tempting to assume that these trigger assembly groups may reveal spatial signatures relevant for the organisation of labour during the crossbows' transportation and placement in the pit. Aspects of the operational chain (chaîne opératoire) employed during the production and transportation of the triggers can be inferred from the spatial patterns of the triggers. Craftspeople and workers were involved in the workshop and stores, and they may also have been involved in creating the battle formations, which would mean that they also worked within the pits. All these factors influence the spatial patterns. However, in reality, the weapon patterns are far more complex than one would expect, and it is therefore important to be contextually sensitive when considering these weapons.

In Chapter 2, I have already discussed the possible distributions that can be inferred from point patterns, ranging from spatial randomness to regularity and clustering, sometimes with the implication that the pattern is driven by particular human activities. In the case of the weapons, a regular or uniform pattern might be considered to reflect a form of intentional arrangement by the workers. Clustering of weapons in the pit may result from a number of factors, but the transportation of the products from different workshops or different units of craftspeople to the pit in bundles would have been influential factors. In contrast, random distributions can be treated as the statistical null-hypothesis, indicating that no specific constraints dictated the particular location of each trigger in the pit. More precisely, they might suggest a mixing up of the spatial distribution in which the weapons were originally stored in the workshop or arsenal at some point before they ended up in the pit.

To avoid using only personal intuition for the assessment of spatial patterns, spatial statistical methods are useful. They provide a way to characterise the point patterns and to examine the degree to which any given assembled trigger group is clustered or regularly spaced, not within the tunnel structure, but within the overall distribution of triggers (also see Bevan et al., 2014b; Li et al., 2014). For this particular project, a pair correlation function using R software, a multiscalar approach to point pattern analysis, has been applied in an attempt to interpret the distribution of the triggers (Baddeley, 2008).

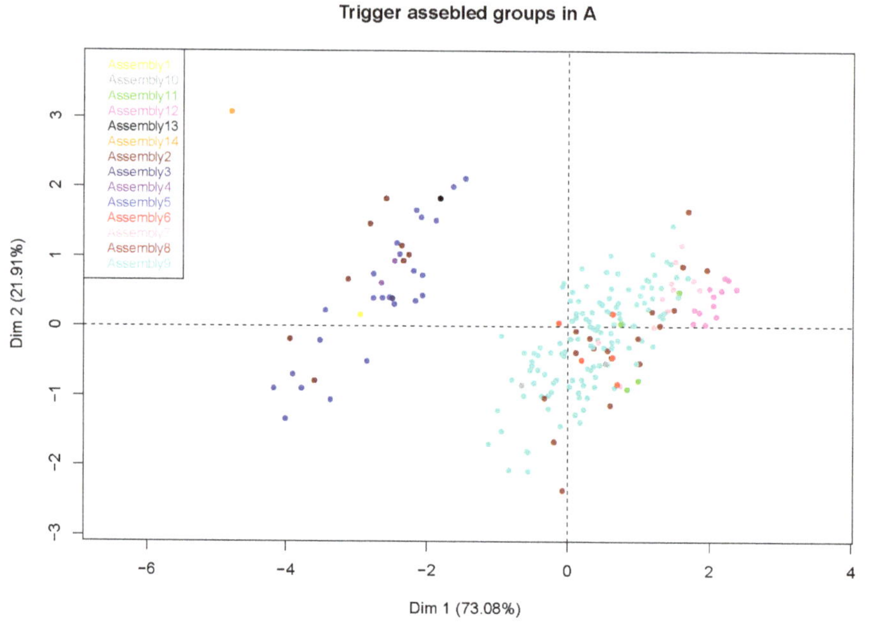

(a)

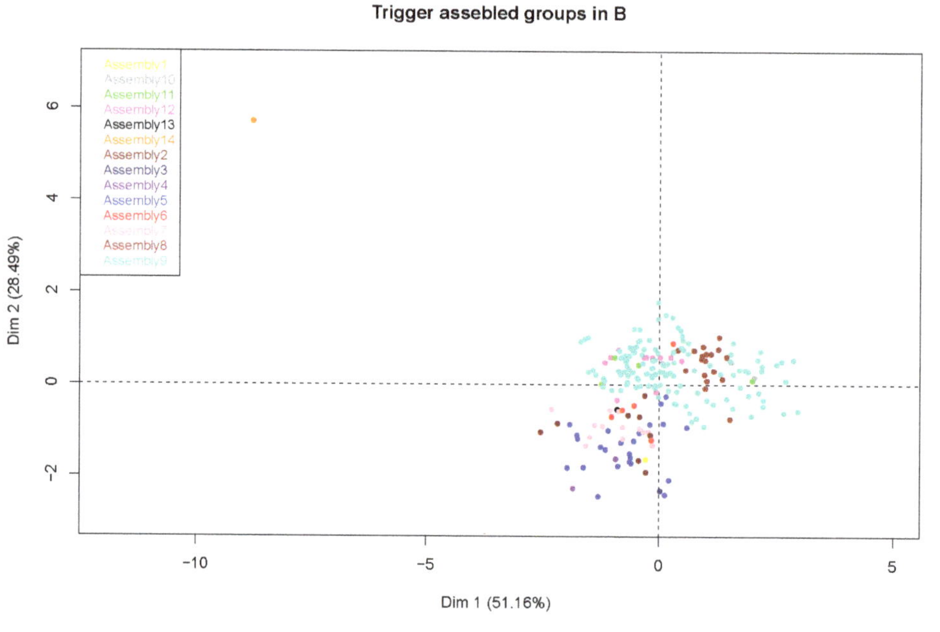

(b)

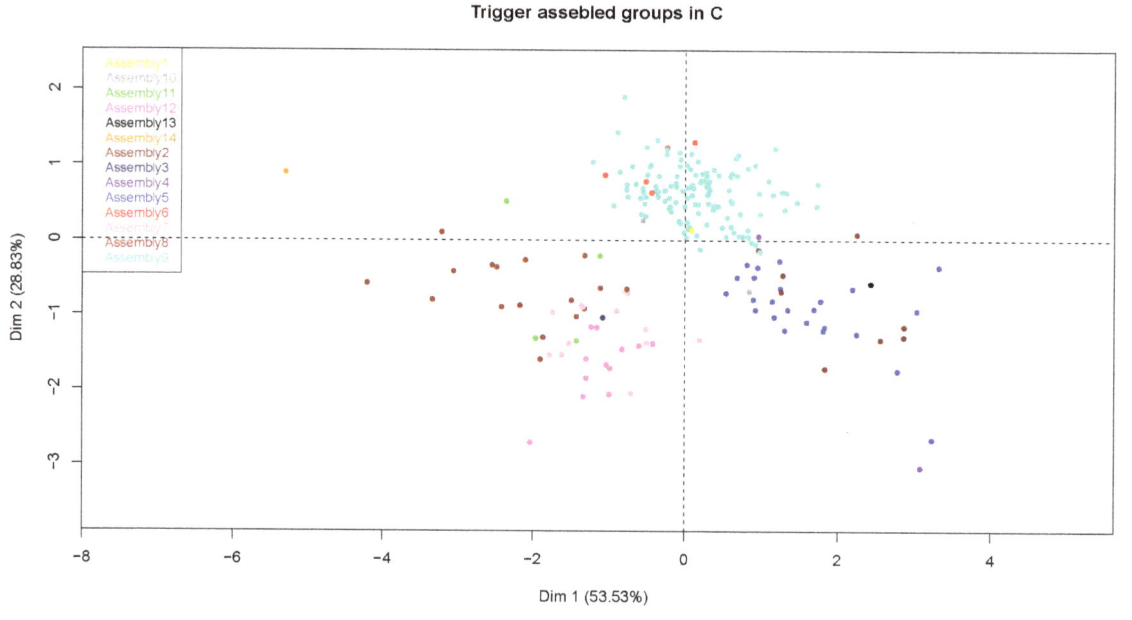

(c)

Fig. 5.22. Concentrated patterns of assembled trigger groups.

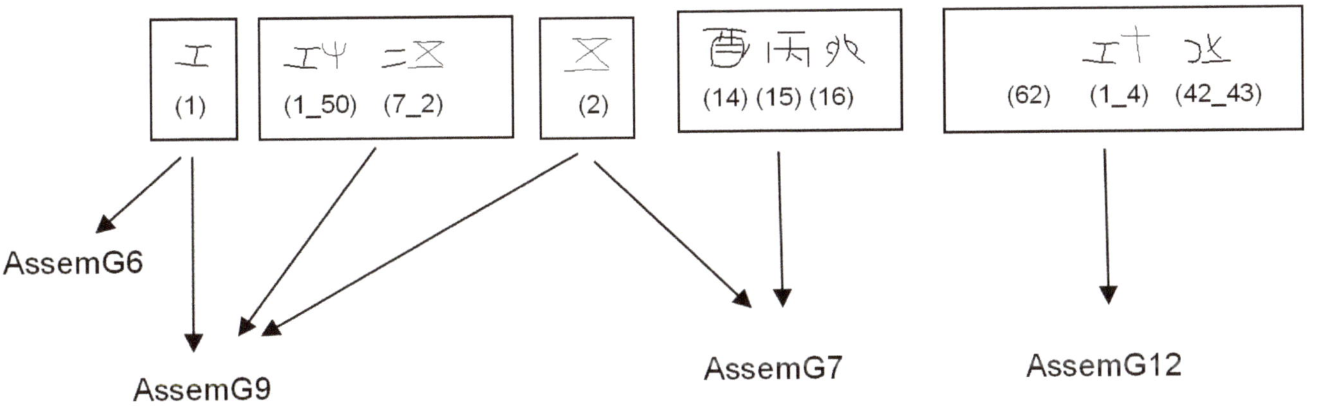

Fig. 5.23. The association between inscriptions at location Ba and assembled trigger groups (inscription codes are shown in brackets).

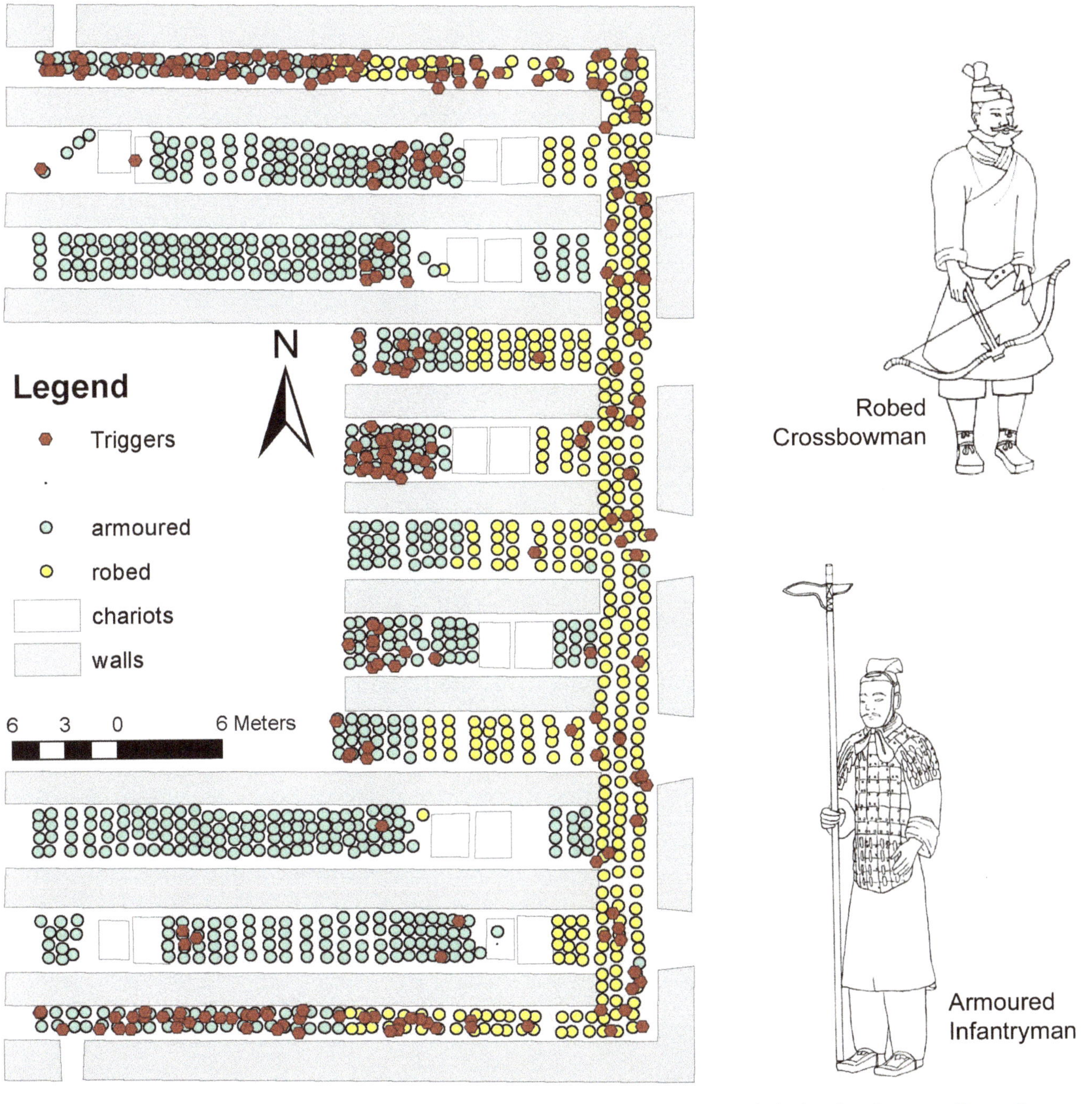

Fig. 5.24. Left: spatial distribution of crossbowmen in battle formation and triggers in Pit 1, showing the generally good correspondence between the two. Right: drawings of a crossbowman carrying a crossbow and an infantryman holding a halberd (courtesy of Ma Yu).

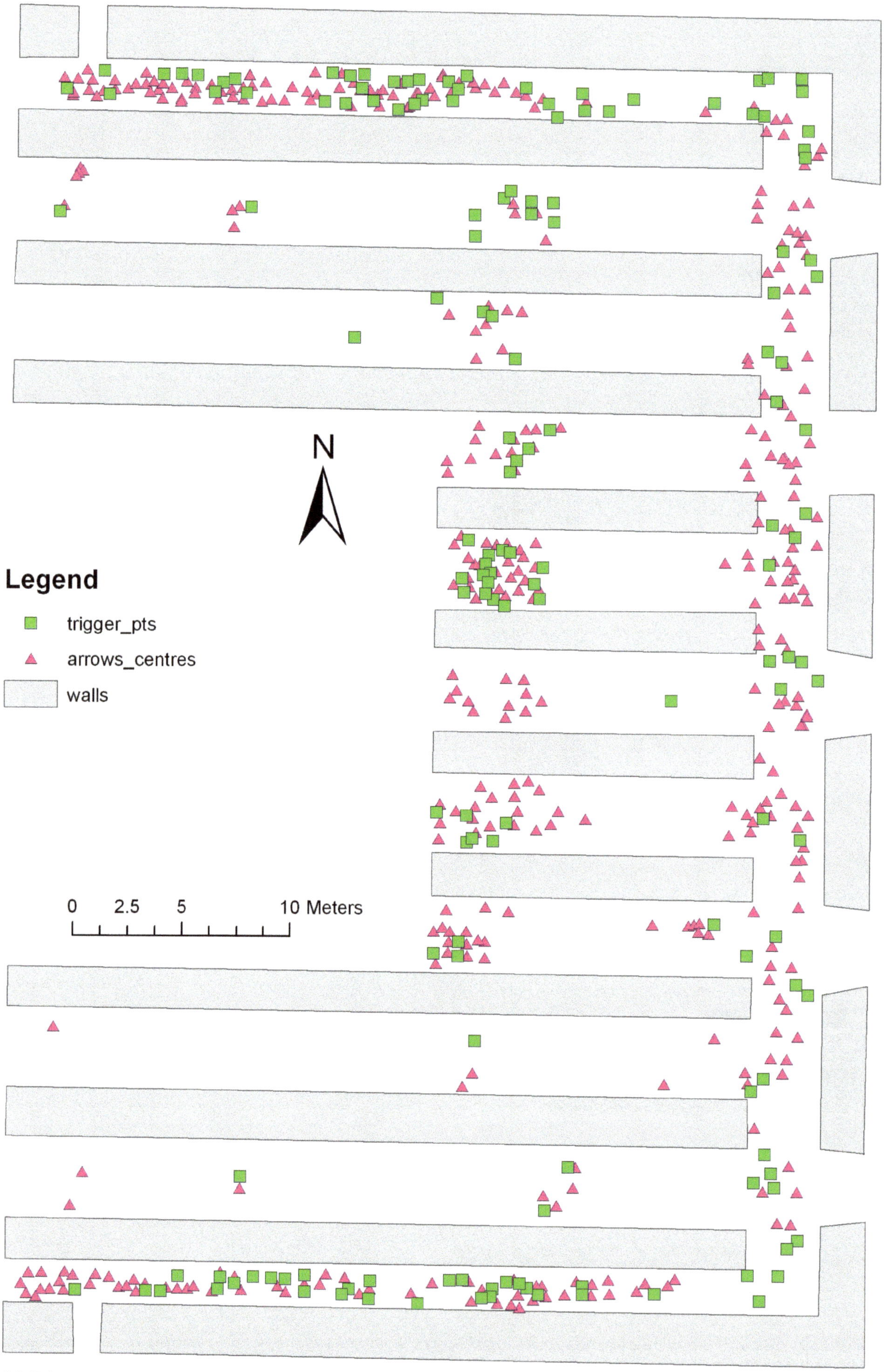

Fig. 5.25. Triggers and arrows in the easternmost five trenches of Pit 1.

5.6.2 A pair correlation function

A large number of statistical tools have been employed in archaeological research to investigate whether the observed points are clustered together, spaced evenly apart, or distributed randomly; examples of such tools include the Nearest Neighbour Index (NNI), the Ripley K and L functions (Ripley, 1977; Orton, 2004; Bevan and Conolly, 2006). The NNI is based on an assessment of the distance between a point and its nearest neighbour, but this only provides a global and shortest-distance measure of clustering, randomness, or regularity in the pattern. It characterises the point pattern at only one scale, when what we would really like to explore are multiple scales and distances for spatial patterning. Ripley's K and L functions are well known and more sophisticated density-based statistical analysis methods employed to measure patterns at a variety of spatial scales, but here the use of a pair correlation function is preferred, which provides a visual image with a combination of the map and statistical summary. Essentially, this function considers each point in the dataset in turn and measures the density of other points found in a series of increasing distance bands away from it.

For this, a form of Monte Carlo simulation method will be employed for testing the spatial patterns of triggers. This is a technique used to approximate the probability of certain outcomes by running multiple runs, and originally named after the city of Monaco with its gambling games and its casinos (Robert and Casella, 2004). It takes advantage of the speed of modern computers and provides a powerful and flexible way of testing spatial patterns. The Monte Carlo simulations are not completely random but are drawn from the overall set of triggers. Therefore, they control for the spatial effects of the battle formation.

The observed results, summarised as an average point density per distance band, would then be compared to sets of expected values derived by the Monte Carlo simulation. If choosing to run 999 Monte Carlo simulations, this allows us to create an envelope of possible values if we assume a random pattern. When the observed values fall above or below the probability envelope generated by these Monte Carlo simulations, a significant pattern of clustering or regular spacing can be assumed (Orton, 2004; Bevan and Conolly, 2006). If the values are higher, the observed data are more clustered, while if the values are lower, they are indicative of more regular groupings.

The pair correlation function was generated in R software on the assembled trigger groups. The parameters were inputted as follows:

- Number of simulation runs = 999
- Minimum distance = 0 metres
- Maximum distance band = 10
- Boundary correction method = none
- Study area = five easternmost trenches of Pit 1
- Points = assembled trigger groups

5.6.3 Results obtained on assembled triggers

The application of the pair correlation function to the main assembled trigger groups presents interesting results. Figure 5.26a shows a small group of assembled triggers, AssemG 2, in the five easternmost trenches of Pit 1. These are notched in part B, curved in part C, and with an Ag1 handle. The spatial pattern of this group is uncertain and difficult to describe as random, clustered, or regularly spaced, justifying the use of spatial statistical methods to assess the scale pattern and to calculate the distances at which the value is present. Figure 5.26b shows the summary pair correlation function for AssemG 2 as a solid black line. The line slightly goes above the grey envelope at 2 and 7 metres. The grey-shaded area marks out a 95% envelope, meaning that if we were to have plotted each of the 999 simulated pair correlation function lines on this chart, 95% of them would have fallen within this envelope. This offers a useful statistical guide for inferring that in cases where the real, observed values are higher than this envelope, the observed pattern is clustered at that distance and scale, or, in cases where they fell below the envelope, they are likely to be regularly spaced (Bevan et al., 2013). The solid line is above the envelope at the points of 2 and 7 metres in Figure 5.26b. It indicates that this group is slightly clustered at about 2 and 7 metres distance.

Figure 5.27 demonstrates a relatively obvious pattern of clustering in the distribution map for AssemG 5, which is also confirmed by the pair correlation function analysis. The solid line in (b) falls above the envelope over the first 3 metres and then drops off, and then goes over again from 5 to 6.5 metres.

The pair correlation function from Figure 5.28 to Figure 5.31 mainly indicates a random pattern. AssemG 6 and 11 possibly present regular spacing over 2 metres, but the sample is too small to have statistical significance. AssemG12 shows slightly clustered at 4.5m, and may indicate the arrangement of triggers in the pit.

Figure 5.32 show strong clustering of triggers AssemG 8 in the first 3 metres. This assembled group includes notched part B and bevelled part C.

The assembled group 9 shows a very strong clustering within 10 metres (Fig. 5.33). These triggers are mainly concentrated on the two side corridors and front. Also, most triggers with '*Gong*' inscriptions are all located in these two flank corridors.

The results of the pair correlation function on the assembled triggers encourage us to build up from the static spatial distribution of the triggers to an understanding of active human behaviour and labour organisation (also see Bevan et al., 2014b; Li et al., 2014). Assembled trigger groups 5 and 8 display a strong clustering over the first 4–8 m. This indicates that these trigger groups were originally produced by different work units and were stored and

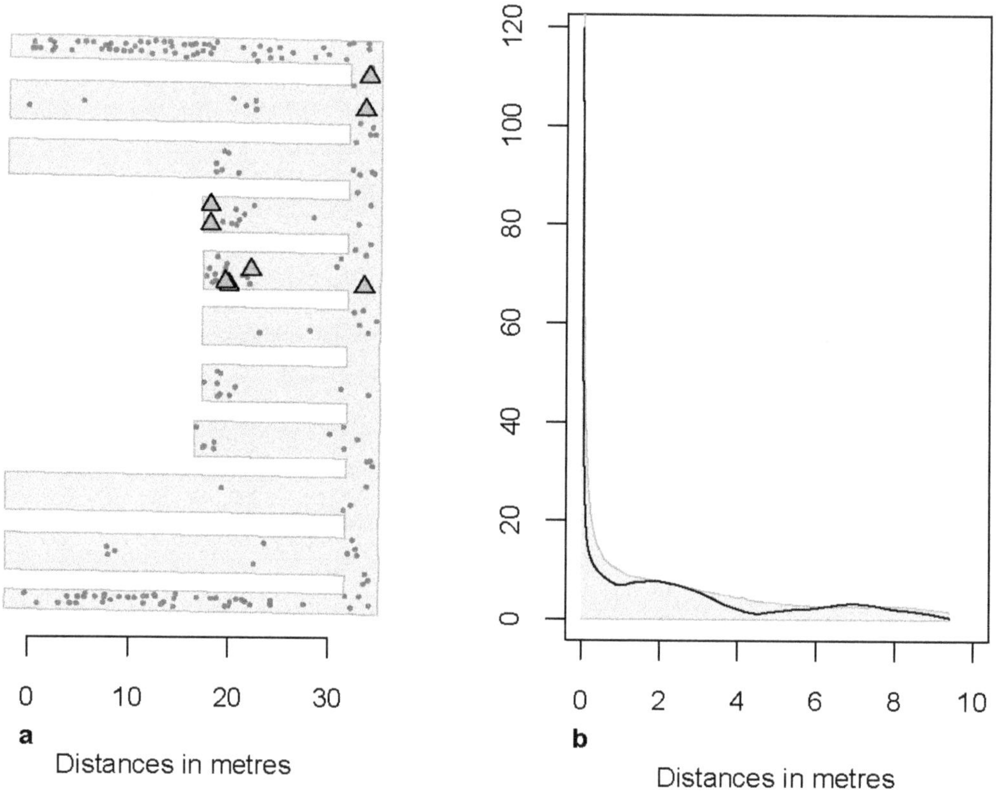

Fig. 5.26 The spatial distribution of Assembly Group 2 triggers (combining parts Ag1-Bg1n-Cg2, and shown as triangles) within the five easternmost trenches of Pit 1 (a), and a pair correlation function considering the spatial pattern over the first 10 metres.

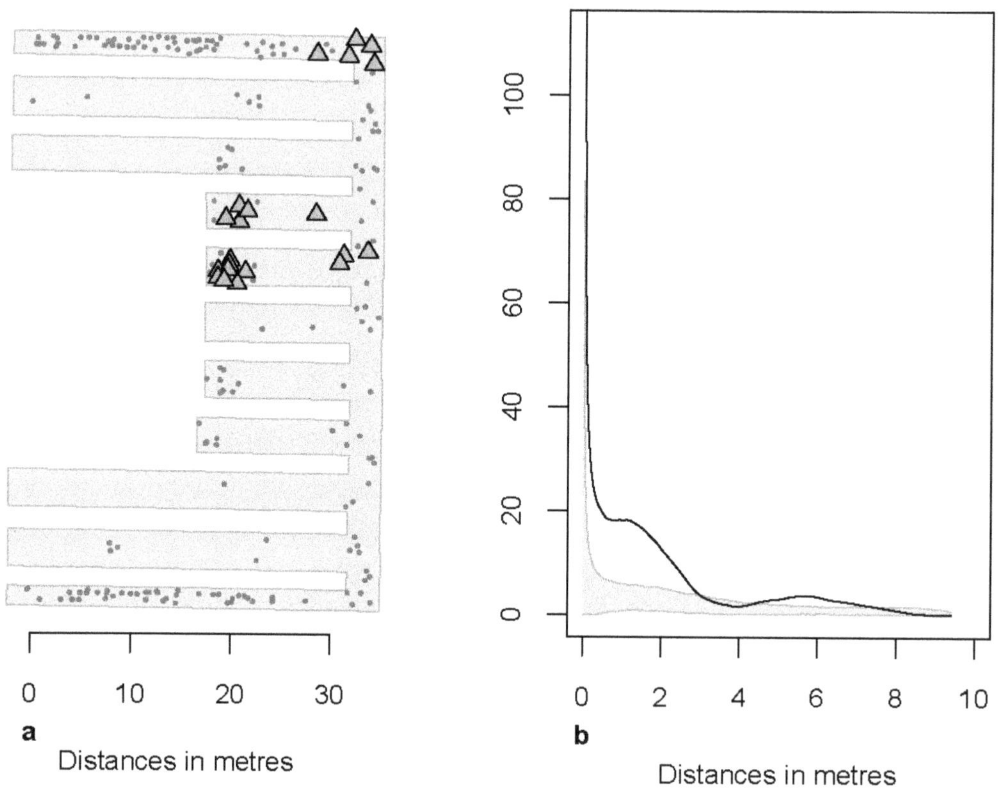

Fig. 5.27. The spatial distribution of Assembly Group 5 triggers (combining parts Ag1-Bg1u-Cg2A, and shown as triangles) within the five easternmost trenches of Pit 1 (a), and (b) a pair correlation function considering the spatial pattern over the first 10 metres.

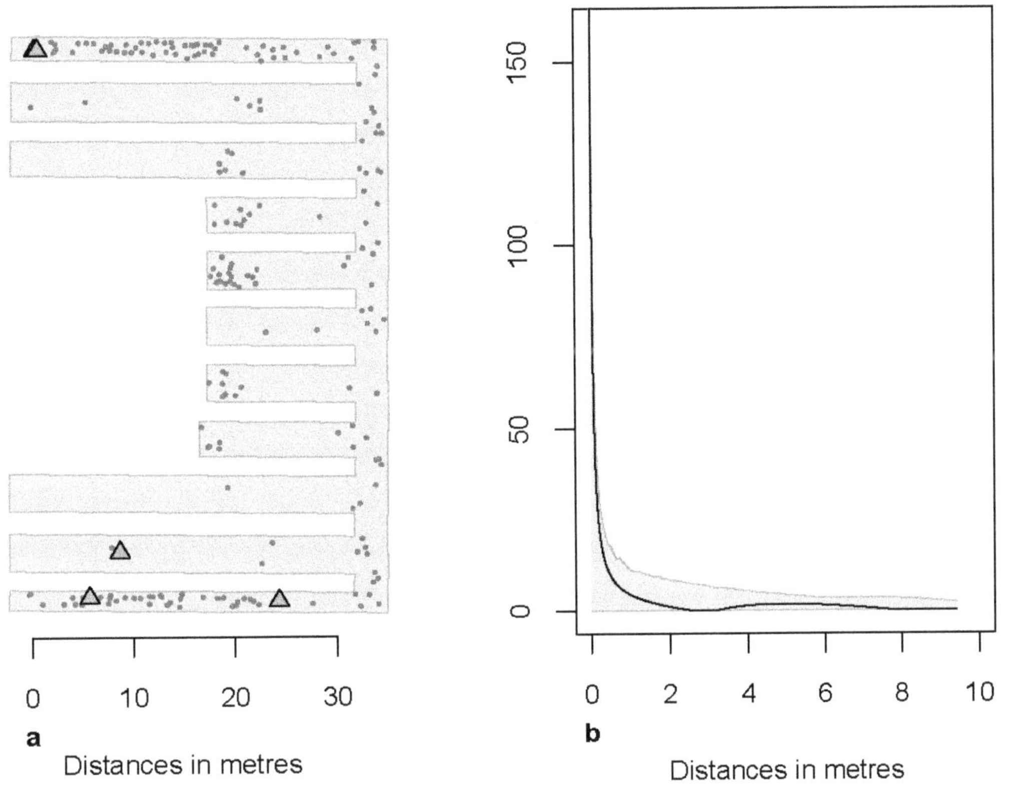

Fig. 5.28. The spatial distribution of Assembly Group 6 triggers (combining parts Ag2-Bg1u-Cg1, and shown as triangles) within the five easternmost trenches of Pit 1 (a), and (b) a pair correlation function considering the spatial pattern over the first 10 metres.

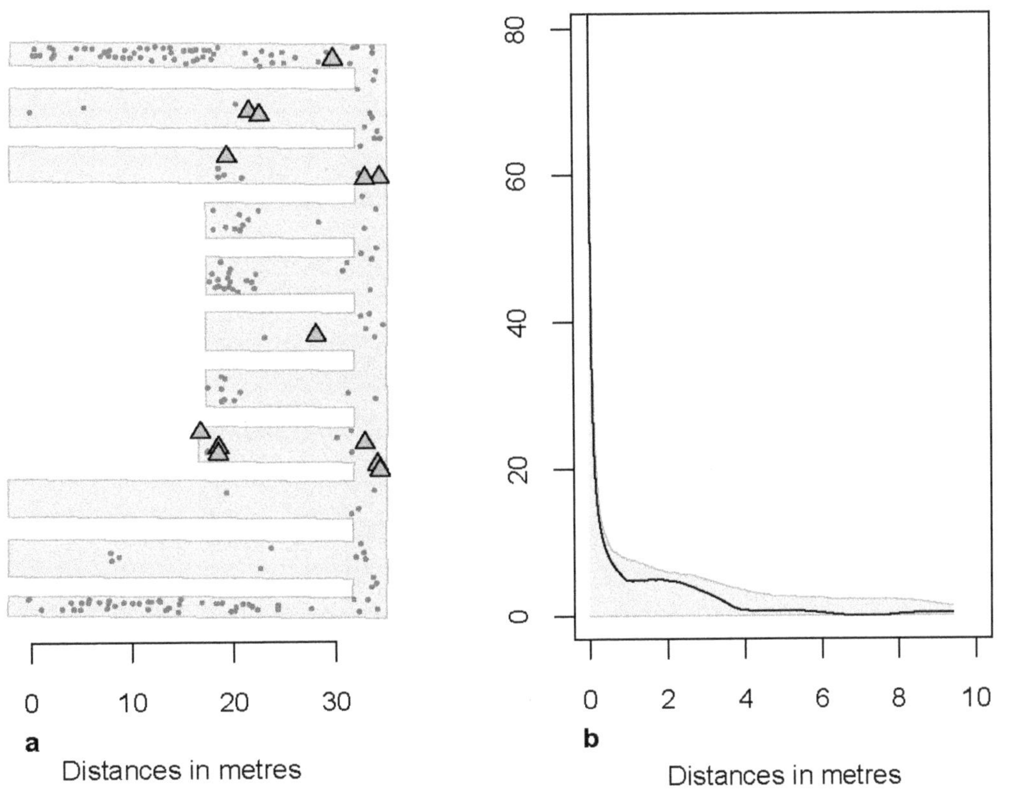

Fig. 5.29. The spatial distribution of Assembly Group 7 triggers (combining parts Ag2-Bg1u-Cg3c, and shown as triangles) within the five easternmost trenches of Pit 1 (a), and (b) a pair correlation function considering the spatial pattern over the first 10 metres.

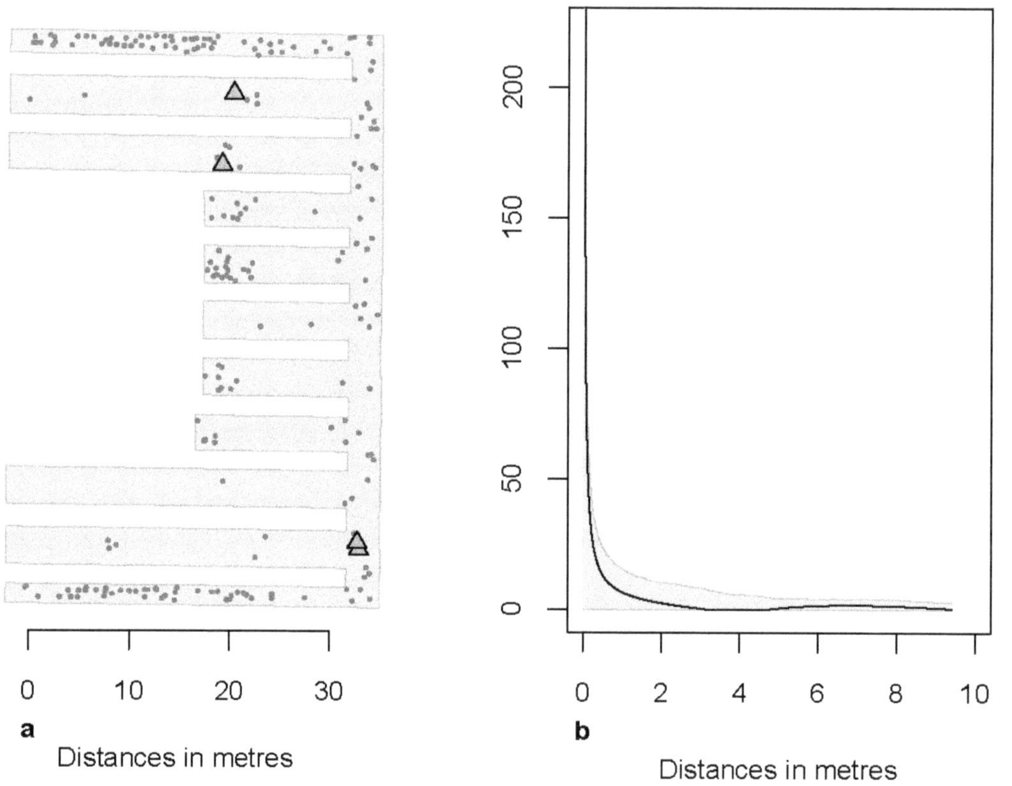

Fig. 5.30. The spatial distribution of Assembly Group 11 triggers (combining parts Ag2-Bg2u-Cg3c, and shown as triangles) within the five easternmost trenches of Pit 1 (a), and (b) a pair correlation function considering the spatial pattern over the first 10 metres.

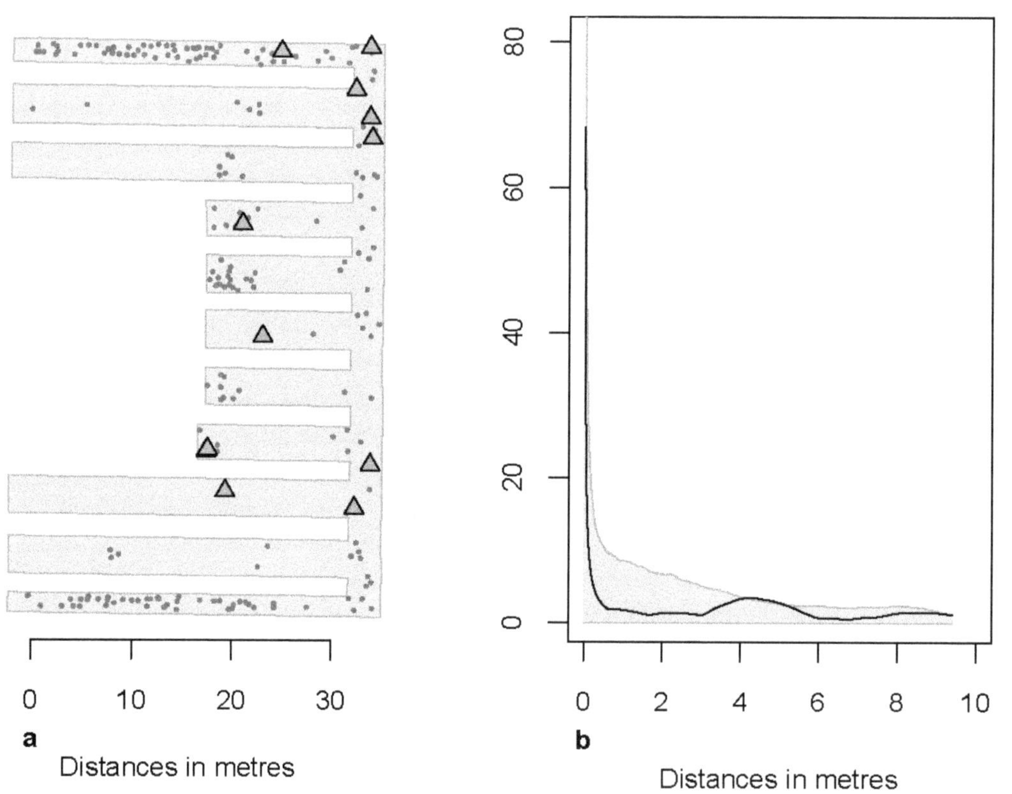

Fig. 5.31. The spatial distribution of Assembly Group 12 triggers (combining parts Ag3-Bg2u-Cg3c, and shown as triangles) within the five easternmost trenches of Pit 1 (a), and (b) a pair correlation function considering the spatial pattern over the first 10 metres.

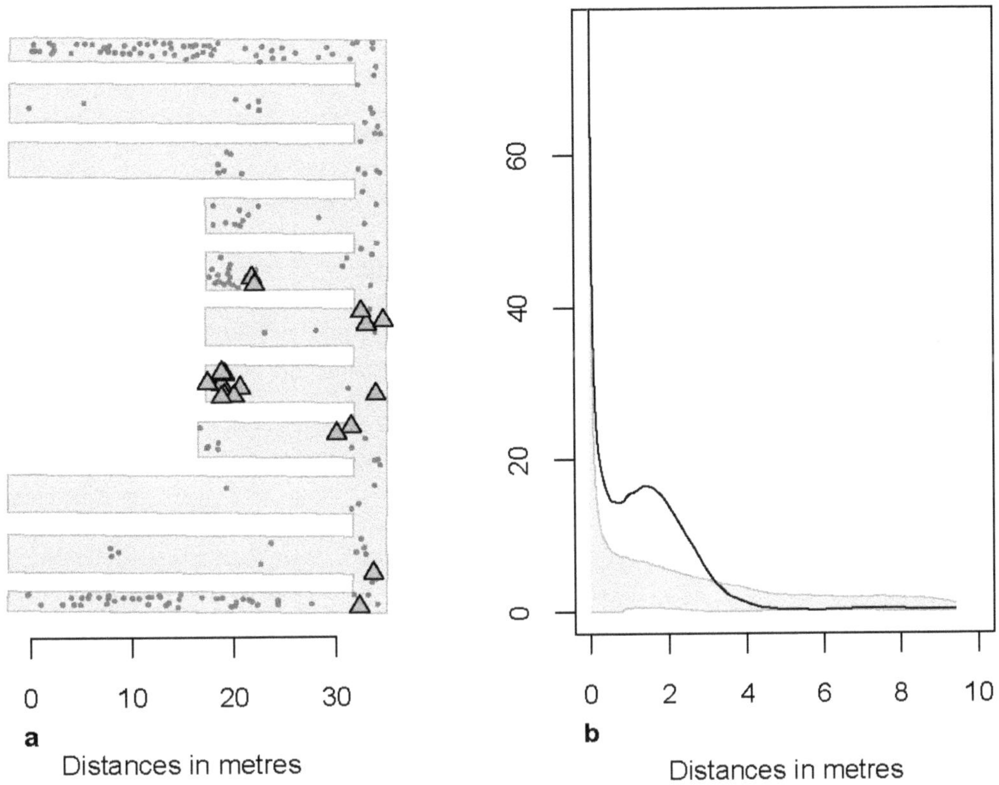

Fig. 5.32. The spatial distribution of Assembly Group 8 triggers (combining parts Ag2-Bg2n-Cg3b, and shown as triangles) within the five easternmost trenches of Pit 1 (a), and (b) a pair correlation function considering the spatial pattern over the first 10 metres.

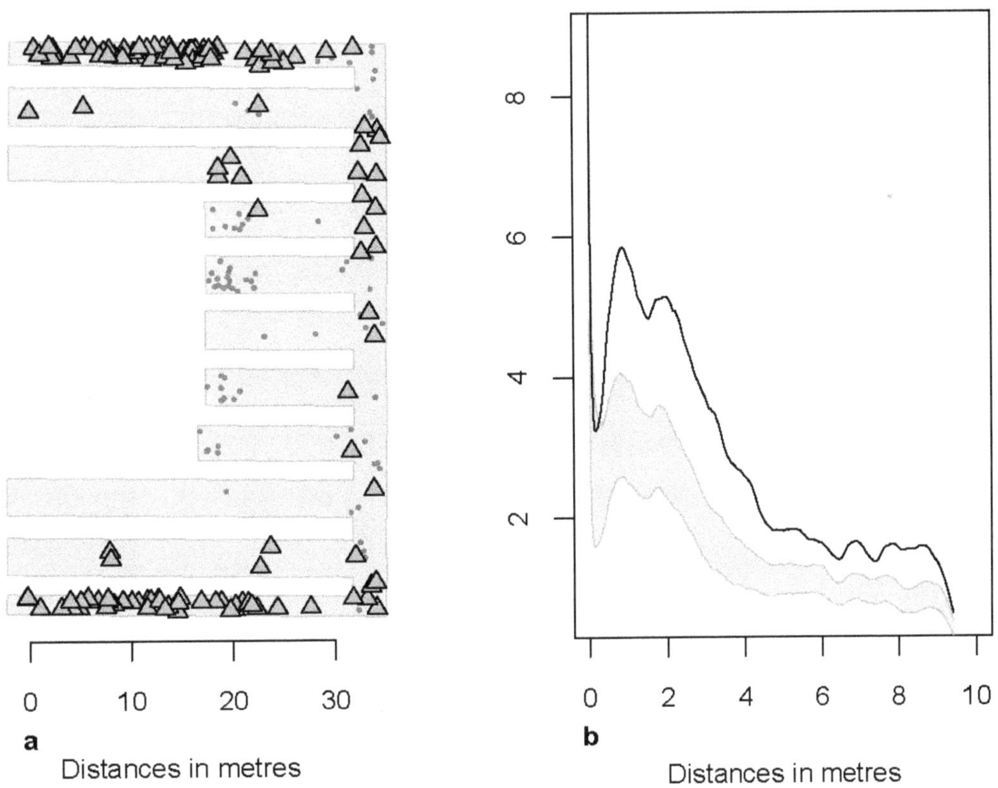

Fig. 5.33. The spatial distribution of Assembly Group 9 triggers (combining parts Ag2-Bg2u-Cg1, and shown as triangles) within the five easternmost trenches of Pit 1 (a), and (b) a pair correlation function considering the spatial pattern over the first 10 metres.

delivered separately to the pit. Their discovery in a clear cluster implies that different batches were generally not mixed when they were collected from the workshop or store (Table 5.6).

Some groups show a more random distribution, and these groups are also the ones that were closer together in terms of the measurement data. As such, it is possible that these groups were produced at the same workshop or working unit, but with many craftspeople involved in their production or many batches, with copy error leading to variation in these triggers. Even if they were produced by different units, their distribution suggests that they would have been mixed within a single storage, before being placed in the pit. This mixing could have occurred when the triggers were assembled and attached to the crossbows, possibly by different specialists.

5.6.4 Activity areas in the pit

The general patterns in the spatial distribution of the assembled triggers attached to the crossbows also allow the identification of what may be different activity areas associated with the placement of the weapons in the pit (Fig. 5.34). AssemG 2 (Ag1-Bg1n-Cg2) and 5 (Ag1-Bg1u-Cg2) are both clustered and overlapped (circled in red) in the north-eastern part of Pit 1, and AssemG 8 (Ag2-Bg2n-Cg3b) is clustered (circled in blue) at the south-eastern corner, while AssemG 9 (Ag2-Bg2u-Cg1) (circled in green) is concentrated on the southern and northern corridors. The two easternmost activity areas show much more trigger group diversity that the two in the northern- and southernmost corridors. The latter two are also the areas with all of the '*Sigong*' inscriptions, which may imply greater standardisation and governmental control for these. It is also possible that each of these areas was placed under the responsibility of a specific group of workers, each with their own provision of crossbows. These activity areas may have functioned in parallel, which would explain the overlaps between the different sections and remain consistent with an overall model of cellular organisation.

5.7 Summary

The typological and metric analysis of the different triggers and their constituent parts has made it possible to identify several subgroups, likely related to specific casting moulds employed by different workshops or production units (also see Li et al., 2014). The trigger parts in each group show a relatively high degree of standardisation. The highest CV values within types are shown to be below 5%, and therefore just above the human ability to perceive the dimensional difference without the help of instruments. However, these trigger parts were all cast with models and moulds. From this perspective, the degree of standardisation should be relatively higher, and CV values should be below 1.7%. Nevertheless, only some types of triggers reach such high degree of standardisation, and most of the CV values are above this point, indicating the limits of standardisation in the production of triggers. Some possible reasons for this were discussed, related to both the organisation of production and human perception (Li et al., 2014).

The generally good correspondence between specific subtypes of parts A, B, and C supports the hypothesis that the finished, assembled triggers were produced in batches in a cellular production model, rather than in a flow line production and assembly model (Li et al., 2014). This impression was ratified by the spatial clustering of some of the assembled trigger groups in the pit of the terracotta warriors, demonstrating a limited degree of mixing between trigger batches. The spatial analysis also revealed potential activity areas in the pit, where different groups of labourers, working with specific assembled crossbows, could have been put in charge of equipping the warriors in a given section of the pit. This production organisation model would have allowed several units to work in parallel, resulting in a more efficacious labour management that would be more adaptable to changes in the master plan.

Table 5.6. Pair correlation results on assembled trigger groups

AssemG	Patterns	Distance	Figure reference
AssemG2 (Ag1-Bg1n-Cg2)	Clustered	2 and 7m	Fig. 5.26
AssemG5 (Ag1-Bg1u-Cg2)	Clustered	0–3 and 5–6.5m	Fig. 5.27
AssemG6 (Ag2-Bg1u-Cg1)	Random		Fig. 5.28
AssemG7 (Ag2-Bg1u-Cg3c)	Random		Fig. 5.29
AssemG8 (Ag2-Bg2n-Cg3b)	Clustered	0–3m	Fig. 5.32
AssemG9 (Ag2-Bg2u-Cg1)	Clustered	0–10m	Fig. 5.33
AssemG11 (Ag2-Bg2u-Cg3c)	Random		Fig. 5.30
AssemG12 (Ag3-Bg2u-Cg3c)	Random clustered		Fig. 5.31

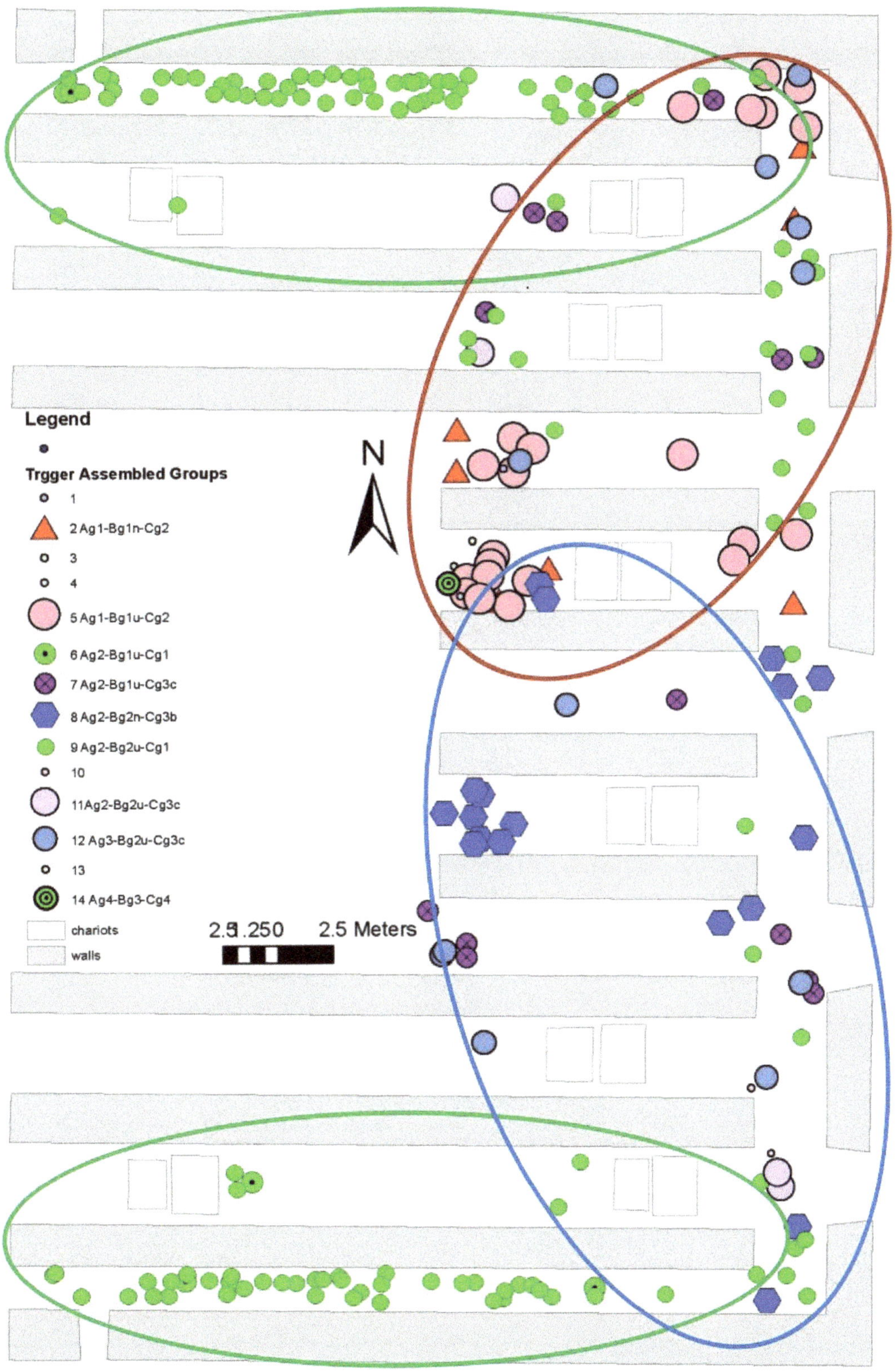

Fig. 5.34. Activity areas in Pit 1.

6

Bronze Arrows

6.1 Introduction

Compared to the mass-produced Qin triggers discussed in the previous chapter, arrows were consumed in larger numbers in ancient warfare, and it is therefore unsurprising that they were often mass-produced as well in past dynasties. However, it is unusual for archaeologists to find such a large assemblage of arrows together in a burial, on a battlefield or in a workshop and thereby have the chance to consider their standardisation and related labour organisation issues pertaining to their production. The Emperor Qin Shihuang's Terracotta Army included many archers and/or crossbowmen in the easternmost vanguard and flanks in Pit 1, and they were buried along with not only a large quantity of crossbows with bronze triggers but also with a much larger quantity of bronze arrows. This assemblage therefore provides a special opportunity for research.

An exceptionally large quantity of arrows, mainly grouped in bundles, was discovered in the different pits within the tomb complex. Altogether, a total of 43,412 arrows (now in storage at the Conservation Department, Museum of Emperor Qin Shihuang's Terracotta Army) have been available for this particular project. These include the arrows unearthed from the five easternmost trenches of Pit 1 excavated in the 1970s; those unearthed from other trenches in Pit 1 during the ongoing archaeological excavations; those discovered in Pit 2 of the terracotta warriors; and those excavated from other sites within and outside the Emperor Qin Shihuang's tomb complex.

Each arrow comprises a bronze head and a bronze tang, as well as a wooden or bamboo shaft fixed around the tang. While most of the wooden and bamboo shafts have rotted away during the intervening 2,000 years, the bronze arrowheads and tangs remain. Previous research has mainly focused on typological analysis, contextualising the development of these arrows within China's long history, in addition to some limited research on their chemical composition and finishing processes (Wang, 1980a, 1980b; Yuan et al., 1981a; Institute and Team, 1988; Yuan, 1990). As such, comparatively little attention has been paid to the systematic analysis of the bronze arrows or to the analysis of the spatial patterns of their distribution in the pit. Thus, such an approach presents a useful opportunity for interpreting how the arrows were standardised and how labour was organised during production. As in previous chapters, this section focuses on the arrows as a source of information about standardisation and labour organisation. The arrowheads and arrow-tangs were carefully measured and used to investigate the accuracy of this mass production and these metric results were then combined with those from instrumental analyses.

6.2 Research review

The quantity and quality of the bronze arrows have attracted much attention from Chinese and Western archaeologists, historians, and archaeometallurgists ever since their discovery (Yang, 1980; Yuan et al., 1981a; Institute and Team, 1988; Yuan, 1990; Research Committee, 1995; Portal, 2007; Shi, 2008). Yuan (1990) and Wang (1983) contributed basic research on their first-hand archaeological data from Pit 1 and from the trial trenches in Pit 2. Further research about the contextualisation of arrows and casting and finishing processes has been carried out in previous papers and monographs (Yuan et al., 1981a; Wang 1994; Research Committee, 1995; Shi, 2008).

Some scientific research on the arrows was also carried out in the 1980s (Wang, 1980a, 1980b; Yuan et al., 1981a) when seven arrows were sampled for their chemical composition analysis. Ongoing work carried out as a part of this project has greatly expanded the number of samples analysed.

Much effort has also been made by Chinese archaeologists to investigate the standardisation of the production of these weapons through limited chemical analyses and studies of the symmetrical shape of the arrows, of finishing processes, and of the assembling of the bronze arrows with wooden or bamboo shafts (Wang, 1980b, 1983; Yuan et al., 1981a; Yuan, 1990). However, all of these studies have so far lacked a rigorous theoretical framework, a large sample size, and statistical methodology. Nevertheless, such publications have influenced many other scholars in contextualising ancient Chinese military tactics from a historical perspective (Research Committee, 1995; Shi, 2008), and the evolution of the arrows' shape has been related to shift from the use of bows to crossbows.

Standardisation and labour organisation are thus not new themes for research on these arrows. However, to date only limited work has attempted to cover a large enough sample enabling observations of statistical value; in addition, spatial patterns were not employed to investigate the labour organisation behind the standardised production, the transportation of the arrows to the mausoleum, and their placement in the pit.

6.3 Arrow data

6.3.1 The archaeological origin of the bronze arrows

Before proceeding to a discussion of the origin of the arrows, it is necessary to clarify how such a large assemblage of arrows as the one studied in this project is described. In order to meet statistical demands and for the purposes of an easy description, the arrows were all grouped in various size bundles for this particular project, instead of using the concept of loose arrows as encountered in the excavation report (280 bundles and 10,895 loose arrows; Institute and Team, 1988). The bundles were divided into two groups: (a) one with an arrow count greater than or equal to 90 (≥90) and (b) one with an arrow count less than 90 (<90), including single arrows (statistical significance, see later Figure 6.3).

In the museum storage, altogether 43,412 arrows were archived, of which 37,348 arrows, in 680 bundles (bundle size varying from 1 to 200 arrows), were discovered in the five easternmost trenches of Pit 1 in the 1970s, with a relatively accurate spatial location. The rest of the arrows were discovered during ongoing archaeological excavations elsewhere in Pit 1 (about 5,312 arrows), in trial trenches in Pit 2 (about 1,227 arrows), and during other archaeological surveys or excavations in the tomb complex (about 40 arrows). There are also 436 arrows with an unclear location, because they were collected from the adjacent field by local farmers, or because the surface soil on which they were found had been disturbed for other reasons (see Table 6.1).

6.3.2 Bronze arrows from the five easternmost trenches of Pit 1

The bronze arrows unearthed during extensive excavations in the 1970s from the five easternmost trenches in Pit 1 constitute the primary focus of this study of dimensional and spatial patterns. These five trenches were completely excavated, with the terracotta warriors and horses from them being totally restored, and the weapons fully documented. Exceptionally, all the bronze arrows found inside these trenches were mapped and archived. These arrows were typically discovered beside the foot plates of the warriors, on the brick floor at the bottom of each corridor (Institute and Team, 1988; Fig. 6.1). As noted above, many of them appear in bundles containing variable numbers of arrows, including single arrows as described in this project. Digital recording was not possible in China in the 1970s, but the map published in the excavation report (Institute and Team, 1988) provided a basic spatial location of arrow bundles, and other archival records in the Conservation Department described the relative locations of the arrows with reference to the warriors' positions. These records can also be used to verify the spatial distribution of the arrows in the pit. This information formed the basis for the digital maps produced for this book.

Table 6.1. The archaeological origin of the bronze arrows.

Sum	Pit 1 main excavation	Pit 1 other excavations	Pit 2	Sites in the tomb complex	Other sites	Not clear
43,412	37,348	4,357	1227	40	4	436

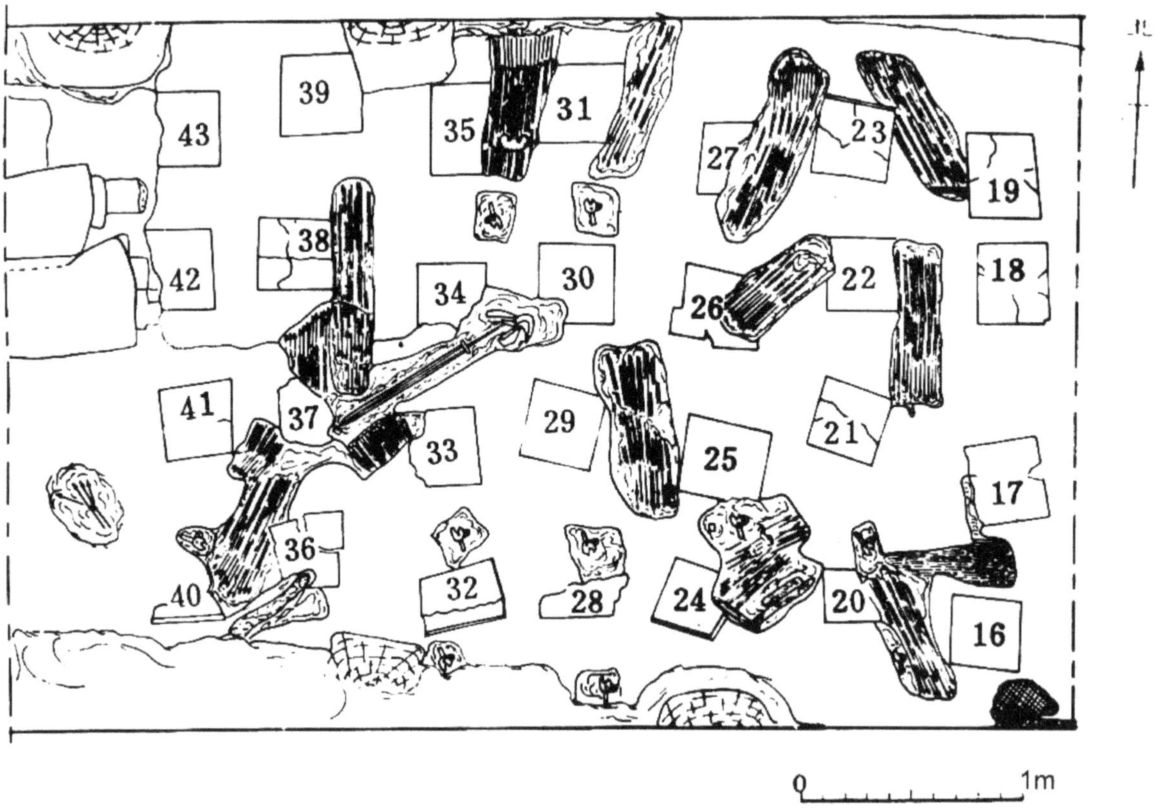

Fig. 6.1. Numbered foot plates of the terracotta warriors and arrow bundles (after figure 186 in the Excavation Report, Institute and Team, 1988: 300).

6.3.2.1 An enumeration of arrows in each corridor

Each corridor contained a different number of arrows. There are 11 corridors marked no. 1 to 11 from south to north, and one more vanguard corridor at the eastern side of Pit 1. Thus, the southern corridor 1 and northern corridor 11 represented the flanks of the army (Yuan, 1990). As mentioned in the previous chapter, the vanguard and flanks usually consisted of archers, who shot arrows at the enemy from a considerable distance (Yuan, 1990; Yates, 2007). This is why these three corridors contain as much as 61.4% of the total number of arrows. In the central corridors, the bronze arrows are fewer, and were mainly found behind the chariots (Table 6.2 and Fig. 6.2a).

The distribution of the arrows (Fig. 6.2a) and triggers (Fig. 6.2b) are corresponding in each corridor except corridor 6, where there is a group of arrow bundles (1,954 arrows) but only one trigger. Figure 6.2c shows that the arrow count per trigger varied in each corridor, from 111 to 422. However, the posture of the warriors in corridor 6 is similar to that of the crossbowmen in the vanguard and flanks. Those warriors may be traditional archers or special arrangements in the battle formation.

6.3.2.2 The size of the arrow bundles

During the excavation, it was obvious that many of the arrowheads appeared consolidated in bundles originating from their decay within quivers (see next section), while large numbers of loose arrowheads or smaller clusters were recorded as well. The excavation report also mentioned that the average number of arrows in each bundle was approximately 100 arrows (Institute and Team, 1988). Figure 6.3 shows a frequency histogram of the number of arrows in each bundle. Although the count of arrows per 'bundle' varies from a single arrow to 200, two main peaks or modes are clearly recognisable, around 1–10 and 91–100. This pattern is further clarified in Figure 6.4, which shows only loose arrows or small groups recorded as bundles of up to 10 arrows, and Figure 6.5, which shows the frequency of bundles containing over 70 arrows. On the one hand, the bundles with 100 arrows seem to constitute a dominate pattern. The fact that some bundles have arrow counts slightly above or below this number is probably due to the loss of the arrows or to the mixing of bundles occurring due to post-depositional processes and during the subsequent recovery. Because of the extremely high density of arrowheads at the site, it is not surprising that not all of the arrows could be conclusively associated to specific bundles. It is likely that the few bundles with unusually large counts may constitute two bundles recorded as a single one during the excavation. On the other hand, the high frequency of single arrows and the tail of the curve from 1 to 10 (Fig. 6.4) suggests that these constituent arrows were detached from large bundles during the course of post-depositional processes, and also that the excavators may have clustered together loose arrows as small 'bundles' simply because they were found together. Overall, it is obvious that the target number of arrows for each bundle was 100.

6.3.2.3 Arrow bundles and quivers

The arrow bundles were originally placed in quivers made of woven hemp that were worn on the backs of the warriors. Traces of rotted quivers could still be identified when they were found in the pit (Institute and Team, 1988; Fig. 6.6), and the reconstruction offered a much clearer image of a Qin arrow quiver (Yuan, 1990; Fig. 6.7).

Table 6.2. The number of arrows and bundles (≥90) in each corridor of Pit 1.

Corridor	Arrow count	Count of all bundles (≥90 and <90 arrows)	Count of bundles with ≥90 arrows
Vanguard corridor	10,194	160	61
Corridor 1	6, 563	104	53
Corridor 2	1, 210	24	8
Corridor 3	702	27	5
Corridor 4	1,657	25	13
Corridor 5	2,117	31	17
Corridor 6	1,954	22	13
Corridor 7	3,003	45	22
Corridor 8	1,733	22	14
Corridor 9	1,103	54	8
Corridor 10	917	24	5
Corridor 11	6,193	142	43
Sum	37,348	680	262

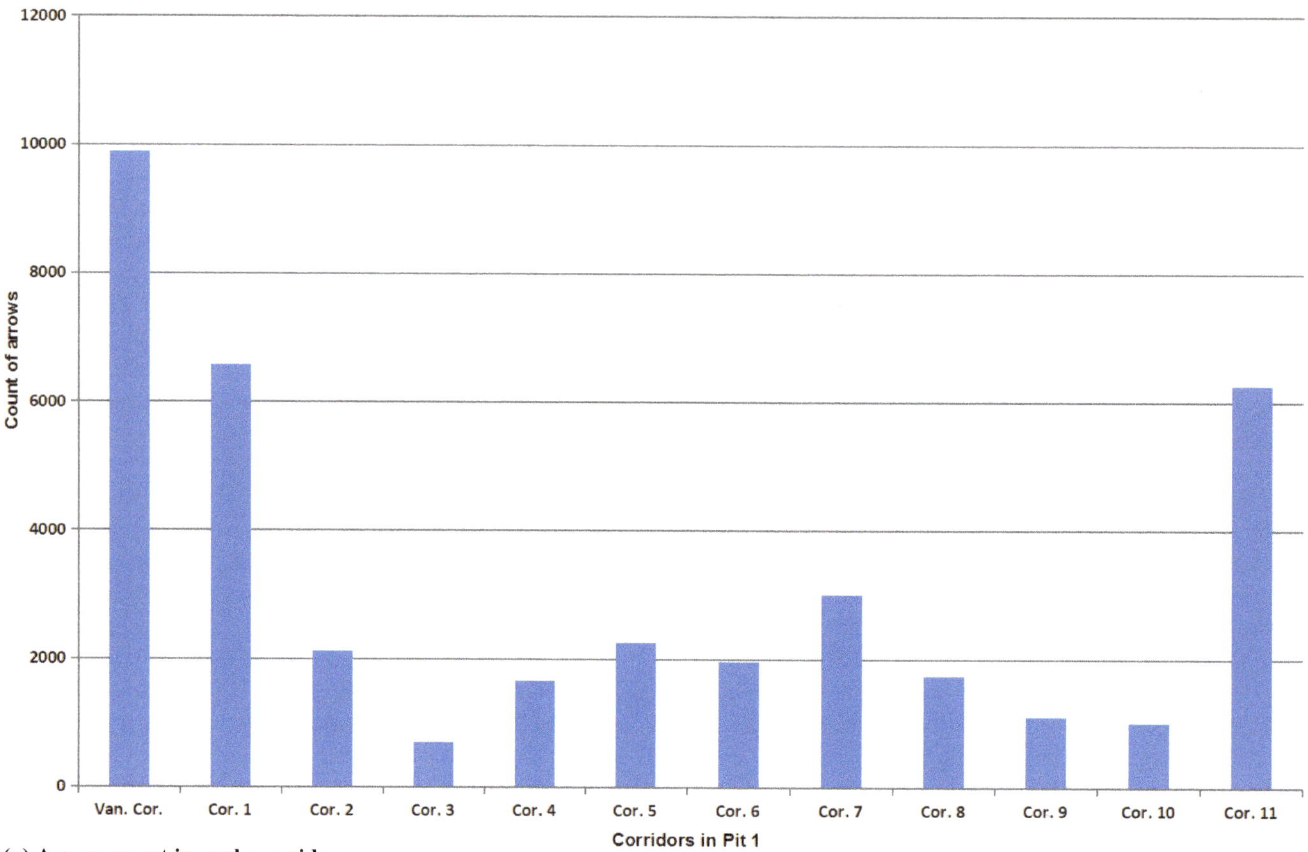

(a) Arrow count in each corridor.

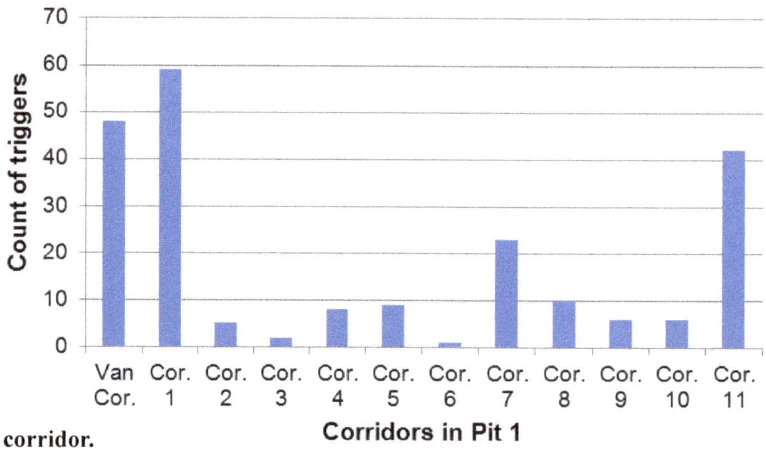

(b) Trigger count in each corridor.

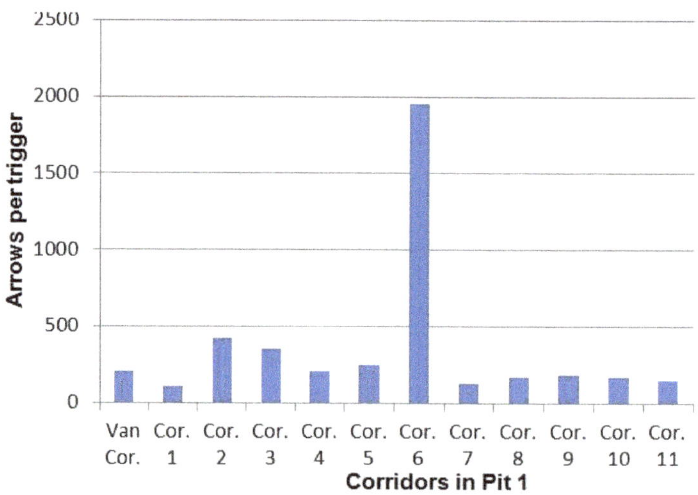

(c) Arrow count per trigger.

Fig. 6.2. Count of the arrows (a) versus triggers (b) in each corridor, and number of arrows per trigger (c).

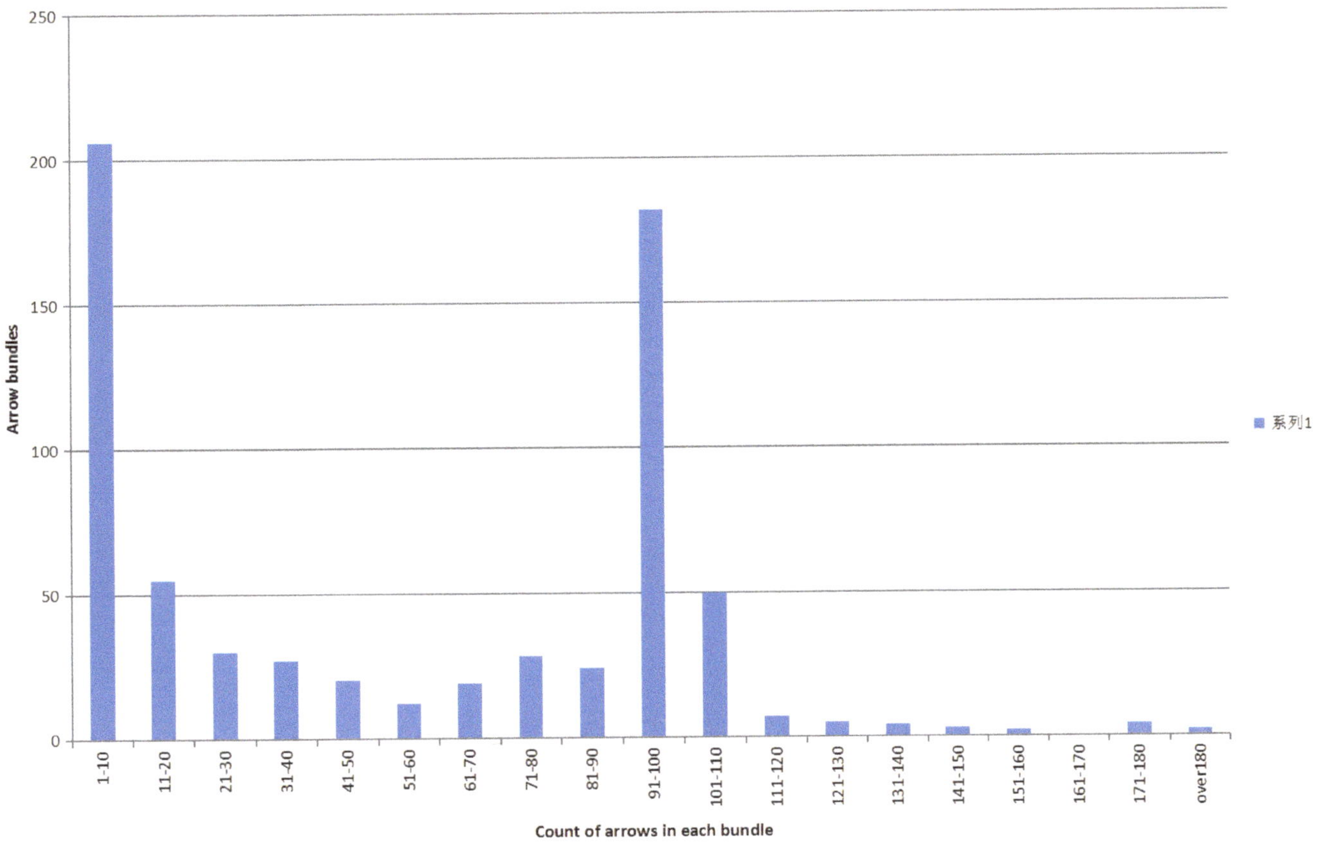

Fig. 6.3 Frequency histogram of the number of arrows and arrow bundles.

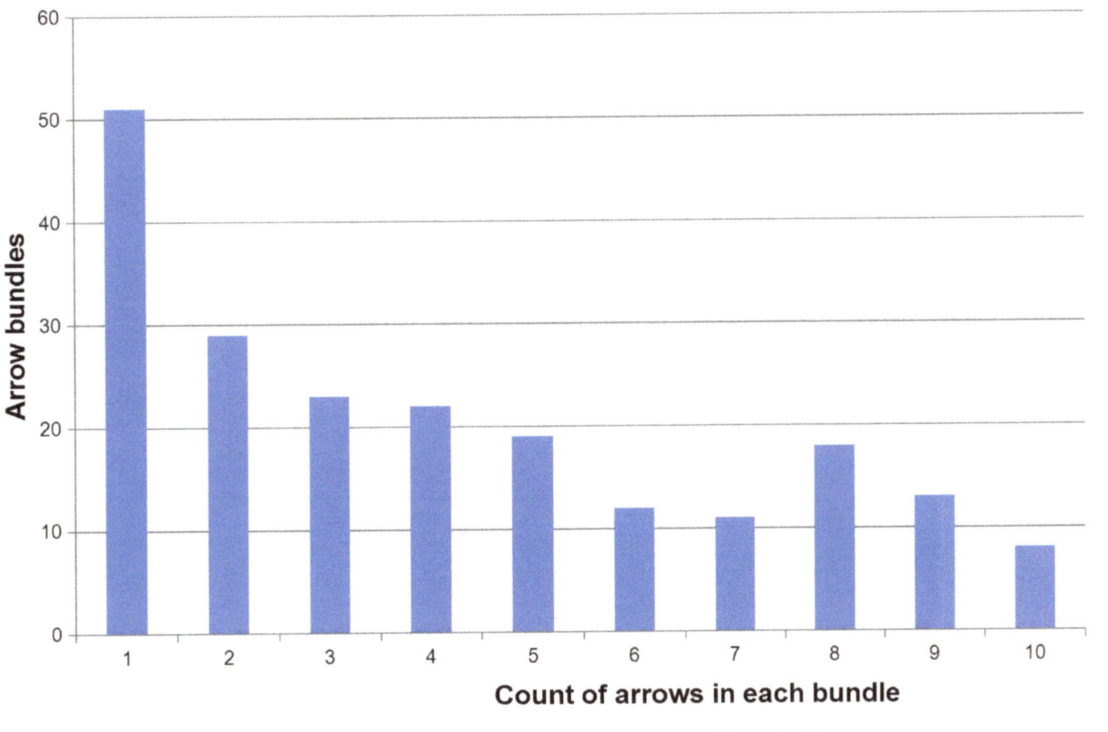

Fig. 6.4 Frequency histogram of the number of loose arrows or small groups with up to 10 arrows.

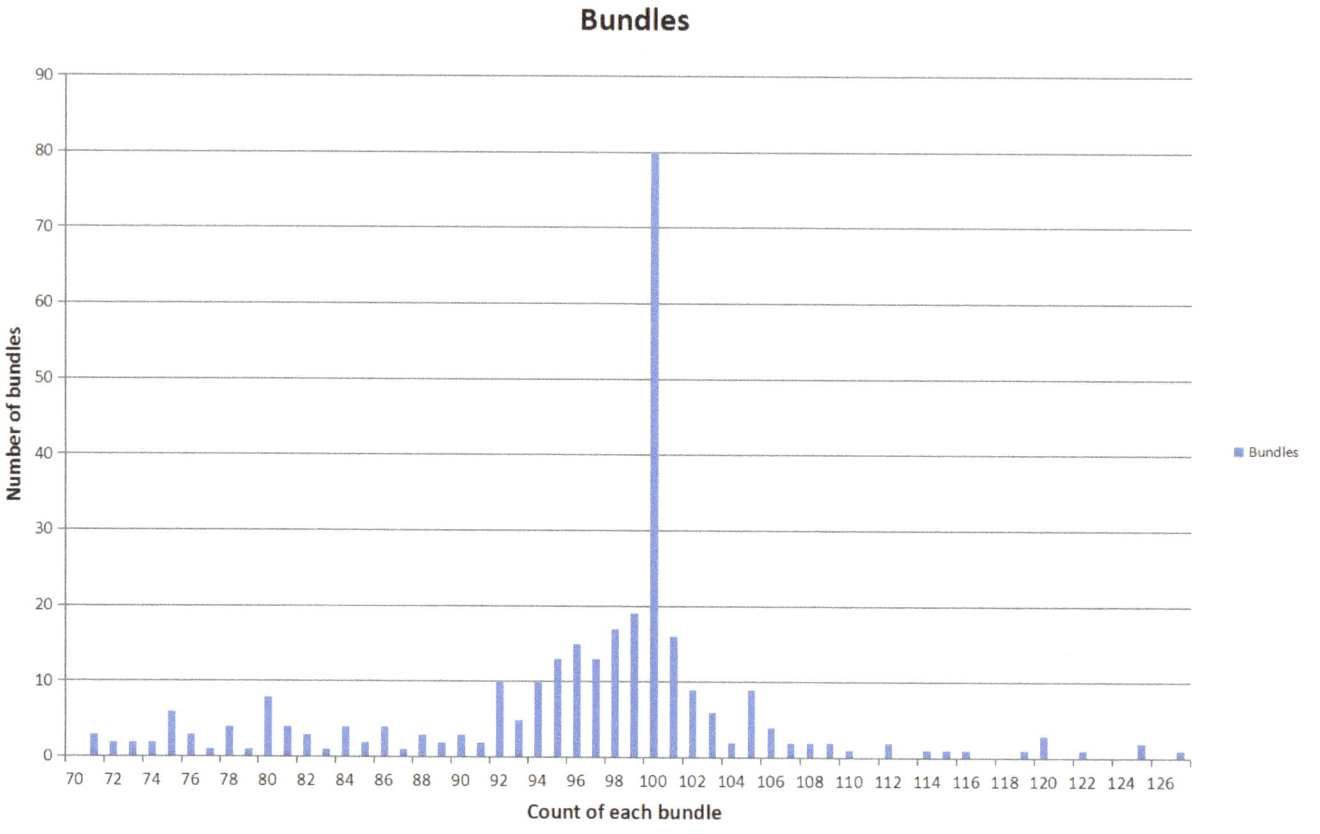

Fig. 6.5. Frequency histogram of the number of arrows in bundles containing between 70 and 120 arrows.

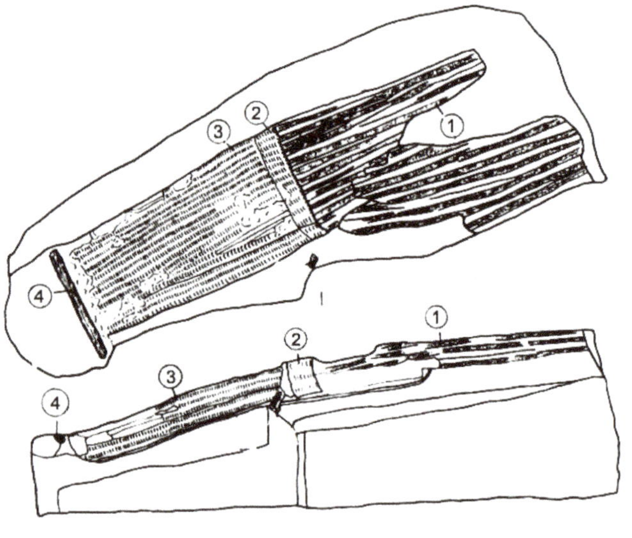

Fig. 6.6. Traces of a quiver with a bundle of arrows in Pit 1: (1) shows the arrows; (2) and (3) show the quiver; (4) refers to the wooden part at the bottom of the quiver (after figure 173 in the Excavation Report, Institute and Team, 1988: 285).

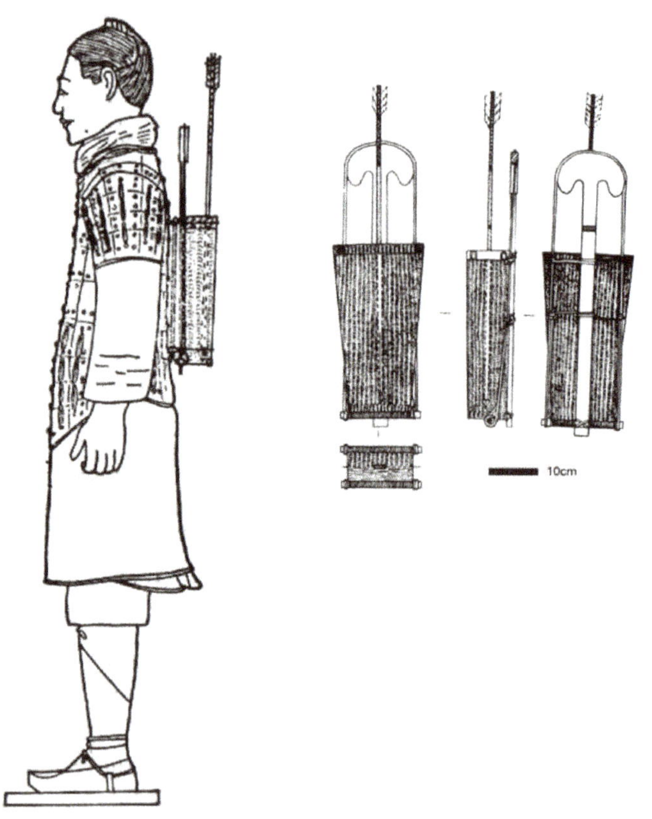

Fig. 6.7. Reconstruction of a Qin arrow quiver (after figure 174 in the Excavation Report, Institute and Team, 1988: 286).

The arrows were put into the quivers with the arrowheads pointing downwards to the bottom and the arrow-tangs with wooden or bamboo shafts and feathers exposed outside. Some traces of the feathers that were originally attached to the ends of shafts could still be seen when they were first discovered in the pit (Institute and Team, 1988; Yuan, 1990). Clay loops with hemp strings were discovered on the back of some warriors, and the quivers hung from these (Yuan, 1990). By examining in detail Figure 6, it appears that the quiver and arrows were not originally placed on the floor, because the soil beneath the quiver of arrows suggests that the arrows must have fallen from a warrior' s back, probably following the collapse of some roofs and walls.

6.3.2.4 Arrow types

A complete Qin arrow normally included an arrowhead, a tang, a longer wooden or bamboo shaft, and, at one end, a feather attachment. Even though the shafts and feathers have rotted away, the traces left in the pit can still be used to reconstruct the arrows and provide us a clear idea of their appearance. Information about complete Qin bronze arrows was also obtained from the two sets of bronze chariots discovered in1980 to the west of the tomb mound (Museum and Institute, 1998; Fig. 6.8), which included arrow reproductions where all the components were made of bronze and have thus survived as complete objects.

Most of the arrowheads show three triangle sides converging in a sharp point. The three sides were all finely ground and polished to obtain a smooth and shiny surface, as well as sharpness. When examined under the SEM, the grinding marks appear as extremely dense parallel lines, and some marks are even transversal on all sides and often continue from one plane to another. It is more likely that a mechanical grinding device, such as a rotary wheel, was employed (Li et al., 2011). However, some exceptional arrowheads were also found, including special arrows, bronze arrowheads with iron tangs, and the truncated arrows on the bronze chariot discovered to the west of the First Emperor's tomb mound (Fig. 6.8). The first two types of heads will be discussed later in this chapter. The truncated arrows, for which there is no evidence in the pit, are believed to have been used for archery practice during the Qin period because they were relatively safe (Museum and Institute, 1998).

Normally, a tapering tang was attached to the arrowhead. The arrow tangs were circular in profile and the filing traces could be clearly identified using a microscope or under the SEM (Li et al., 2011). They were originally inserted into a wooden or bamboo shaft when produced and assembled, and, to date, there are still some bamboo traces and linen strings visible on the arrows (Figures 6.9 and 6.10).

As shown in a previous archaeometallurgical study on the arrows from the pit, the tangs were cast first and the arrowheads subsequently cast on (Yuan, 1990). However, evidence from assembled arrows from the bronze chariot indicates that the arrowhead and tang were cast separately and joined together (Fig. 6.11; Museum and Institute, 1998). In many cases, small and thin sheets of bronze, henceforth referred to as 'necks', were found between the head and tang (Fig. 6.12). In my personal view, it is likely that these necks were leftovers from the second cast when the arrowhead was subsequently cast on to the tang. The details for this linking-cast require further laboratory observation and analysis by other experts. Even if the tang had the function of balancing the flying arrow, it needed to be combined with the wooden or bamboo shaft as a whole. The traces of the shafts found in Pit 1 were marked with black and red pigments into three sections (Institute and Team, 1988; Yuan, 1990; Wang, 1994). The black section is an approximately 12 to 14 cm section just after the arrowhead, and is followed by the red section (about 38 cm) and by another black section of approximately 17 cm (Yuan, 1990). A few shafts have alternative sections, 1/3

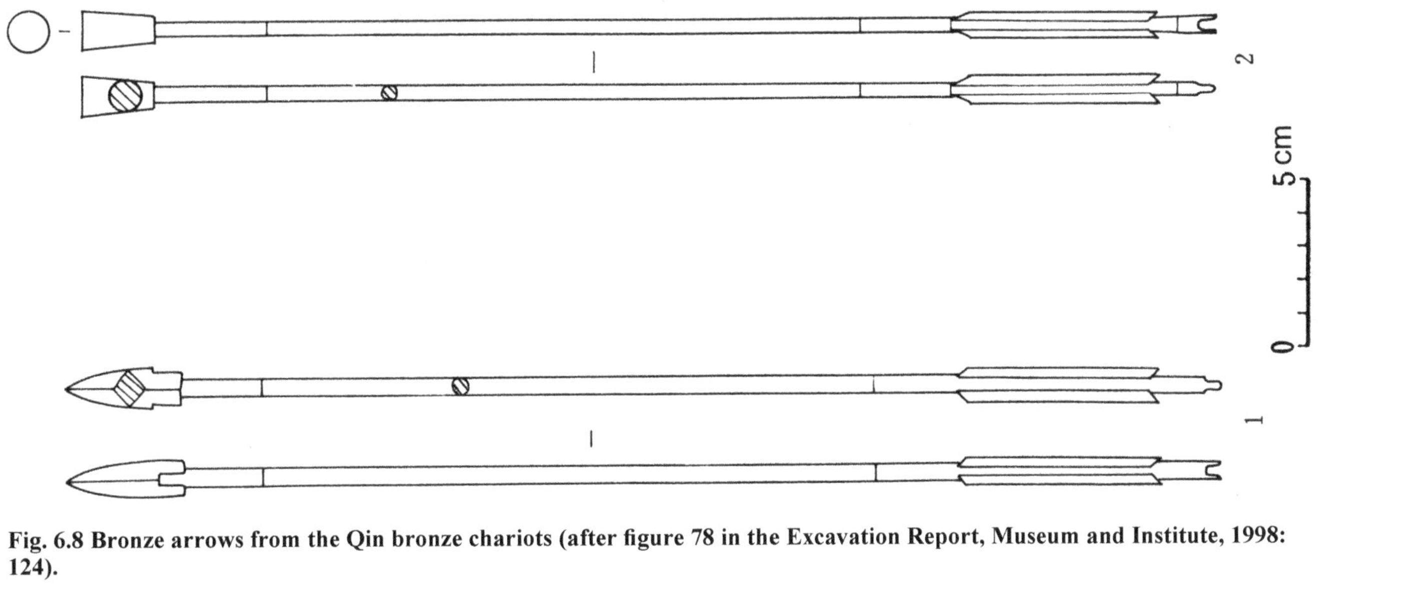

Fig. 6.8 Bronze arrows from the Qin bronze chariots (after figure 78 in the Excavation Report, Museum and Institute, 1998: 124).

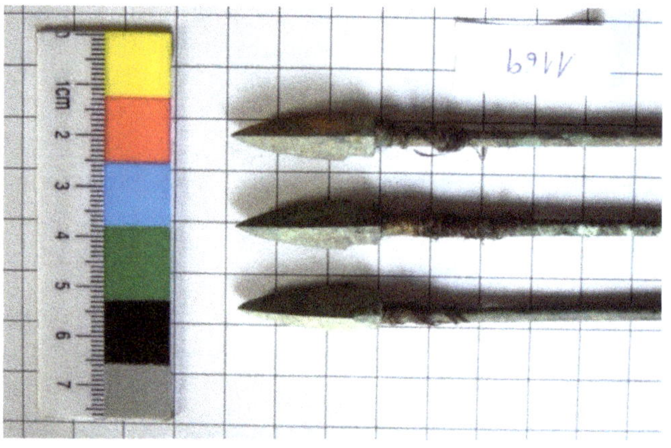

Fig. 6.9. Bronze arrows with linen strings coiled around them.

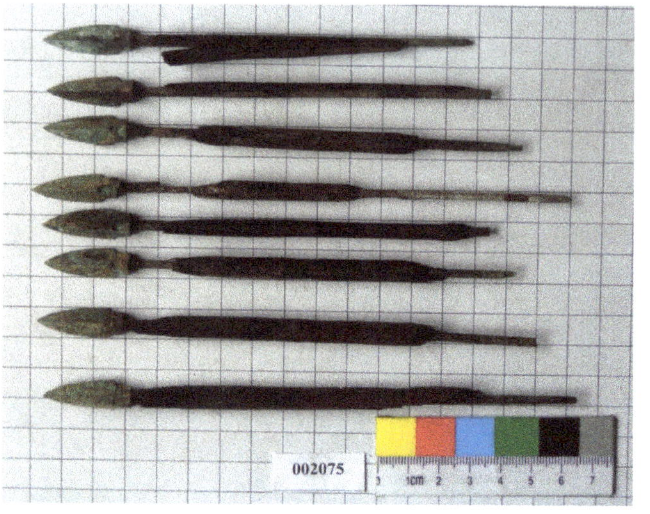

Fig. 6.10. Bronze arrows showing bamboo remains of shaft.

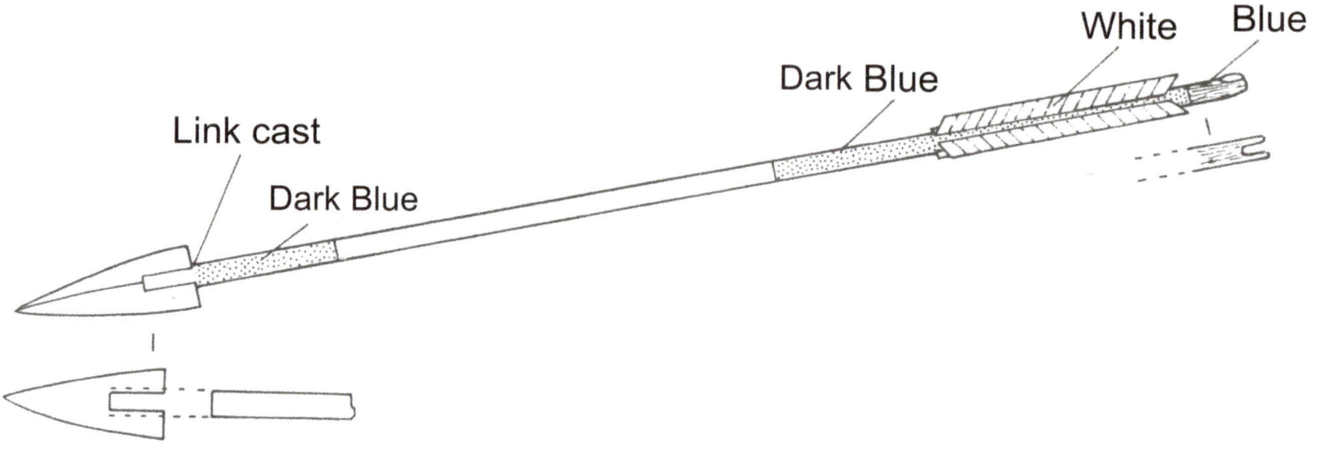

Fig. 6.11. Linking between an arrowhead and tang from the bronze chariot (after figure 188 in the Excavation Report, Museum and Institute, 1998: 311).

Fig. 6.12. The neck between the arrowhead and tang.

in black and 2/3 in red (Yuan, 1990), adding up to a total arrow length of about 70 cm. The arrow tangs found on the bronze chariot are also divided into four sections, and are marked with dark blue, red, dark blue, and light blue (Fig. 6.11) (Museum and Institute, 1998). The instructions of how to assemble the arrowhead and shaft, as well as the length sections, were recorded in an ancient document, *Zhouli·kaogongji*, originally written about the rituals of the Zhou Dynasty (1050–221 BC). According to this source, if you divided a shaft into three, the tang should be 1/3 of the length (Yuan, 1990). This length, expressed as a percentage of the arrow shaft, would allow arrows to balance better when in flight and be targeted accurately.

Based on their lengths, the bulk of arrow tangs can be divided into two main subtypes: 1) with a comparatively long tang; and 2) with a comparatively short tang (Fig. 6.13). The longer tangs range mainly from 9.5 to 18 cm in length, while the shorter ones are mainly 7–9.4 cm. The detailed measurements will be discussed in the following sections.

In addition to the bronze arrows with bronze tangs, two bronze arrowheads with iron tangs were unearthed from Pit 1 (see detailed spatial location in Fig. 6.34). These two bronze arrowheads are morphologically different from the other arrows mentioned above, and the iron tangs are both seriously corroded, with one broken into several pieces (Fig. 6.14). To date, no iron arrowheads or arrows made entirely of iron have been discovered in the pits of the terracotta warriors.

Another exceptional type of arrow, referred to as a 'special arrow', was represented by seven examples in Pit 1. These have a head like a Chinese writing brush and a long tapering stick (Fig. 6.15). The function of this type of arrow remains unclear (Institute and Team, 1988).

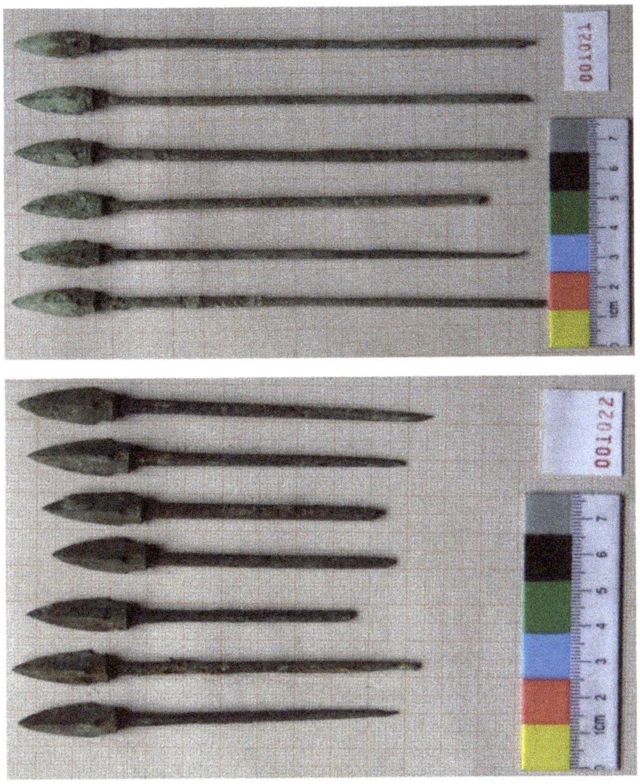

Fig. 6.13. Shorter and longer arrow tangs.

Fig. 6.14. Bronze arrowhead with iron tang.

6.3.3 A comparison between the arrows from Pit 1 and those from Pit 2 and other sites

Even though this project is mainly concerned with the arrows from the five easternmost trenches of Pit 1, arrows from the other trenches in Pit 1, Pit 2, and other sites are worth mentioning before proceeding to further discussion of the former. In addition to the arrows (normally 7–18 cm in length and about 9–19 g each in weight) found in Pit 1, bolts or quarrels (about 32.6 cm in length and about 42.5 g each in weight from the museum archive) with much longer tangs were discovered in Pit 2. The bolts can be divided into two types: A) one where the arrow tang is straight, with traces of string wrapped around (Fig. 16); and B) one where the arrowhead is narrower and the tang has two sections, thicker at the topside and thinner at the lower side (Fig. 6.17). These are related to the special units of more heavily armed crossbowmen in Pit 2, different from the light crossbowmen in Pit 1. The light crossbowmen in Pit 1 would have given the Qin battle formation speed and mobility, while the heavy crossbows in Pit 2 would have permitted them to fire heavier bolts or quarrels, more accurately, with greater force and penetrating power, and at a longer distance (Yates, 2007). The heavier bolts were almost certainly far slower to load; however, by training the crossbowmen to act in unison, the unity of their firing action would have more than compensated for their slower loading times. While the front line of crossbowmen was firing together, the line behind would be loading their bolts or quarrels. Once the front line had fired, the second line would move forward to replace them. This type of manoeuvre seems to be what is represented by the terracotta warriors in Pit 2. The crossbowmen formation in the north-east corner of Pit 2 comprises kneeling crossbowmen positioned to the front and centre, with standing crossbowmen behind them. The kneeling and the standing crossbowmen would have fired bolts at the enemy in turn (Yuan, 1990; Yates, 2007).

While the arrows from other sites in the tomb complex are mainly of the general type described for Pit 1 above, arrows from earlier Qin sites collected by the museum generally show a very different morphology, with wings on each side (Fig. 6.18). The longer development of arrow technology from earlier periods through to the Qin dynasty has been discussed by many researchers (Yang, 1980; Research Committee, 1995; Shi, 2008), and the different morphology may be due to technological changes taking place during the earlier Qin period or driven by the archery development from bows to crossbows. The Qin arrows from Pit 1 lack the wings of the older style, and are finely polished on three sides of the arrowheads to enhance sharpness and penetrating power.

Fig. 6.15. A special arrow from Pit 1 (after figure 168–13 in the Excavation Report, Institute and Team, 1988: 278).

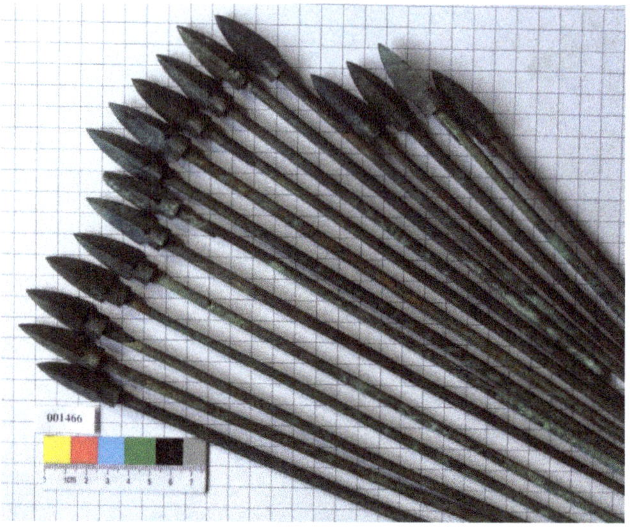

Fig. 6.16. Type A of the bronze bolts from Pit 2.

Fig. 6.17. Type B of the bronze bolts from Pit 2.

6.4 Measurements of the bronze arrows

6.4.1 Sampling strategy

As there are almost 40,000 bronze arrows to be considered, devising a sampling strategy was essential for this part of the research (Fig. 6.19). Decisions had to be made as to how many arrows had to be studied to answer the research questions effectively and how many could be feasibly studied within the time frame of this project. Altogether, the 277 arrows bundles with 90 arrows or more were considered as the main target samples, since they were deemed to be more coherent and informative than the more fragmented bundles of arrows. Among these bundles, 262 originate from the five easternmost trenches of Pit 1, having specific spatial locations recorded, while the other bundles are from other ongoing trenches of Pit 1 and Pit 2, not considered for statistical and spatial analysis in this book.

As a first stage, a pilot survey was undertaken to examine the arrow bundles. Theoretically, if this pilot survey showed there were arrows with similar characteristics within each bundle, then a systematic sampling procedure could be applied to a subset of the arrows that would be deemed as representative in each bundle. For example, the same number of arrows would be systematically selected for measurement from each bundle. However, if the pilot survey showed measurable differences within each bundle, larger than the differences between bundles, then some spatial cluster sampling would be more appropriate.

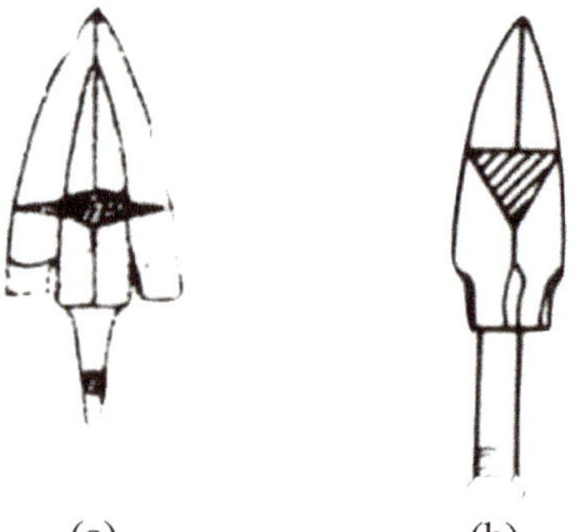

Fig. 6.18. (a) An arrow from another Qin site, and (b) one from Pit 1 of the terracotta warriors (after figure 168–5 in the Excavation Report, Institute and Team, 1988: 278).

The first impression from handling the bundles was that, notwithstanding many exceptions, there was a certain degree of internal coherence for most of them, most noticeably in terms of the length, thickness, hardness, and straightness of the tangs (as the heads looked indistinguishable within and between bundles). XRF analyses of a large number of arrowheads and tangs (conducted by Marcos Martinón-Torres, and to be reported in detail in section 6.5.1) demonstrated that there was a significant degree of chemical consistency within the bundles and variability between bundles. On this basis, the bundles were systematically sampled. Six arrows were randomly sampled from each bundle for detailed measurements on tangs and arrowheads.

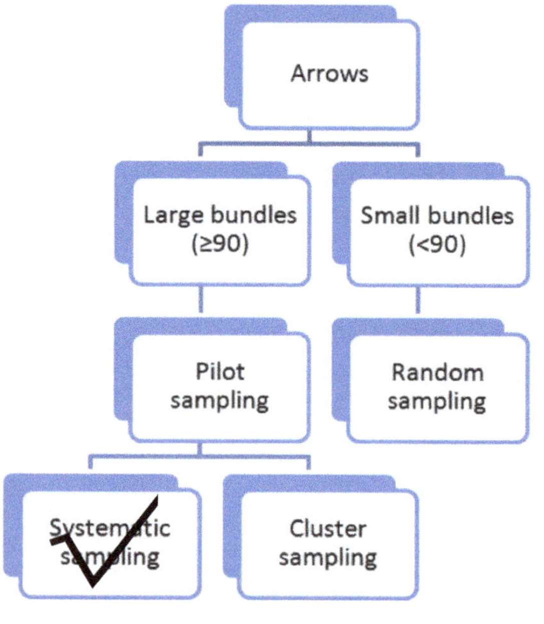

Fig. 6.19. Sampling strategy used for the bronze arrows.

6.4.2 Methodology of the measurements

Measuring these arrows involved a slightly different procedure from the one employed for the triggers: 1) taking photographs in the Conservation Department at the Museum of Emperor Qin Shihuang's Terracotta Army in China using millimetric paper as background; 2) measuring the arrows digitally in an image analysis package (Adobe Photoshop, see Chapter 3); 3) recording the measurements in a spreadsheet; and 4) digitising the spatial data in a GIS (ESRI's ArcGIS).

For the photography, the same method employed to study the triggers was used: the camera was set to level, metric paper was placed on a board, and the arrows were placed at the centre. The museum identification number of each arrow or arrow bundle was clearly marked on the paper. Typically, two pictures were taken, one for the whole bundle of arrows and the other for six arrows random sampled for detailed measurement (Fig. 6.20).

When these digital photographs were transferred to a computer, the measurements were carried out in an image analysis package. The arrow tangs were measured along their entire length, while the heads were measured across their width and length, respectively. The dimensions were recorded on a spreadsheet for further analysis. The means, standard deviations, and coefficient of variation were then calculated for the measurements of the six arrows in each bundle, as well as for various series of arrows across bundles.

6.4.3 Patterns in the dimensions of the arrows

6.4.3.1 Measurements of the arrow tangs

The detailed measurements of the arrow tangs provided a clearer picture of their patterns (see Fig. 6.21). These patterns generally matched the macroscopic division mentioned above (see Fig. 6.13), between a shorter and a longer type of tang. However, the length distribution histograms also suggested a further subdivision.

In Figure 6.21, three histograms representing the frequency distributions of the measurements of the arrow tangs are displayed. Figures 6.21a and 6.21b are examples of the measurements on the first and second arrows in each bundle, respectively, while 6.21c shows the mean length of all six arrows sampled. From these results, the length of the arrow tangs can be tentatively divided into 2 clusters based on the three histograms: 1) short tang, from 5 to 9.4 cm; and 2) longer tang, from 9.5 to 18 cm. Figure 6.22 presents the photographs of arrows belonging to the two groups.

A further way of visualising these data is shown in Figure 6.23, but the overall pattern appears to remain the same.

Some inaccuracy should be taken into consideration when interpreting the clusters showing the division of the arrows according to tang length, particularly as some of the arrow tangs were seriously damaged due to a fire shortly after the pit was built, or affected by the collapse of the earth supports. In addition, all the objects in the pit have been subject to long term corrosion. All of these factors could influence the final length of the arrow tangs, but do not dramatically change the frequency of the clusters. These factors should also be considered when discussing the spatial patterns based on the length of the arrow tangs.

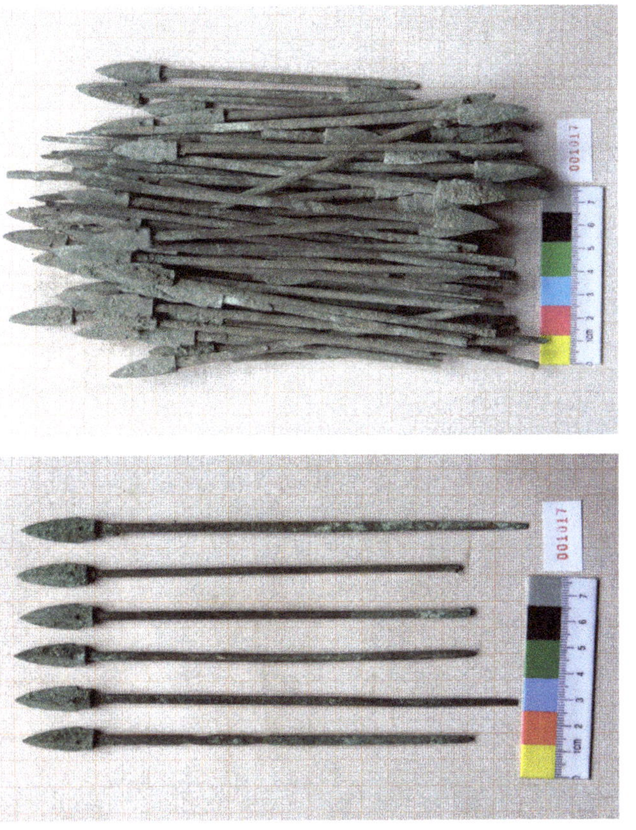

Fig. 6.20. The arrow bundle (ID001017) and the six sampled arrows.

6.4.3.2 Measurements of the arrowheads

The bronze arrows were all carefully ground and polished after casting, leaving extremely fine marks that can be examined under the scanning electron microscope (Li et al., 2011, 2012). The grinding and polishing processes not only removed the excess metal from casting seams, but also conferred sharpness and penetrating power to the arrowheads. The production of such a large quantity of weapons entailed processes of mass production and a highly organised labour force. Measurements of these weapons demonstrate the scale of this production in a new light. The measurements of the length of arrowheads (Fig. 6.24) indicate a single peak in the histogram, concentrated at 2.7 cm. Most arrowhead lengths concentrate at 2.6–2.8 cm, and the two tails are generally symmetrical, with a slight skew to the right.

The six arrowheads selected from each bundle also show remarkable consistency in their width. The peak of the

(a) Length: measurement 1st sampled arrow in the bundle

(b) Length: measurement 2nd sampled arrow in the bundle

(c) Mean length: measurement of 1st -6th sampled arrows

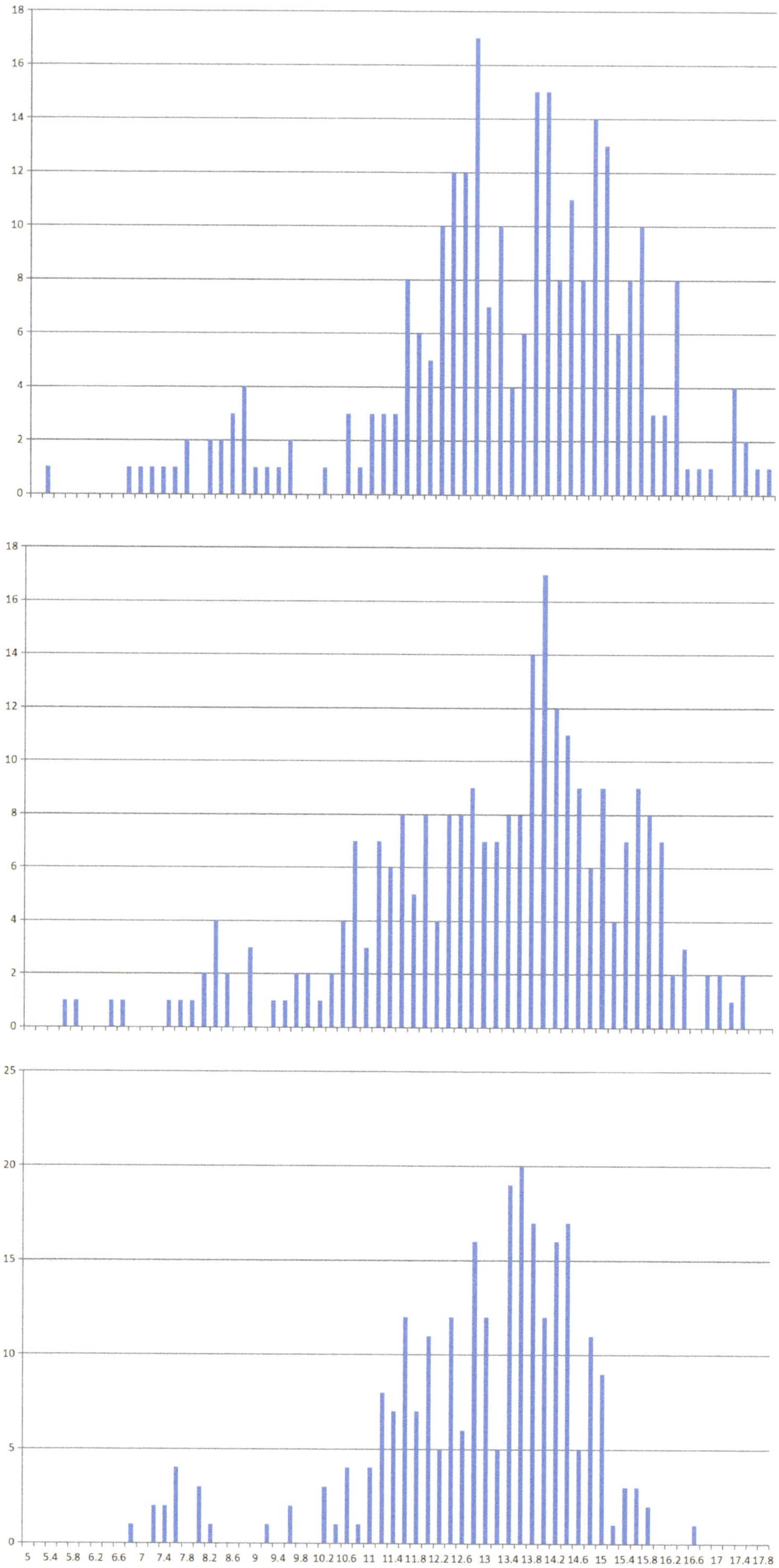

Fig. 6.21. Histogram of the frequency distribution observed on the arrow tang length measurements.

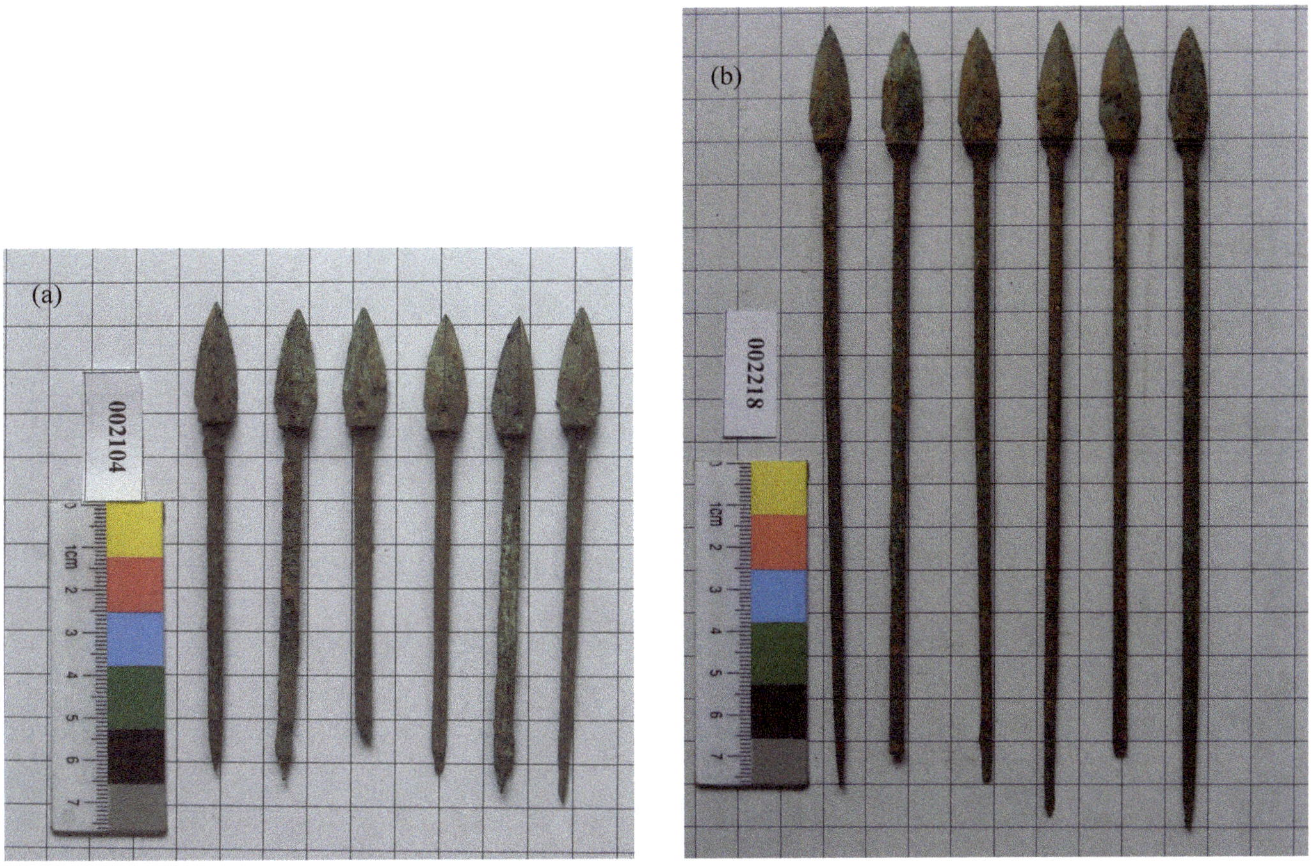

Fig. 6.22. Images of examples of the two subgroups of arrow tangs: (a) short and (b) long.

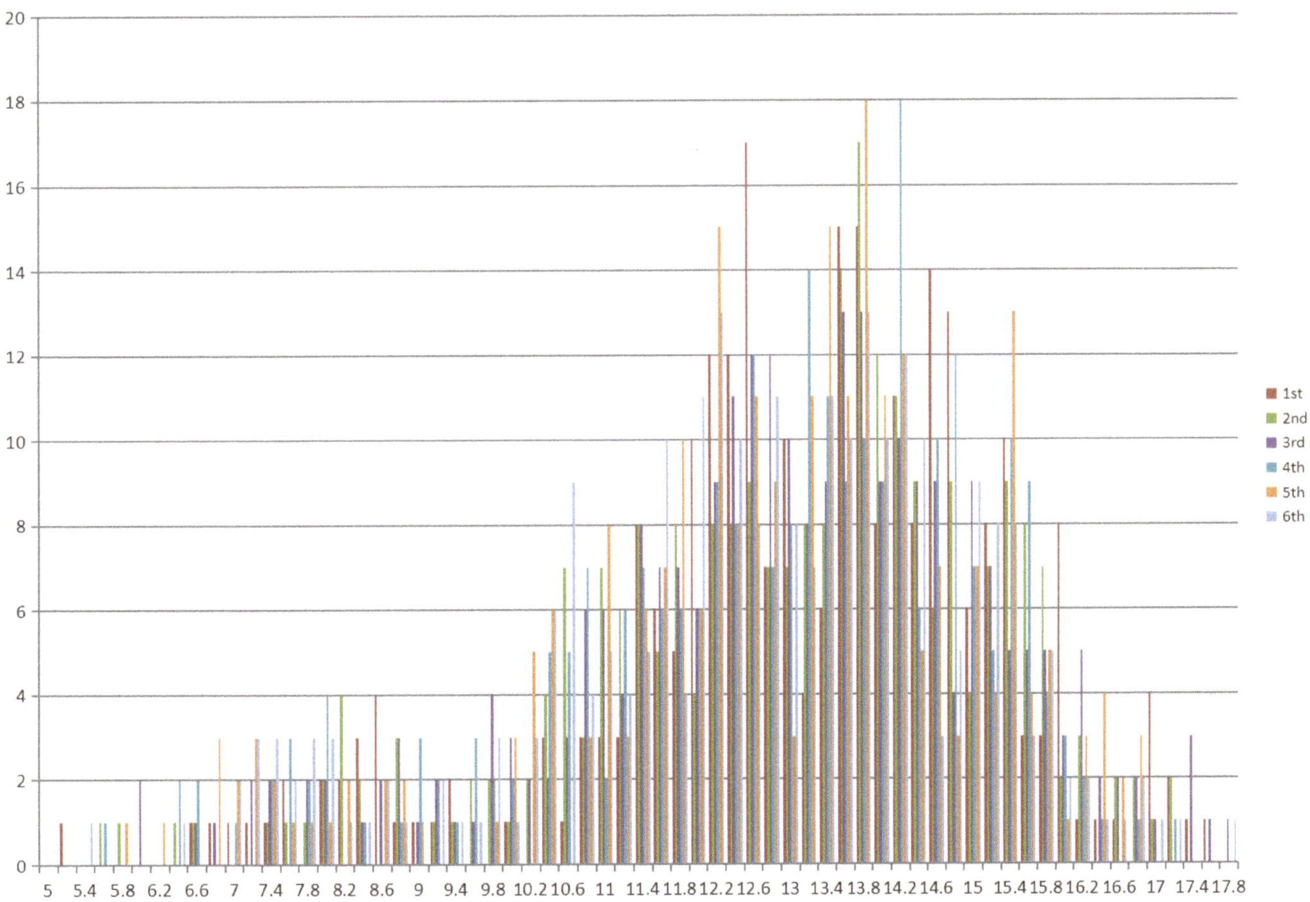

Fig. 6.23. Histogram of the frequency distribution observed on the arrow tang length measurements (each colour represents one arrow randomly selected from each bundle).

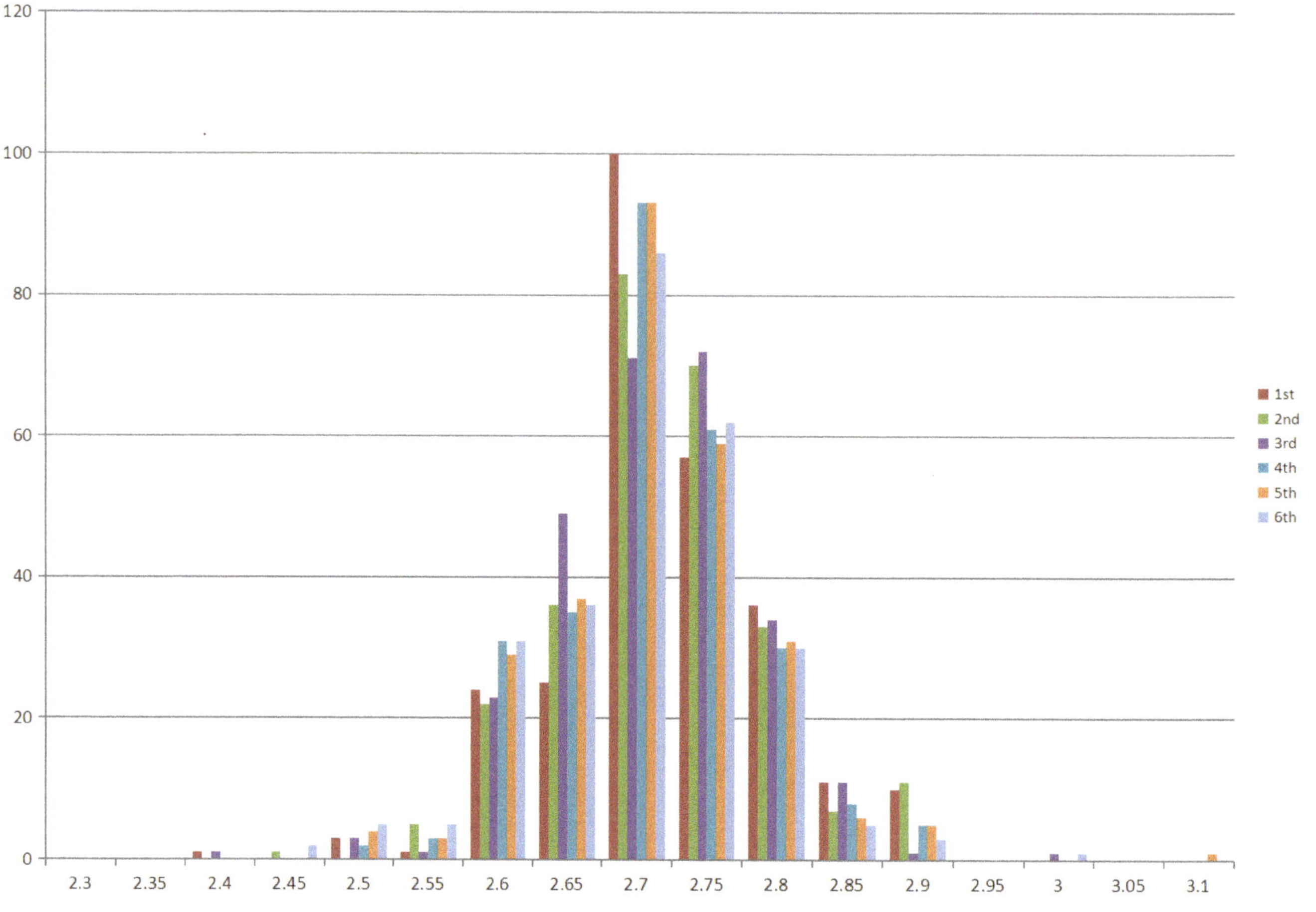

Fig. 6.24. Histogram of the frequency distribution of six length measurements of the arrowheads. Each colour represents one arrow randomly selected from each bundle.

histogram is at 0.95–1 cm (Fig. 6.25), and the overall distribution is symmetrical, with very little variation around the mean (most calculations of standard deviation are zero). This suggests a high degree of standardisation and production control.

6.4.4 Coefficient of variation values and degree of standardisation

As with previous weapons, coefficient of variation values have been employed to assess and compare the degree of standardisation of the arrows. A coefficient of variation was calculated for each dimension (head length, head width, tang length) of the six arrows randomly selected from each bundle.

6.4.4.1 CV values for the arrow tangs

Figure 6.26 shows the mean tang length for the sampled arrows in each bundle, and their respective CV values. The CV values generally illustrate a relatively high variation of tang length (Table 6.3). As noted above, there are two main modes within arrow tangs. Group 1 contains a relatively large number of arrows, and the tang length is over 9.5 cm. The CV values for this group range from 1.7% to over 20%. Group 2 contains only a relatively small number of arrows, and the tang length is less than 9.4 cm. The CV values for this group are from 5% to 15%. At the top of Group 1 and Group 2, there are some further outlying points with significantly higher CV values. Closer inspection of these outliers demonstrates that they typically include: i) some broken or seriously corroded arrows; and ii) a mixture of longer and shorter arrows, possibly as a result of mixing during the archaeological recovery, given the high density of arrows at the site (Fig. 6.27).

Even when these values are left aside, the tang lengths of the arrows in Group 1 and Group 2 still suggest a relatively low degree of standardisation. CVs for tang length are in most cases above 5%, often much higher. As detailed in Chapter 2, the Weber fraction for the perception of the length of a line is about 3% (Teghtsoonian, 1971). Moreover, Eerkens and Bettinger (2001) have further explained that CV values in the range of 2.5–4.5% are typically indicative of the minimum error attainable by manual production without the use of external rulers, and thus reflect the threshold of variability that falls below what human senses can perceive (Longacre, 1999; Eerkens, 2000). However, most of the CV values for the two groups of arrow tangs are above 3%, and would be clearly discernible to the naked eye as they show a rather high variability.

Most of the tangs were cast with two-piece moulds and the seam lines between the mould valves can be seen on the surface of tangs (Yuan, 1990) in spite of the remarkable efforts often made to file away and grind the excess

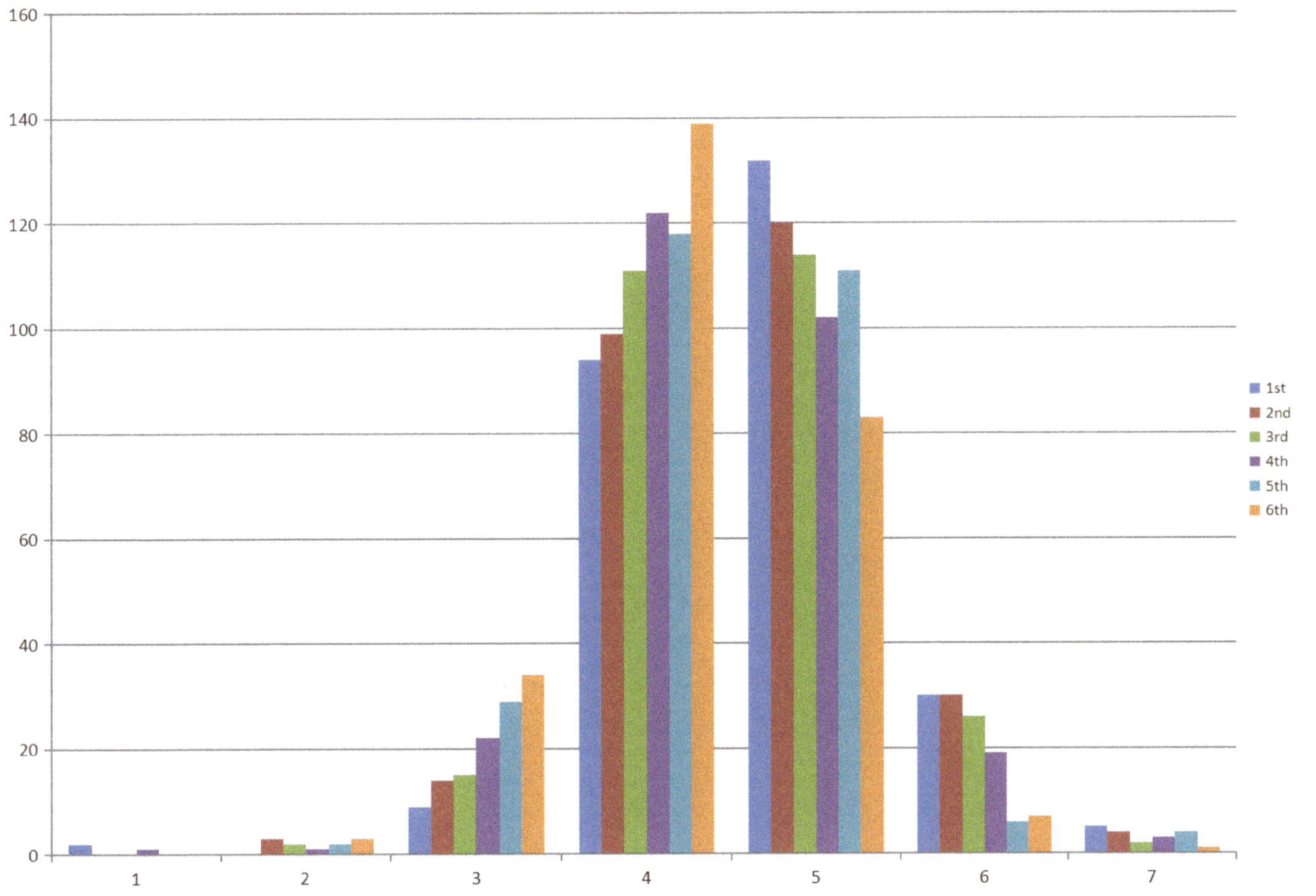

Fig. 6.25. Histogram of the frequency distribution of 6 width measurements of the arrowheads. Each colour represents one arrow randomly selected from each bundle.

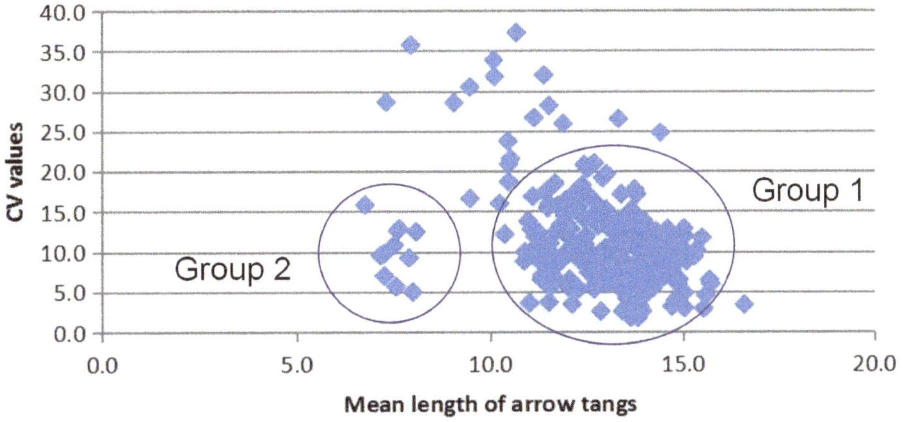

Fig. 6.26. Mean length of arrow tangs versus their CV values from each bundle.

metal (Li et al., 2011). It is likely that, in some cases at least, relatively long rods of metals were cast before being cut into shorter segments to be used as tangs, as indicated by the cut-marks apparent at the extremes of some tangs (Martinón-Torres et al., 2014b). While the variable tang diameters and rigidity noted macroscopically already suggested that a range of moulds (and, possibly, craftspeople or workshops) was engaged in the production of tangs, the variation in their length also indicates that this dimension was not very tightly controlled. This is in stark contrast with the standardisation in arrowhead dimensions noted below.

For the purposes of individual assessment, Figure 6.28 maps the arrow bundles with their museum identifiers and Appendix 7 offers a summary version of the measurements taken on each bundle. Further details and the full set of measurements are provided in Appendix 7.

6.4.4.2 CV values for the arrowheads

In comparison with the arrow tangs, the arrowheads are much more standardised. Figure 6.29 presents the mean length and the mean width of arrowheads in each bundle. In most of the bundles, the CV values for the width and

Table 6.3. Pair correlation results on shorter and longer tang arrows.

Arrows	Patterns	Distance	Figure reference
Shorter tang arrows	Clustered	5–7 m	Fig. 6.36
Longer tang arrows	Clustered	1–4 and 8–9 m	Fig. 6.37

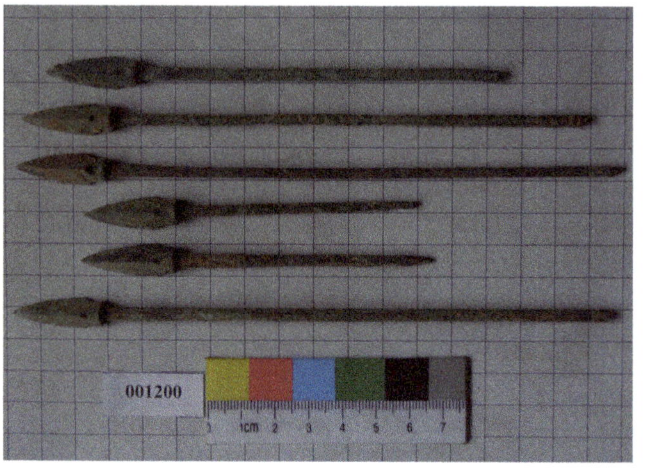

Fig. 6.27. Examples of the longer and shorter tangs in an arrow bundle.

length of six sampled arrows within each bundle are close to zero.

Figure 6.30 shows the CV values of the arrowheads' width and length across the different bundles. Six CVs on length were calculated taking one arrow from each of the bundles and the results are all below 3.5%; the same CV calculations on width were taken and the results are all below 4.5%. The CV values for the length of the arrowheads are 2.5–3%, and those for the width are 4–4.5%. The slightly higher variability in width may be due to the application of rotary devices to intensively grind and polish transversally the surfaces of the arrowheads (Li et al., 2011). While this process would have removed any casting imperfections and homogenised their overall appearance, sharpness, and sheen, it may have increased differences in width between some of the arrowheads. In any case, the most important aspect to highlight is that, theoretically, human eyes have difficulty perceiving dimensional differences if the CV values are below 3%. In other words, and assuming that the sample of over 1,500 arrows analysed is representative of the entire population, all 40,000 arrowheads buried with the terracotta warriors can thus be considered to be generally identical.

Such an extraordinary degree of standardisation attests to the very tight control that must have existed over the production of these weapons, particularly on those parts of the arrows that, unlike the tangs, would be visible once the arrows were finished. At first sight, such typological and dimensional uniformity could be taken as a strong indication that all of these arrowheads would have been produced in a single workshop or, at least, in very close proximity and under tight central supervision and quality control. However, how was this central supervision and quality control organised during the arrows' production and assembly processes?

6.5 Arrow production, assembly and labour organisation

In the previous chapter, it was argued that production of triggers was arranged as a cellular production, with little room for batch mixing. In contrast, the typological uniformity and high degree of dimensional standardisation of the arrowheads makes it virtually impossible to divide them into subgroups or batches based on metric differences. Even though the length of the arrow tangs is less standardised than that of the arrowheads as a whole, it was not possible to subdivide that heterogeneity further based on metric or visual features. As such, the initial impression is that of a prescriptive model or flow line production: one could imagine one specialised workshop unit producing arrowheads more or less constantly, with the next one producing tangs, another one bamboo shafts, and so on. The different parts would then reach another workshop unit, where the arrow-heads and tangs would be filed and polished before being fitted together and subsequently assembled to shafts and feathers (presumably produced in different specialised units), before bundling the arrows together in groups of 100.

However, X-ray fluorescence (XRF) and scanning electron microscope (SEM) analysis carried out by our cooperative team suggests a different model for the labour organisation of the production of these bronze arrows. The first interesting aspect of the chemical analyses was the differences noted between the arrowheads and arrow tangs (Martinón-Torres et al., 2014a, 2014b; Xia et al.,

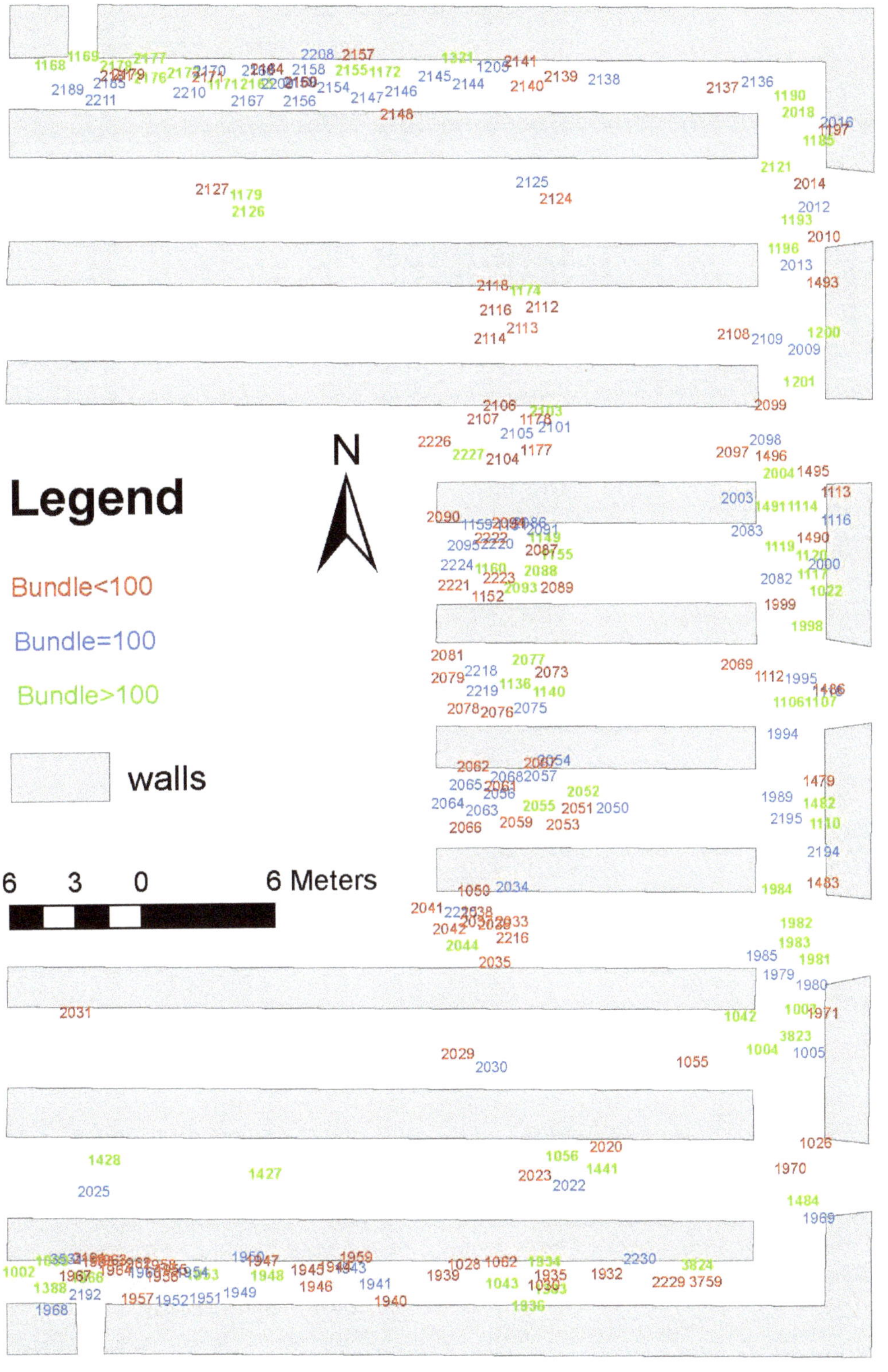

Fig. 6.28. Arrows labelled with their museum IDs (due to overcrowding, these labels are slightly offset from the objects' real locations).

2014), with the heads being systematically richer in tin. High tin would make arrowheads harder and enhance their penetrating power, but at the expense of a higher brittleness. Conversely, the tangs would be made of a lower-tin bronze that would ensure a higher toughness, reducing the risk of fracture when inserted in the bamboo shaft and perhaps allowing for a certain degree of flexibility for its oscillation during the arrow's flight (Martinón-Torres et al., 2014a; Xia et al., 2014).

Of more relevance to the reconstruction of the labour organisation, the XRF results indicate that the arrows and bundles were assembled near the location where the heads and tangs were cast. This impression is supported by the internal chemical coherence of the heads and tangs within each bundle, which most likely derives from individual metal casts or 'batches' that produced all of the arrows to be joined in a single bundle, with little room for the mixing of different batches during assembly or transport.

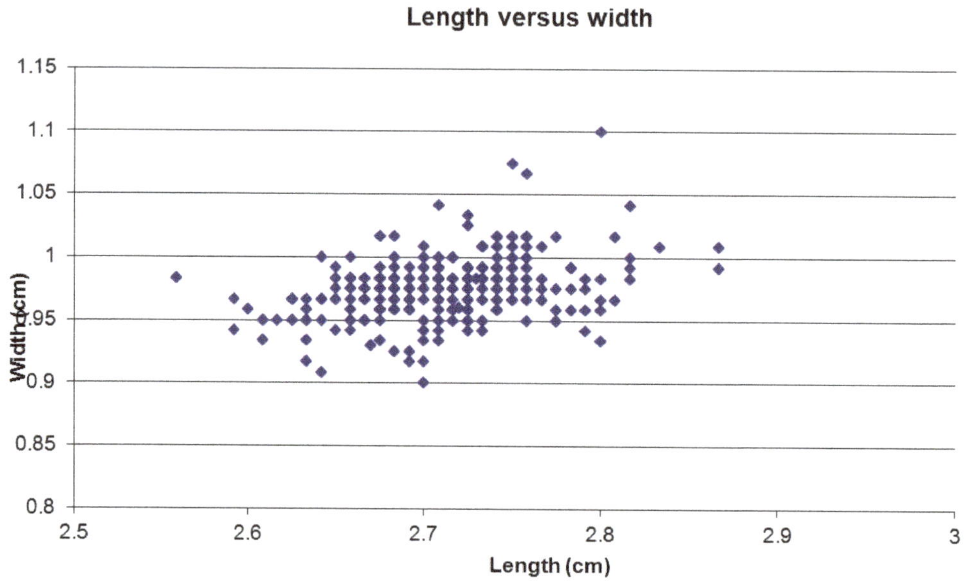

Fig. 6.29. Mean length versus mean width of the six sampled arrowheads from each bundle.

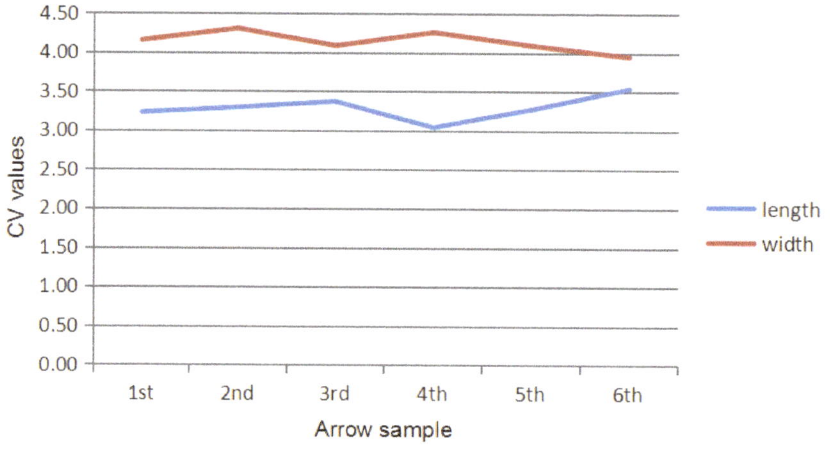

Fig. 6.30. CV values of arrowheads length and width in six series.

6.5.1 Consistency and variability of arrow bundles

As previously noted, the visual observation of the arrows conveys a sense of homogeneity within bundles, especially with regards to the tangs, in their length (roughly as a group), thickness, hardness, and weight. The XRF analyses on 18 bundles from different corridors in the pit confirm that the chemical composition of the arrows is internally coherent within each bundle, with small but significant variability between bundles.

We can illustrate this aspect by focusing on two bundles (Fig. 6.31). Figures 6.31a and 6.31c show the bundle ID 1152, from the 7th corridor in the middle of the easternmost trenches of Pit 1, comprising 92 arrows, (a) as a whole bundle and (c) for 9 selected arrows. From the images, one can observe the similar quality of the arrows within the bundle, a little bit bent and degraded. The length of tangs appears less standardised, although this is likely to be exacerbated by the fact that some are broken, due to their remarkable thinness. Figure 6.31b and 6.31d show comparable pictures of another bundle, ID 1985, from the easternmost crossing corridor of Pit 1, containing 100 arrows. The arrows within this bundle show visual similarity in their quality, straightness, hardness, and even the length of the tangs. Overall, in spite of the obvious similarities, these two bundles appear slightly different from each other: generally, arrows in ID1152 are bent and corroded, while arrows in ID1985 are straighter and tougher in comparison.

When the chemical composition of a sample of arrows from these bundles are presented in a scatter plot, each bundle forms a clear cluster, with the two clusters separated from each other: the straighter and tougher bundle with relative higher lead content versus the bent and corroded bundle with lower lead content. Both bundles also show higher tin in the arrowheads than in the tangs (Fig. 6.32). Moreover, some of the variability that exists within bundles can be explained as resulting from the lower accuracy of the surface XRF analyses when carried out on corroded surfaces: analyses focused on less corroded arrows and

(a) Bundle ID1152

(b) Bundle ID1985

(c) Bundle ID1152

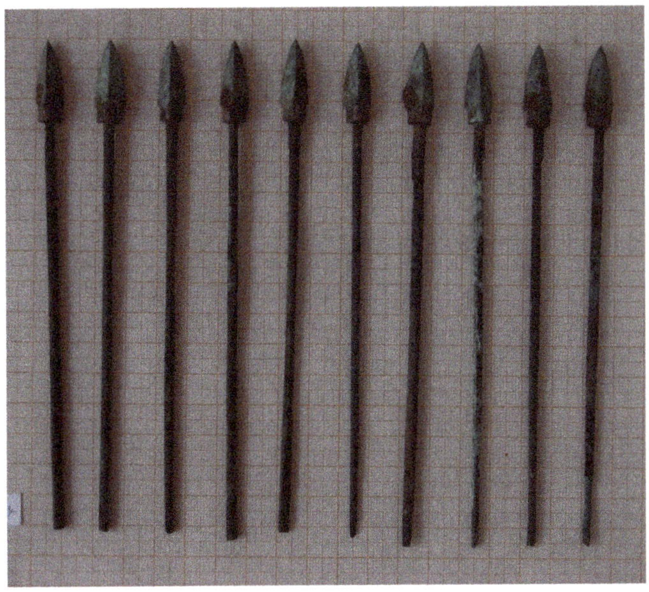

(d) Bundle ID1985

Fig. 6.31. Two arrow bundles selected for comparison.

selected cross-sections by XRF and SEM-EDS show even tighter clusters when the effects of corrosion are neglected.

When 18 selected bundles are plotted on the chart, this pattern is obscured by the many compositional overlaps among bundles, but it is still observable that each bundle tends to form a relatively coherent chemical cluster (Martinón-Torres et al., 2014a; Xia et al., 2014). This strongly suggests that all the arrowheads and tangs within a given bundle represent individual 'batches' of metal coming from a single crucible or furnace load – one for the heads, and another one for the tangs. As such, based on the chemical analyses on sampled arrow bundles, the production organisation of the arrows appears to be the same as that of the triggers', and it can also be argued that it was organised in semi-autonomous, multi-skilled cells of labourers rather than as a single production line. In other words, one group of craftspeople working as a cell would produce arrowheads and arrow tangs, then polished and fit them together, assembled them with bamboo shafts, and attached feathers to them. These arrows were bundled before being stored in the arsenal or sent to the pit. The entire production process for each bundle seems to be finished by one group of craftspeople or one working unit, perhaps with many cells working in parallel in the governmental workshop or chronological sequence producing the weapons for the terracotta warriors. In the alternative hypothesis, a flow line production, it would be much more likely that different metal batches would be mixed, as the different units would be mass-producing specific parts and then piling heads, tangs, and shafts together as they proceeded through to the assembly line.

6.5.2 Batch or cell production organisation

Based on the above results, I tend to argue in favour of a model of cellular organisation of labour for the production of arrows. In a cellular production, one or more workshop units would produce complete, finished bundles from the beginning to the end, responding to demand and quality

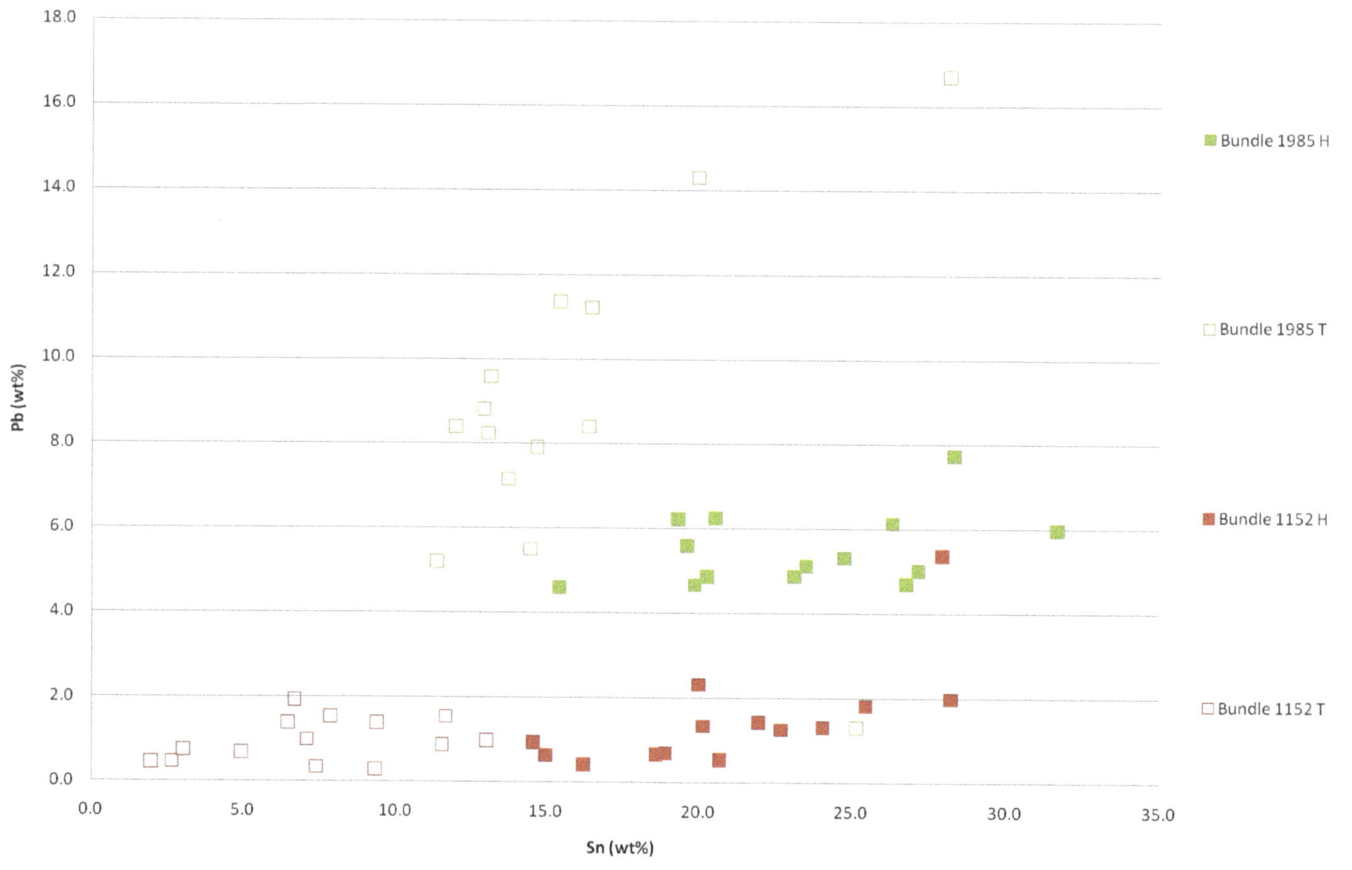

Fig. 6.32. Chemical composition of arrows in bundles ID 1152 and ID 1985 (image courtesy of Marcos Martinón-Torres).

control. A significant challenge for this type of production organisation would be to ensure that the different cells worked in harmony and to the same high standards, effectively producing items that were identical to each other. For this purpose, models, moulds, and standards were needed, but so was centralised supervision. We can obtain an insight into the system of supervision from the inscriptions on the longer weapons, such as lances and halberds, recording the names of up to four hierarchical levels of supervisors, workshop officials, and individual workers (Yuan, 1984; Li et al., 2011; see Chapters 4 and 7). Such supervisory organisation structures would be consistent with such a production model.

The next section presents the analysis of the spatial distribution of the arrows in Pit 1, which was carried out with the hope that it would shed further light on the organisation of the activities involved in the production, transportation, and arrangement of these arrows from the workshop or storage to the pit. However, considering the extreme standardisation of the arrowheads, which has prevented the identification of any typological subgroups, the potential of such spatial analysis will be explored mainly on groups identified according to arrow tangs.

6.6 Spatial data

The source of the digital map of the arrows is the same as that of the triggers and other weapons, and is based on a combination of the original distribution map published in the Excavation Report (Institute and Team, 1988) and the archives in the Conservation Department of the Terracotta Army Museum. The arrows were all digitised as vector data, and these location records were integrated within a GIS environment (specifically in ESRI's ArcGIS package; see also Chapter 3, section 3.2) in the local coordinate system.

Altogether, the location of 680 bundles or loose arrows, including 262 large bundles (bundle size ≥90 arrows) and 418 small bundles (size < 90 arrows), including single loose arrows were digitised as vector point locations (Fig. 6.33, (a) all arrows; (b) large bundles versus small bundles). Most of these bundles consist of bronze arrows. However, there are two exceptional cases, which are made of an iron tang with a bronze arrowhead (Fig. 6.34). One was found on the chariot in corridor 9, and the other was in the east vanguard corridor. Their significance will require further discussion in a separate paper. In addition, there are seven special arrows of peculiar shapes in the pit (Fig. 6.35), and they are evenly spaced across the battle formation. Given the shape of the heads of these arrows, and possible later analogies, they may have been used as signal arrows, but this possibility still requires further investigation.

6.6.1 Parameters affecting the spatial patterns

As with the bronze triggers and other long weapons mentioned in previous chapters, the spatial distribution of the arrows

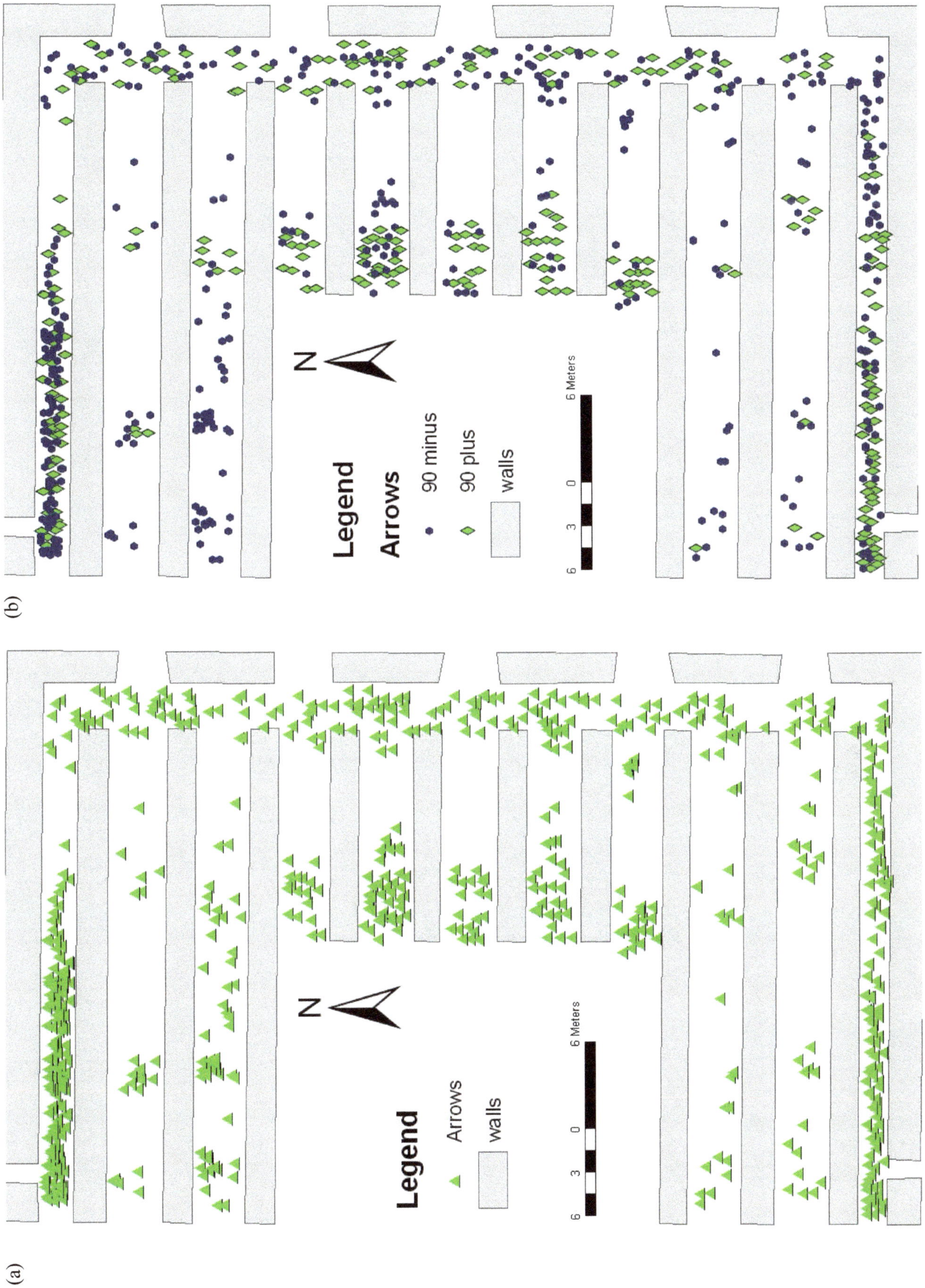

Fig. 6.33. Spatial distribution of arrows (a) and arrow bundles (b).

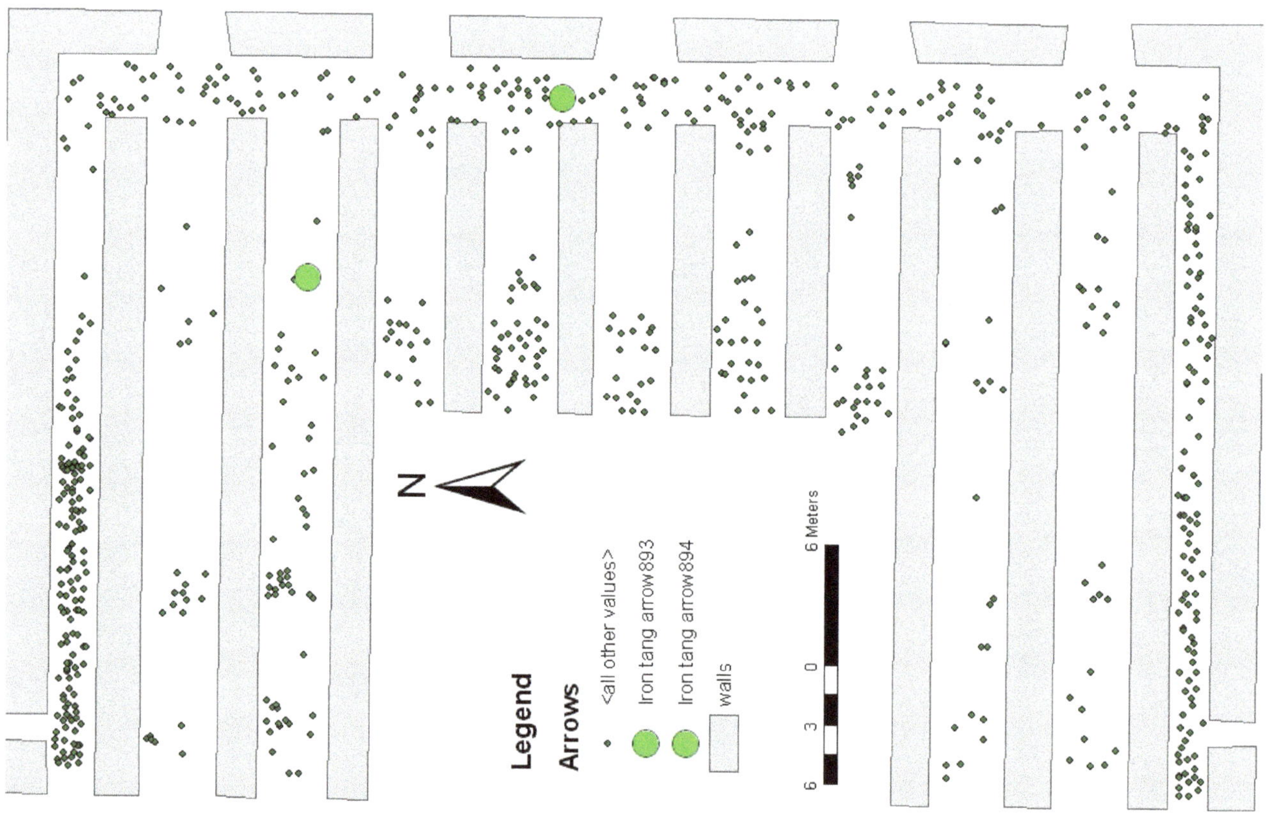

Fig. 6.34. Location of the arrows showing iron tangs assembled with bronze arrowheads.

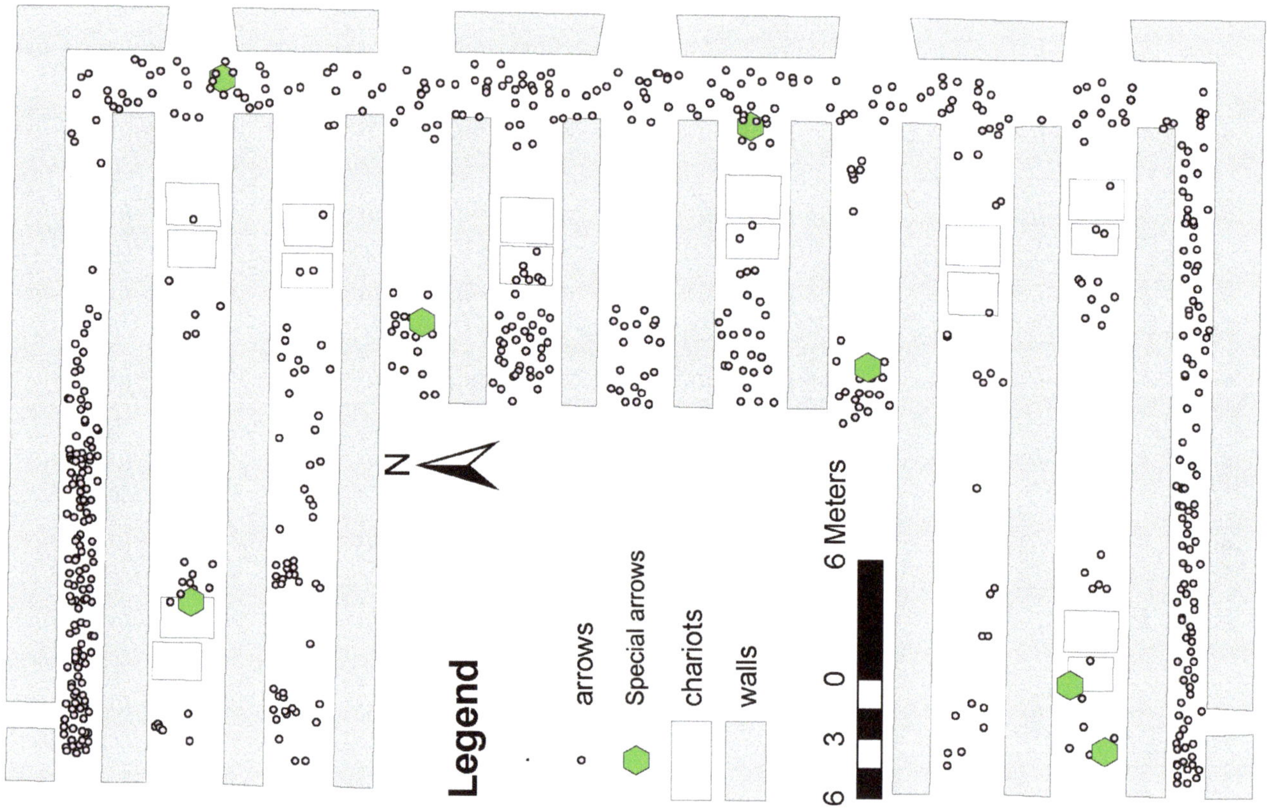

Fig. 6.35. Location of the special bronze arrows.

is affected by two main factors. One is their relationship to the terracotta crossbowmen arranged in a battle formation within the earth-wooden tunnel structure; the other is the organisation of the labour force during the production of the arrows, storage in the arsenal, delivery, and their placement in the pit. Furthermore, some other factors need to be considered as well, such as looting, post-depositional alteration, and/or effects of the archaeology itself.

The general spatial distribution of the arrows is tightly correlated with that of the bronze triggers to match the crossbowmen in the battle formation. The arrows are densely distributed in the two flank corridors (south and north side of the pit) and in the front vanguard corridor (east), which were normally the predominant locations for crossbowmen (Fig. 6.36). However, some of the infantrymen were also equipped with bundles. For example, at the back of corridors 3 and 9, most warriors appear to be holding long weapons from their postures, but some arrows are still dispersed in these corridors. Alternatively, in corridor 6 there are some warriors with crossbowmen postures but without arrows.

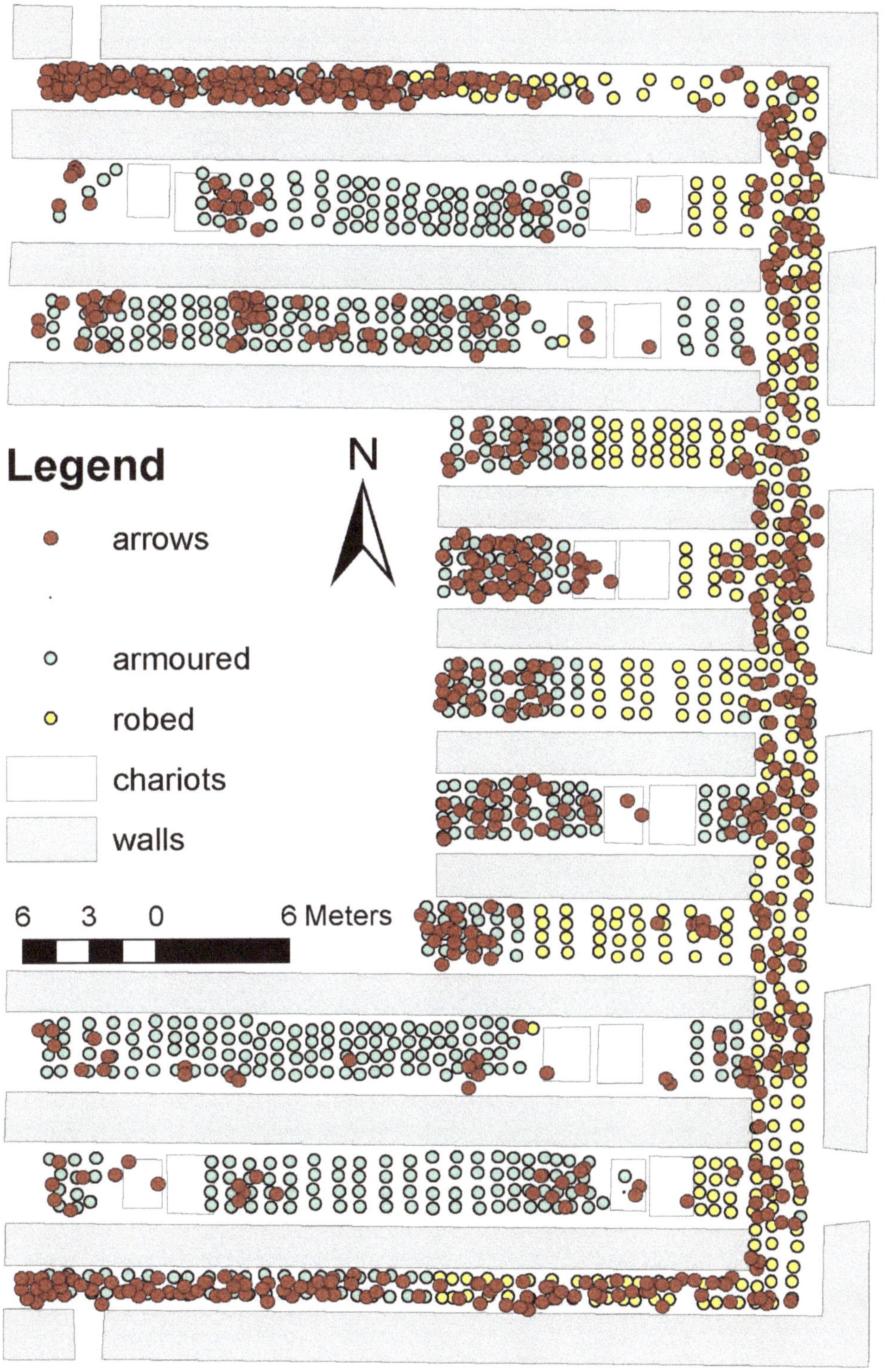

Fig. 6.36. The location of bronze arrows matched to the crossbowmen in the pit.

In addition to the relationship with the crossbowmen in the battle formation, we argued that the distribution of arrows with particular morphological or compositional characteristics, *within the overall pattern of arrows*, was driven by the model employed for the weapons' production organisation, transportation, and placement into the pit. As mentioned above, these arrows present rather tight consistency within the bundles and variability between the bundles. In addition to supporting a model of cellular production as discussed above, this pattern confirms that the arrows had never been used in the battlefield before being placed in the pit. Ideally, the spatial patterns of the arrows in the pit still bear information pertaining to workshop activity and arrow delivery.

The pair correlation function analysis of the arrows will mainly focus on the arrow tangs, given the very high degree of standardisation observed on all arrowheads and the identification of no subgroups from their measurement. Even though the chemical composition analysis by XRF indicated variability between bundles in both arrow tangs and arrowheads, specific subgroups of bundles cannot be identified due to the relatively small sample size and the low analytical accuracy. Arrow tangs were tentatively divided into two subgroups based on their average length (see section 6.3.3.1). Such differences might be driven by several factors, such as different workshops and different working units and/or cells working under the same roof. The spatial patterns (regular, clustered, or random) of the subgroups are theoretically related to the workers' behaviour and the underlying labour organisation. A regular or uniform spatial pattern of arrow tangs or analytically similar subgroups may reflect a form of intentional arrangement by the workers, perhaps linked to special kinds of workshop or storage, or to certain warrior types. The clustering of arrows may result from a number of factors, but the existence of different workshops or different units of craftspeople would seem to have been influential factors. In contrast, random distributions offer a useful statistical null-hypothesis.

6.6.2 A pair correlation function

A pair correlation function was generated using R software on the two subgroups of arrows defined by the average tang lengths. Figure 6.37a demonstrates the spatial patterning for short arrow tang bundles (see Fig. 6.22 as well) with standardised arrowheads. We can visually note that the possible patterning of some arrows concentrate in the middle corridors, but we should confirm that this is so, over the distribution of arrow bundles as a whole, as well as to calculate the distance at which this cluster is really present. Figure 6.37b shows a pair correlation function providing a summary spatial statistics for the group of short tang arrow bundles. From the x-axis measures, the solid curve line of the observed value is higher over the envelope at 5–7 metres (Table 6.3), a distance at which the clustering of the short tang arrow bundles is demonstrated. As discussed for the triggers in the previous chapter, there are interesting processes that may be behind such a clustered distribution of the shorter tang bundles in the pit. A plausible interpretation of the clustering may be that this results from specific activity areas where designated workers equipped warriors with these arrows

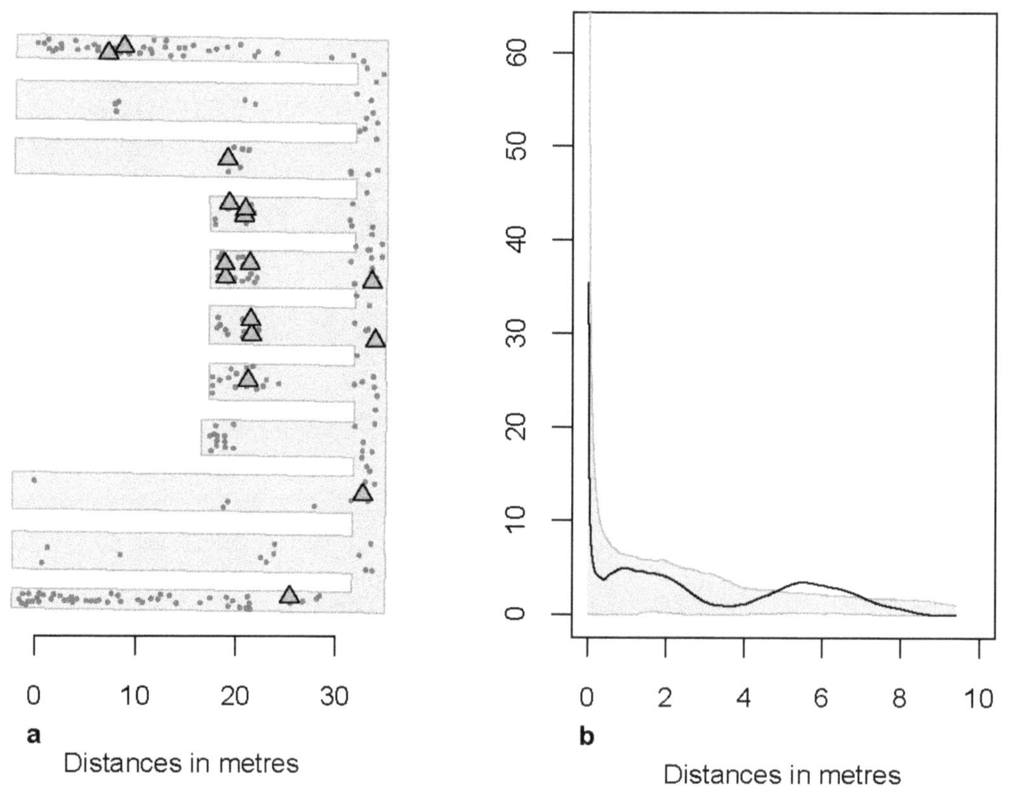

Fig. 6.37. The spatial distribution of short tang arrows (shown as triangles) within the five easternmost trenches of Pit 1 (a), and (b) a pair correlation function considering the spatial pattern over the first 10 metres.

in the pit. The short tang arrow bundles would have been produced by a group of craftspeople or a cell, and stored in a storage or an arsenal, and then transported to the Qin First Emperor's tomb complex to supply workers in this specific activity area.

Figure 6.38 shows a pair correlation function for the long tang (9.5–18 cm) arrow bundles. The observed value is above enveloping at the distance of 1–4 and 8–9 metres (Table 6.3), which indicates clustering, and may again represent the activity areas related to the placement of arrows into the pit. However, it should be remembered that this group was difficult to define neatly, as it fell between the shorter and longer tang groups without a clear demarcation. Considering the relatively large variability noted for the arrow tangs, it is also possible that there are various subgroups here, which cannot be discerned with certainty.

Overall, the spatial analysis of the distribution of arrow bundles in the pit is perhaps not as conclusive as that of the triggers, in terms of the information we may obtain regarding labour organisation. As noted above, the high standardisation of the arrows, together with the high variability of the tangs, made the identification of subgroups very difficult. As such, without a clear estimate of the number of moulds or work units involved, the identification of potential activity areas becomes much harder. The patterns exhibited in Figures 6.37 and 6.38 appear consistent with the presence of activity areas for the placement of arrow bundles in the pit, but these results should be regarded as provisional until we can subdivide the bundles further – perhaps using chemical data. In the meantime, the relatively internal coherence between bundles, both in tang length and in chemical composition, strongly points to a cellular model of labour organisation for the production of the arrow bundles.

6.7 Summary

The large quantity of bronze arrows recovered in Pit 1 demanded a pilot survey be undertaken followed by a systematic sampling strategy in order to achieve meaningful results. The detailed measurement of 6 arrows from each of the 290 bundles containing more than 90 arrows provided valuable data that makes it possible to develop our reconstruction of the standardisation and labour organisation in the production and placement of weapons into the funeral pit.

The assessment of CVs revealed a very high degree of standardisation for the arrowheads, with most of the arrows measured being identical to the millimetre. The arrow tangs were much less standardised, perhaps because they were less functional than the arrowheads, and also because their higher variability would be concealed when they were inserted in the longer shafts. Based on the measurements of tang lengths, the bundles could be provisionally divided into two groups, short (5–9.4 cm) and long (9.5–18 cm). These could be related to different working units in the workshop or small cells, but it has to be acknowledged that these subdivisions, particularly the latter one, may well hide further subdivisions that we simply cannot discern. As such, the implications of the

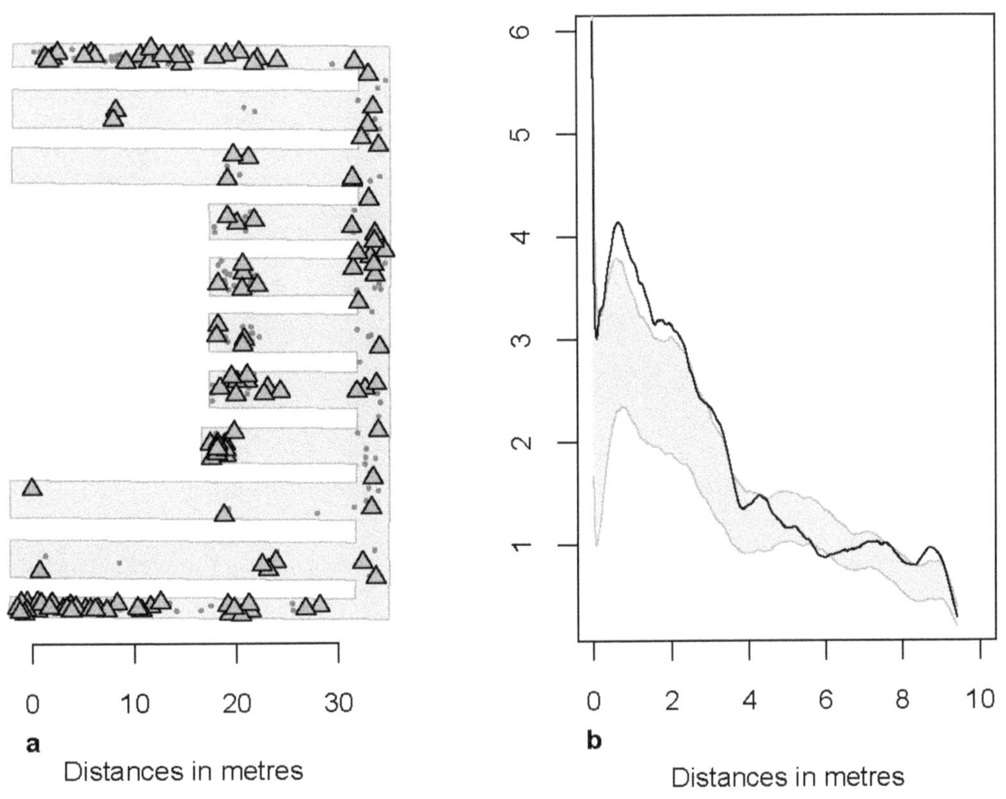

Fig. 6.38. The spatial distribution of the longer tang arrows (shown as triangles) within the five easternmost trenches of Pit 1 (a), and (b) a pair correlation function considering the spatial pattern over the first 10 metres.

typological and spatial analysis remain more limited than in the case of the triggers.

Macroscopic examination indicated a certain degree of internal coherence within the bundles, particularly noticeable in the tang shape and length. This was corroborated by XRF analyses showing slightly different chemical compositions between each of the arrow bundles. These differences were driven by both technological and organisational factors. The size of the crucible might have provided only sufficient liquid bronze for one bundle of arrowheads, and another crucible might have been only enough for one bundle of arrow tangs. The technological factors and labour organisation mutually reinforced each other. However, it is important to note that different batches of metal were rarely mixed, thus confirming that each bundle of 100 complete arrows was completed by a single production cell, rather than by arranging the different parts in a flow production and assembly line. A combination of shared models and moulds with a central governmental monitoring and hierarchical levels of supervisors would have ensured that the cells or units manufactured standardised products of acceptable quality.

The spatial patterns and spatial statistics attempted an assessment of how labour was organised in the placement of arrows into the pit. Besides matching the battle formation, the distribution of arrow bundles in the pit suggests the presence of activity areas, thus informing about how the workers were involved in the arrangement of the arrows. However, these patterns will have to remain provisional, given the difficulties encountered when trying to identify arrow subgroups based on their typology or metric data.

7

Bronze Ferrules and Long Weapons

7.1 Introduction

The weapons unearthed from the pits of the terracotta warriors are divided into three categories (for details, see Chapter 3): short weapons such as swords and hooks, long distance weapons such as crossbows, and long weapons such as spears, lances, and halberds. The long weapons comprise three parts: a bronze blade (found on dagger-axes, spears, lances and halberds, etc.), a wooden shaft (rotten), and a terminal ferrule. The term 'ferrule' refers to the bronze cap that was fixed to one end of the wooden shaft in long weapons. The ferrules are cylindrical in shape, and some of them are decorated with three central bands (Fig. 7.1). The function of a ferrule was both to decorate and to protect the end of the wooden shaft of a long weapon, because the shafts were especially sensitive to water damage. If water was constantly allowed to seep into the wood from contact with the wet ground, the wood would become weak and split. Therefore, protection from dampness was essential to the good maintenance of these weapons. They could also be employed as back-up spearheads in case the main head got blunted or fell off, or, sometimes, they made it easy to stick a spear in the ground. Ferrules are still used today on walking sticks and on some wooden handles, and they perform a similar function as in the past.

Even though the ferrules are not actually weapons themselves, they are so closely linked to the bronze long weapons that they merit examination. Figure 7.2 shows the intact linkage of a halberd, wooden shaft, and ferrule discovered in Pit 1. It was originally to be held upright by a terracotta warrior (Fig. 7.3) and measured approximately 287 cm in length. Some of the ferrules unearthed from the pit have a '*Sigong*' inscription on the surface, the same inscription recorded on other bronze weapons (Institute

Fig. 7.1. Examples of ferrules unearthed from the pits of the terracotta warriors (image courtesy of Xia Juxian).

Fig. 7.2. Ferrule, wooden shaft and halberd unearthed from Pit 1 (images courtesy of Xia Juxian).

and Team, 1988: 272; see Chapter 4 above). These inscriptions indicate that the ferrules were produced in the same workshop and were subject to the same quality control as the other bronze weapons.

The quantity and the quality of the bronze ferrules discovered in the pit provides a rare opportunity to study patterns of standardisation and labour organisation

Fig. 7.3. Reconstruction drawing of a warrior holding a long weapon (image courtesy of Ma Yu).

involved in their mass production during the Qin period, as well as their relationship to the long weapons. Overall, 126 bronze ferrules were examined for this particular project, and, among these, there are 80 ferrules from the five easternmost trenches of Pit 1 excavated during the 1970s, which are thus associated with a specific spatial location. The other ferrules were unearthed from ongoing excavations in the other trenches of Pit 1 and Pit 2 and their spatial patterning is too patchy to be considered here. These ferrules are all well preserved, despite being buried in the pit for more than 2,000 years. Apart from two of them that are deformed, the rest are all intact, and can therefore be accurately measured and analysed in order to assess the procedures and organisation of their production.

The ferrules have also been analysed typologically and spatially in a systematic manner. In addition, this chapter also considers the lances, spears, dagger-axes, and halberds that were presumably attached to the long wooden handles, and which were mostly discovered beside the ferrules (Institute and Team, 1988). There is value in assessing the extent to which the location of these long weapons matched that of the ferrules, and to understanding why more ferrules were unearthed from the pit than the actual blade-ends of long weapons.

7.2 Research review

7.2.1 Ferrules

Long weapons played a very important role in ancient Chinese warfare and they were recorded in ancient historical documents. Therefore, the related archaeological finds normally attracted much attention (Institute and Team, 1988; Yates, 2007). When the actual blade-ends of long weapons were found in Pit 1 of the terracotta warriors, much more attention was dedicated to them than to the ferrules (Wang, 1980b; Liu, 1982; Qin and Zhang, 1983;

Yuan, 1990). Whereas the long weapons found during these excavations were immediately analysed, the ferrules were generally neglected, and, apart from the initial typological analysis (Institute and Team, 1988), little systematic analysis has been conducted on them by archaeologists. Thus, a review of the Qin ferrules unearthed from Pit 1 and other archaeological sites, as well as related information, appears necessary for the purpose of the present research.

Basic information about the ferrules unearthed from Pit 1 is contained in the Excavation Report (Institute and Team, 1988). According to the excavation report, the 80 ferrules recovered from this area can be divided into three types (I, II, and III) based on their general shape, and this basic division is maintained below, while also identifying further subdivisions within them. In this report, the focus was only on linking the ferrules to their original wooden shafts, but no proper interpretation about the ferrules themselves was provided. Some bamboo traces were found inside the ferrules, suggesting that bamboo handles might have been used in addition to wooden ones. Most of the wooden or bamboo shafts had, however, rotted away during the 2,000 years of their burial, except for the halberd trace preserved intact (Institute and Team, 1988).

In addition to the ferrules found in the pits of the terracotta warriors, some other ferrules produced during the Qin period have been unearthed from other archaeological sites. One of the earliest Qin inscribed ferrules, discovered in 1995 from a Qin tomb at Ta'erpo, Xianyang City, has attracted much attention, because it contains a long inscription on the surface. The inscription reads '*in the 19th year (343 BC), Shang Yang, as Da Liang Zhao, supervised the making of this ferrule*' (十九年大良造庶长鞅之造殳犛郑; Wang, 1999; Xianyang Cultural Relics Bureau, 2002; Yates, 2007; Fig. 7.4). Another ferrule was recorded in the ancient document *Shuang Jian Ji Jin Tu Lu* (《双剑吉金图录》下), bearing the inscription '*in the 16th year (346 BC), Shang Yang, as Da Liang Zhao, supervised the making of this ferrule*' (十六年，大良造鞅之造，雍，矛; Yuan, 1984). These two ferrules provide a clear date

Fig. 7.4. Ferrule with inscription referring to *Shang Yang* (image courtesy of Zhao Zhen).

of production, and the supervisor for this production was, Shang Yang, an advisor to the duke of Qin, and also known as a key reformer during that period (Yuan, 1984; Yates, 2007). In contrast to these long inscriptions, some ferrules from the pit of the terracotta warriors only bear short '*Sigong*' inscriptions, while others bear no inscription at all.

Along with the broader political reforms initiated by Shang Yang, mentioned in Chapter 1, it seems that changes also took place at this time in the control of the weapons' production. Apart from the ferrules mentioned above with Shang Yang inscriptions, a halberd with the same inscription was also recorded in an ancient document entitled *San Dai Ji Jin Wen Chun* (《三代吉金文存》20，21，1; Yuan, 1984). The halberd and ferrules were carved with the name of the supervisor who controlled the quality of the production. Shang Yang was in the position of *Da Liang Zhao* (大良造) during the period from the 10th to the 24th year of the reign of Duke Xiao (秦孝公)，which corresponds to 352–338 BC (Yuan, 1984). It was precisely during this period that carved inscriptions bearing the supervisor's name began to appear on the weapons, reflecting the tighter control of the production of weapons by the Qin central government. The 16th- and 19th-year ferrules with Shang Yang's name are good examples of this. The fact that both halberds and ferrules were carved with inscriptions bearing the name of Shang Yang also indicates that the ferrule was assigned similar importance to the halberd blades during the early Qin production.

From then on, some Qin weapons began to bear inscriptions containing the name of the chancellor who also served as supervisor of the production, and such inscriptions can be found on some of the halberds and lances in the pits of the terracotta warriors. However, the chancellor's name is Lu Buwei rather than Shang Yang.

Ferrules were also found in a metallurgical workshop at Xianyang, the ancient Qin capital. According to the excavation report, the workshop was seriously disturbed, but eight intact ferrules were discovered (Shaanxi Institute of Archaeology, 2004). These ferrules were divided into three types, of which two are comparable to type I and type II ferrules from the pits of the terracotta warriors. Of the ferrules from this workshop, the one that corresponded to the type I ferrules under consideration here measured 13.8 cm in length, 3.4 cm in diameter, and 0.2 cm in thickness, and the example corresponding to type II measured 12.8 cm in length and 4.6 cm in diameter. In addition to these ferrules from the aforementioned workshop, one type I ferrule (12 cm in length, 3.5 cm in diameter, and 0.2 cm in thickness) was also discovered in a Qin tomb at Hang Jiagong, Shaanxi province, in the 1970s (Shaanxi Institute of Archaeology, 2004). For all these ferrules discovered outside of the tomb complex of the Qin's First Emperor, only the typological analysis was mentioned in the excavation report, and, since then, no more attention has been devoted to them.

Even though the long weapons, such as spears, dagger-axes, halberds, and lances, from the pits have attracted more attention than the ferrules, the present study offers a more complete analysis that integrates these data with those obtained on the ferrules.

7.2.2 Bronze blades of long weapons

A wooden or bamboo shaft with a ferrule at the one end and a blade weapon at the other formed a long weapon. The blade weapons included lances, spears, dagger-axes, and halberds. In Qin period China, spears and dagger-axes were sometimes used separately, and sometimes fixed together to function as a halberd. The lances had another Chinese name – *a long spear* (长矛) to distinguish from the more common spears (Qin and Zhang, 1983; Wang, 1983; Institute and Team, 1988).

Overall, 6 spears (5 bronze and 1 iron), 1 dagger-axe, 4 halberds, and 16 lances (all made of bronze) were found in the five easternmost trenches during the excavations carried out in the 1970s (Institute and Team, 1988). The discrepancy between the large number of ferrules and the small number of lances and halberds needs to be addressed. While it is also possible that some ferrules did not belong to weapons but were placed at the end of flags or other ensigns in the battle formation, the contrast may also relate to different levels of preservation and requires further investigation, as presented below.

Past research on these bronze long weapons mainly focused on their origin, development, and functions within the battle formation, as well as on identifying their names based on information from ancient Chinese documents (Qin, 1975; Yang, 1980; Dang, 1987; Research Committee, 1995). For example, historical research has focused on when, how, and why spears began to be assembled with dagger-axes to form halberds. Interestingly, one of the earliest pieces of evidence of a bronze halberd was discovered in a Shang Dynasty (1600–1050 BC) tomb at Gao Cheng, Hebei Province (Hebei Museum, 1977), but it was not until the East Zhou period (770–221 BC) that bronze halberds started to be widely used by the infantrymen or warriors on the chariots (Yang, 1980). Some examples show that the spear and dagger-axe were cast together before being attached to the wooden shaft, while others were separately cast and attached to the shaft. The latter combination presented advantages for chariots warriors and infantrymen. The dagger-axe could be used to drag the enemy closer and the spear to pierce them. Lances, other important weapons in ancient China, are also recorded in many documents (Qin and Zhang, 1983), but there had been no idea of their exact form until the discovery of the 16 lances from Pit 1 of the terracotta warriors. Generally, data pertaining to long weapons and ferrules had not been previously used to address issues of standardisation and labour organisation. In Chapter 4, I presented details of the information regarding the supervisory structures pertaining to the production of these weapons that can be obtained from inscriptions. The focus will be devoted to furthering this information through metric and spatial analyses.

7.3 Measurements of the bronze ferrules

7.3.1 Sources of data

As with the triggers and arrows, all the ferrules and related bronze spears, dagger-axes, halberds, and lances discovered from Pit 1 of the terracotta warriors have been moved from their original location to the store in the Conservation Department of the Museum of the Terracotta Army. Contextual data (1976–1984 excavation) has been obtained from the excavation report (Institute and Team, 1988), available in Chinese, and from the museum database in the Conservation Department, which also contains information on the continuing archaeological excavations of the pits.

A total of 126 ferrules have been studied, photographed, and measured for this particular project. Among these, 3 ferrules originate from the trial trenches in Pit 2, 80 from the five easternmost trenches in Pit 1, and a further 43 ferrules from the other trenches identified during ongoing archaeological excavations in Pit 1. All 126 ferrules were measured, but the spatial analysis will only focus on the 80 ferrules from the five easternmost trenches. Summary information is presented in the following table (Table 7.1):

Table 7.1. The types of ferrules, their locations, and the research methods employed.

Locations	Count of Types			Sum	Methods involved
	I	II	III		
Pit 1 The five easternmost trenches	26	4	50	80	Measurement; Multivariate analysis; and spatial analysis.
Pit 1 On-going archaeological excavations	11	1	31	43	Measurement and multivariate analysis
Pit 2	1		2	3	Measurement and multivariate analysis
Total	126			126	

7.3.2 Methodology for the measurements

In comparison to the triggers' complicated assembled parts and to the large number of arrows, the ferrules are relatively straightforward and easy to measure, as there are not as many parts involved and the shape is fairly simple. Thus, a digital calliper was employed to measure the selected dimensions. The following measurements were taken consistently across the assemblage (Fig. 7.5).

Length, the diameters of the top and bottom, and thickness were the dimensions selected for digital measurements. The top and bottom of the ferrules are elliptical, and so the parameters include: T1-diameter (the longest diameter on the top), T2-diameter (the shortest diameter on the top), and B1-diameter (the longest diameter on the bottom), and B2-diameter (the shortest diameter on the bottom) (Fig. 7.5). All the measured data, including length, T1-diameter, T2-diameter, B1-diameter, B2-diameter, and thickness, were directly transferred to the computer as a spreadsheet.

For the purposes of individual assessment, Figure 7.6 maps the ferrules with their museum identifiers and the following Tables 7.2, 7.3, and 7.4 offer a summary version of the measurements taken on each type. Further details and the full set of measurements are provided in Appendix 8.

Fig. 7.5. Dimensions measured for the ferrules.

7.3.3 Measuring patterns in the ferrules

Both scatterplot (Fig. 7.7) and principal component analysis (PCA; Fig. 7.8) were employed for further research into these measurements. The scatterplot offers a relatively straightforward first impression of variability in ferrule measurements. However, it can only show two variables at a time on its horizontal and vertical axes, and a principal component analysis is ultimately more useful for considering multidimensional patterns in the ferrules' measurements.

Both Figures 7.7 and 7.8 present the patterns of the ferrule measurements in different but complementary ways. Figure 7.7 is a scatter plot of the length against the T1-diameter of the ferrules. It clearly displays the concentration of type I and type III ferrules, but each type includes one or two exceptions that are isolated from the others. The five ferrules in type II are relatively dispersed; three of them are close to each other, and the other two are separated from each other. Figure 7.8 provides the patterns concerning all six variables, which includes length, T1-diameter, T2-diameter, B1-diameter, B2-diameter, and thickness. Table 7.5 provides the correlations between the six original variables of ferrules and the five principal components. In the PCA plot, it is possible to discern a subdivision within the type III ferrules. This will be a starting point for further analysis regarding the function of the ferrules in the battle formation and their standardisation in the workshop.

Excluding the singular ferrules noted above, which fall into the same broad typological groups but evidently represent different types, the three main types of ferrules clearly present distinctive features (Fig. 7.9). Type I ferrules, including a total of 38 samples, show three characteristic bands on their surface, and account for about 30% of the total number of ferrules; type II, containing only 5 ferrules, show a plain cylinder shape without any decoration; type III, numbering 83 examples in total (about 65% of the total number of ferrules), are shorter and wider than type II, and some of them bear '*Sigong*' inscriptions (Fig. 7.10; Table 7.6).

A type I ferrule normally has a hole on one side, which is assumed to have been used to fasten the wooden shaft of the long weapons. Some of them still have the pin left inside (Fig. 7.11). Conversely, type III ferrules have two pins inside, and traces of them can still be seen on the surface of this type of ferrule (Fig. 7.12). There is no evidence of holes and pins on type II ferrules.

From both the examples where the full weapons are preserved (see Fig. 7.2) and the spatial link between the bronze blades and ferrules in the pit (see Fig. 7.26 and Fig. 7.27), we know that the spears, dagger-axes, and halberds were normally attached with type I and type II ferrules, while the lances are all linked with type III ferrules (Institute and Team 1988: 270; Fig. 7.13). However, only 5 spears, 1 dagger-axe, and 4 halberds were found in the eastern trenches of Pit 1, as compared to 26 type I and

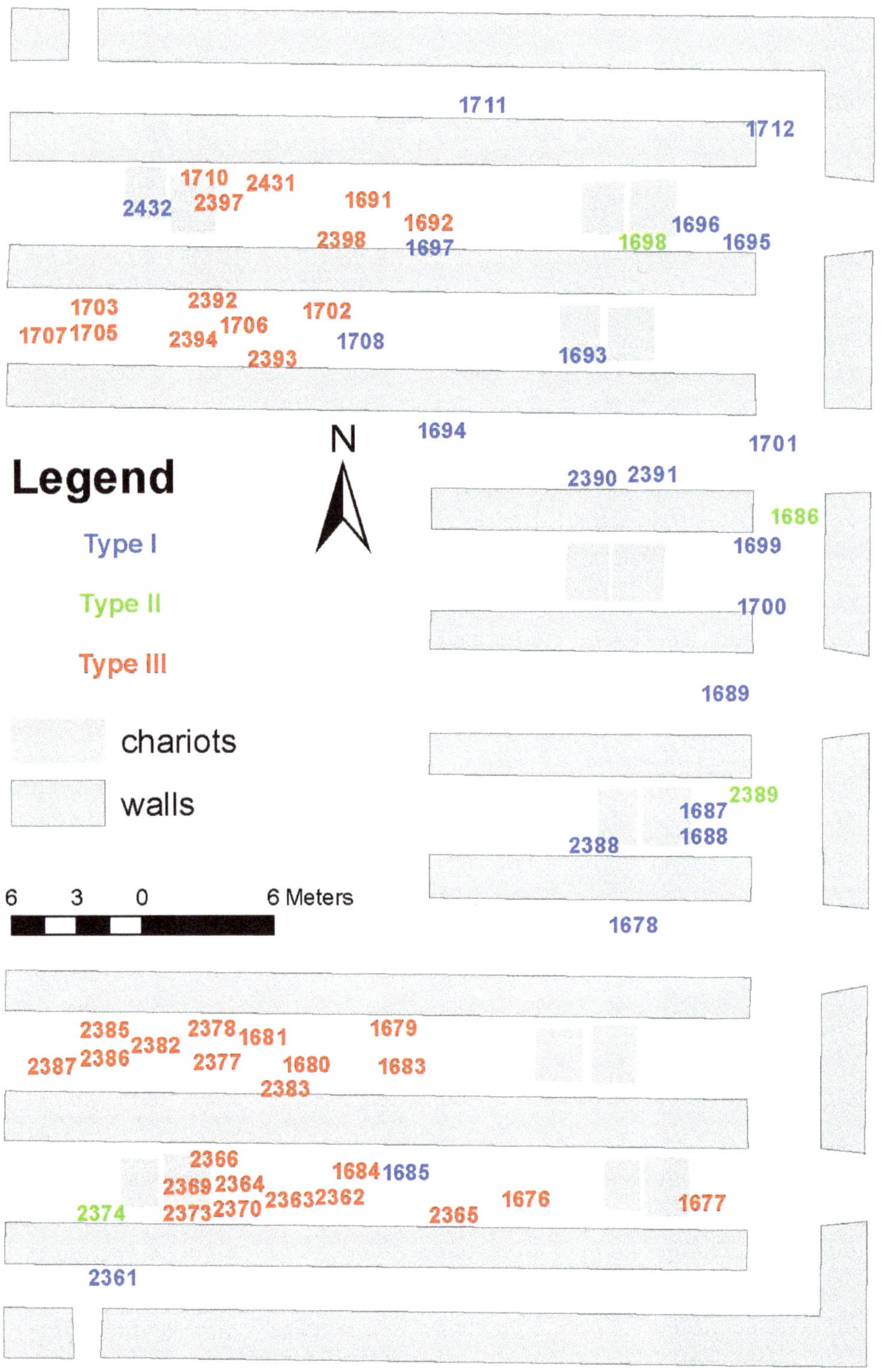

Fig. 7.6. Ferrules from the five easternmost trenches of pit 1, labelled with their museum IDs.

4 type II ferrules; likewise, 16 lances versus 50 type III ferrules is not a balanced proportion. There are several possibilities that might explain this numerical discrepancy. First, the halberds and lances might have been looted during a farmers' rebellion shortly after the pit was built (Yuan, 1990). Second, the archaeological remains might have been disturbed by agriculture, which did damage to the long weapons, as their shafts were very long and any collapse of the pit might have caused the tops of the bronze weapons to emerge out of the roof.

The shafts were mostly made of wood, but two type I ferrules with bamboo traces inside were observed (Institute and Team, 1988: 274). There is no clear evidence to indicate that these two ferrules were attached to any weapons, and they were possibly linked with other ensigns, such as flags.

7.3.4 Analysis of the patterns of each type of ferrules

The pattern of the ferrule measurements provides basic information about their production and represents a

Table 7.2. Measurements taken on type I ferrules.

Ferrule ID	Type	Inscription	Length	Diameter T1	Diameter T2	Diameter B1	Diameter T2	Thickness	Upper length
001675	I	0	11.32	4.09	3.29	3.73	3.02	0.21	4.53
001678	I	0	11.22	4.05	3.46	3.69	3.09	0.30	4.57
001685	I	0	11.21	4.01	3.45	3.63	3.21	0.20	4.52
001687	I	0	11.28	4.08	3.42	3.72	3.03	0.19	4.49
001688	I	0	11.25	4.15	3.37	3.82	3.01	0.18	4.49
001689	I	0	11.27	4.09	3.23	3.87	3.07	0.13	4.54
001693	I	0	11.22	3.97	3.40	3.60	3.13	0.16	4.53
001694	I	0	11.53	4.29	3.42	3.85	3.19	0.16	4.85
001695	I	0	11.19	4.01	3.49	3.62	3.14	0.16	4.58
001696	I	0	11.19	3.96	3.33	3.63	3.19	0.21	4.52
001697	I	0	11.55	4.15	3.29	4.13	3.23	0.17	
001699	I	0	11.21	4.00	3.43	3.55	3.06	0.14	4.43
001700	I	0	11.19	4.00	3.36	3.59	3.10	0.16	4.55
001701	I	0	11.21	3.95	3.39	3.66	3.24	0.13	4.49
001708	I	0	11.24	4.00	3.39	3.66	3.09	0.19	4.52
001711	I	0	11.45	3.88	3.24	3.83	2.99	0.18	4.77
001712	I	0	11.18	3.98	3.35	3.61	3.01	0.22	4.45
002361	I	0	11.30	4.10	3.36	3.83	3.12	0.18	4.50
002367	I	0	11.55	4.14	3.45	3.88	3.17	0.19	5.04
002368	I	0	11.20	4.08	3.39	3.68	3.06	0.18	4.53
002372	I	0	11.46	3.94	3.16	3.86	3.02	0.13	4.87
002388	I	0	11.19	4.00	3.37	3.69	3.05	0.15	4.65
002390	I	0	11.22	4.03	3.41	3.67	3.12	0.19	4.46
002391	I	0	11.27	3.90	3.51	3.62	3.09	0.17	4.56
002432	I	0	11.22	3.94	3.41	3.59	3.06	0.13	4.59
003806	I	0	11.18	3.97	3.38	3.64	3.16	0.19	4.53

Table 7.3. Measurements taken on type II ferrules.

Ferrule ID	Type	Inscription?	Length	Diameter T1	Diameter T2	Diameter B1	Diameter T2	Thickness
001686	II	0	10.07	3.27	3.19	3.13	3.12	0.27
001698	II	0	10.25	3.31	3.26	3.22	3.18	0.24
002374	II	0	10.06	3.20	3.19	3.13	3.11	0.19

Table 7.4. Measurements taken on type III ferrules.

Ferrule ID	Type	Inscription	Length	Diameter T1	Diameter T2	Diameter B1	Diameter T2	Thickness
001676	III	0	3.50	3.77	3.19	3.76	3.18	0.35
001677	III	0	3.48	3.81	3.19	3.78	3.23	0.20
001679	III	0	3.49	3.79	3.14	3.79	3.19	0.20
001680	III	1	3.43	3.77	3.36	3.80	3.38	0.26
001681	III	1	3.52	3.83	3.68	3.85	3.66	0.25
001682	III	0	3.42	3.80	3.18	3.79	3.21	0.26
001683	III	0	3.47	3.80	3.18	3.78	3.22	0.23
001684	III	1	3.48	3.82	3.22	3.76	3.20	0.30
001690	III	0	3.15	2.83	2.79	2.83	2.79	0.23
001691	III	0	3.51	3.87	3.42	3.83	3.40	0.21
001692	III	0	3.51	3.80	3.25	3.76	3.23	0.21
001702	III	1	3.53	3.90	3.66	3.88	3.67	0.20
001703	III	1	3.52	3.87	3.67	3.84	3.66	0.25
001704	III	0	3.49	3.81	3.24	3.81	3.22	0.20
001705	III	1	3.48	3.82	3.43	3.82	3.40	0.25
001706	III	1	3.43	3.80	3.35	3.79	3.35	0.24
001707	III	1	3.44	3.81	3.32	3.82	3.36	0.20
001709	III	0	3.46	3.84	3.21	3.83	3.23	0.22
001710	III	0	3.46	3.85	3.16	3.82	3.28	0.16
002363	III	1	3.47	3.85	3.16	3.86	3.25	0.22
002364	III	1	3.46	3.85	3.25	3.83	3.24	0.11
002365	III	0	3.50	3.84	3.21	3.84	3.24	0.24
002366	III	1	3.52	3.80	3.23	3.85	3.24	0.18
002369	III	0	3.51	3.81	3.25	3.84	3.23	0.20
002370	III	1	3.50	3.90	3.66	3.89	3.69	0.25
002371	III	1	3.51	3.78	3.19	3.81	3.24	0.17
002373	III	0	3.48	3.87	3.22	3.85	3.24	0.16
002375	III	0	3.50	3.81	3.21	3.81	3.25	0.17
002376	III	0	3.51	3.84	3.24	3.82	3.27	0.19
002377	III	1	3.46	3.81	3.41	3.87	3.41	0.28
002378	III	0	3.44	3.83	3.28	3.84	3.25	0.28
002379	III	0	3.51	3.83	3.21	3.82	3.22	0.22
002380	III	1	3.53	3.83	3.61	3.84	3.64	0.18
002381	III	1	3.52	3.87	3.58	3.87	3.64	0.16
002382	III	0	3.49	3.81	3.25	3.86	3.25	0.24
002383	III	0	3.50	3.87	3.63	3.90	3.66	0.21

Table 7.4. Measurements taken on type III ferrules (*cont.*).

002384	III	1	3.55	3.86	3.61	3.86	3.64	0.24
002385	III	0	3.42	3.82	3.19	3.84	3.24	0.17
002386	III	1	3.50	3.78	3.20	3.78	3.20	0.20
002387	III	1	3.45	3.80	3.41	3.81	3.40	0.22
002392	III	1	3.52	3.86	3.64	3.86	3.73	0.18
002393	III	1	3.49	3.83	3.23	3.82	3.28	0.15
002394	III	0	3.49	n/a	n/a	3.81	3.34	0.20
002395	III	1	3.51	3.91	3.68	3.87	3.68	0.16
002396	III	1	3.53	3.82	3.71	3.90	3.70	0.19
002397	III	1	3.50	3.87	3.65	3.87	3.69	0.18
002398	III	1	3.46	3.77	3.37	3.81	3.41	0.19
002431	III	0	3.50	3.83	3.20	3.83	3.21	0.25
002434	III	0	3.52	3.80	3.26	3.81	3.22	0.16

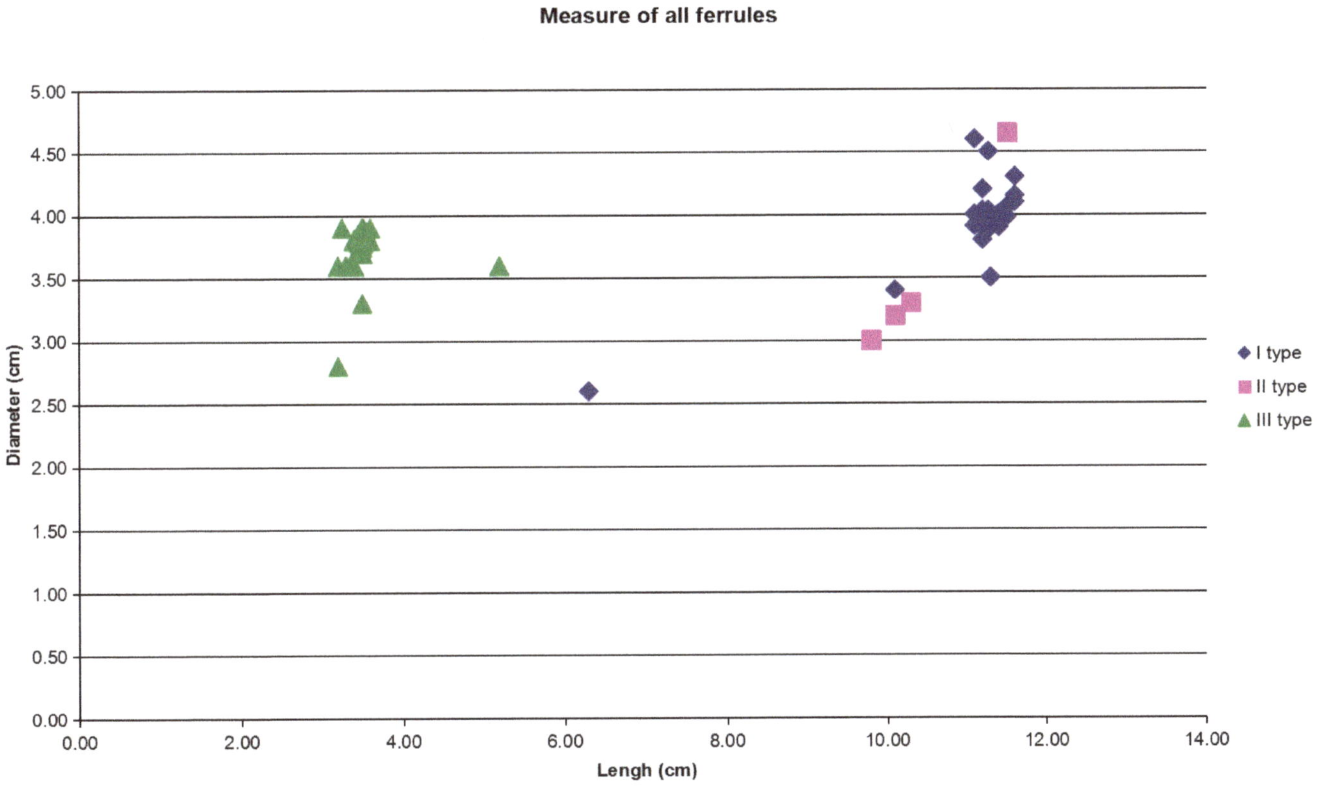

Fig. 7.7. Scatterplot of the ferrule measurements: length versus T1-diameter.

starting point for analysing the standardisation and labour organisation behind their production. Type II is represented by only three ferrules that are fairly dispersed in their dimensions. They therefore have only limited statistical potential. However, the type I and type III ferrules include a sufficiently large number of examples to allow statistical analysis for the purpose of examining the standardisation of production methods and the organisation of craftspeople.

7.3.4.1 Subgroups within Type I ferrules

Out of the 38 type I ferrules considered here, 26 are from the five easternmost trenches in Pit 1, 11 more are from the ongoing excavation trenches elsewhere in Pit 1, and 1 is from Pit 2 (see Table 7.1). In order to cooperate with the following spatial analysis, PCA was employed only on 26 examples with specific locations in the five easternmost trenches in Pit 1.

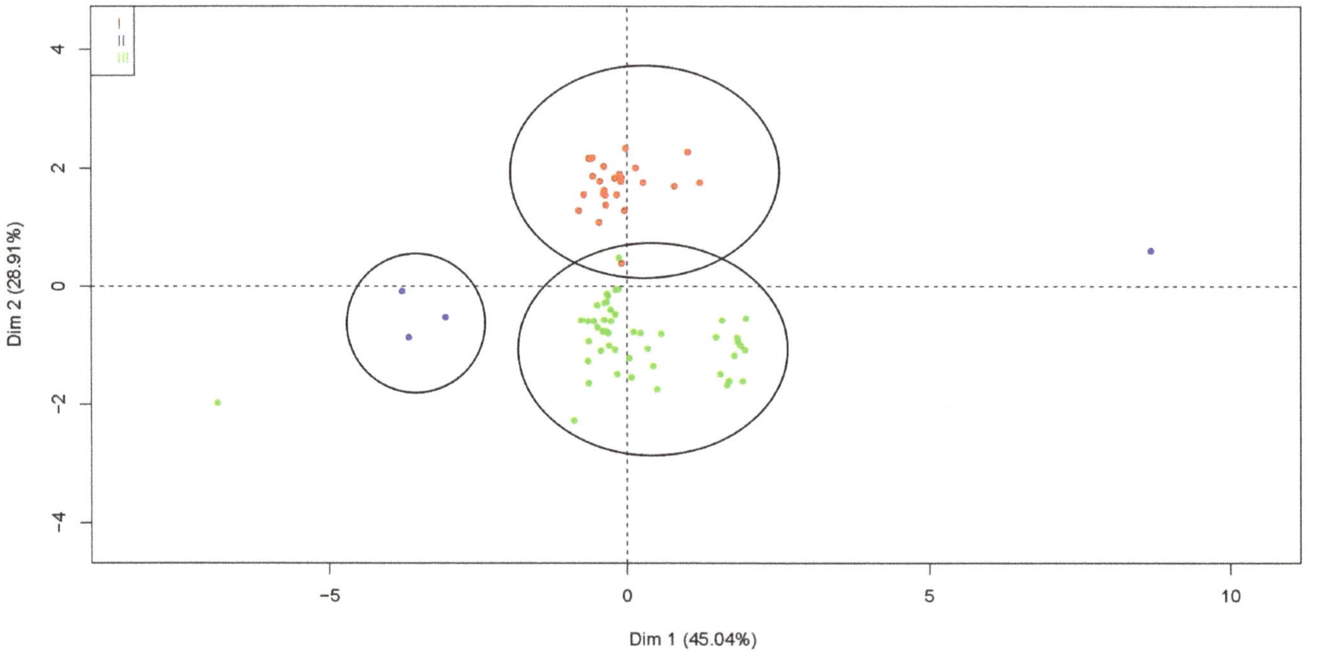

Fig. 7.8. A plot of the first two PCA components of the ferrules' measurements including all variables recorded.

Table 7.5. The correlations between the six original variables of ferrules and the five principal components.

Ferrules	Dim.1	Dim.2	Dim.3	Dim.4	Dim.5
Length	-0.058	0.886	0.440	-0.005	0.130
DiameterT1	0.753	0.543	-0.021	0.344	-0.125
DiameterT2	0.867	0.0515	0.370	-0.313	-0.065
DiameterB1	0.850	-0.066	-0.356	0.359	0.121
DiameterB2	0.809	-0.464	0.0616	-0.337	0.068
Thickness	-0.045	-0.656	0.598	0.457	0.006

Fig. 7.9. Type I, type II, and type III ferrules.

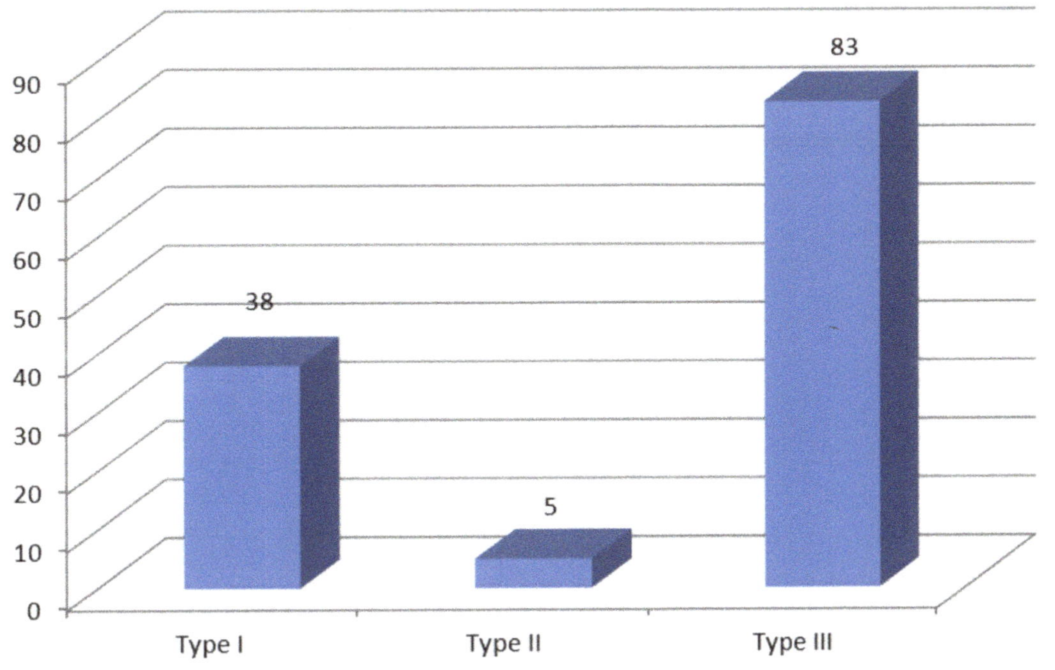

Fig. 7.10. Proportions of the three types of ferrules within the assemblage.

Table 7.6. Type I, type II, and type III ferrules.

	Type I	Type II	Type III
Shape	Slim cylinder	Slim cylinder	Shorter and wider
Cross-section	Elliptical	Round	Round
Decoration	Three bands	Plain	Inscriptions (some)
Length	10.05–11.55 cm	6.63–11.51 cm	3.15–3.58 cm
T1-diameter	3.43–4.29 cm	3.2–4.71 cm	2.83–3.95 cm
T2-diameter	3.16–3.51 cm	3.19–4.68 cm	2.79–3.71 cm
B1-diameter	3.10–4.13 cm	2.51–4.43 cm	2.83–3.91 cm
B2-diameter	2.99–3.24 cm	2.49–4.36 cm	2.79–3.73 cm
Thickness	0.13–0.30 cm	0.19–0.28 cm	0.11–0.35 cm

Figure 7.14 presents the multivariate patterning or subgroups of type I ferrule measurements using the first two components of a PCA. Table 7.7 shows the correlations between the six variables of type I ferrules and the five principal components. On this basis, the type I ferrules can be tentatively divided into three subgroups: I-A, I-B, and I-C, and the details of these three subgroups are listed in the following table (Table 7.8).

These three subgroups may be related to the time period, different casting moulds, different craftspeople working in the workshop, and/or different spatial distributions in the pit. As such, further spatial analysis will be carried out below on the samples from the five easternmost trenches to consider these issues.

7.3.4.2 Subgroups within type III ferrules

The type III ferrules include 83 examples, of which 50 come from the five easternmost trenches of Pit 1, 31 from the other trenches in Pit 1, and 2 from Pit 2 (see Table 7.1). In order to coincide neatly with the spatial analysis, the PCA only focused on the samples from the five easternmost trenches, and considered 48 out of a possible total of 50 ferrules (two ferrules could not be measured because their shape has been heavily deformed).

A plot of the first two PCA components of measurements of type III ferrules is presented in Figure 7.15 and suggests the existence of 2 subgroups, III-A (17 samples) and III-B (30 samples). One more III-C is considered separately.

Fig. 7.11. A small hole and a pin shown on type I ferrules (ferrule ID 1678 and ID 1771).

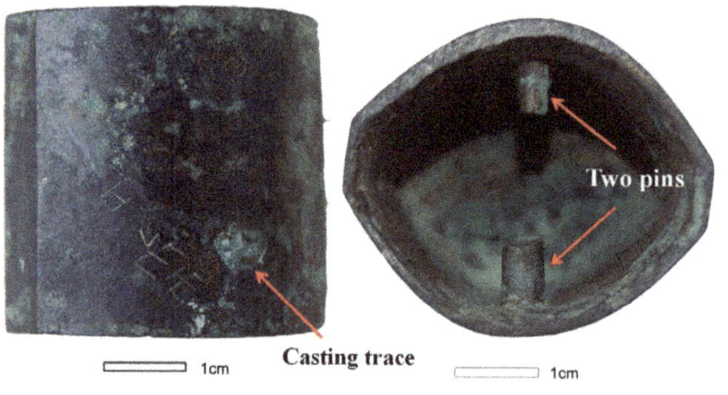

Fig. 7.12. Two inside pins and a surface trace on a type III ferrule (ferrule ID 1707 from side and top).

Table 7.9 shows the correlations between the six original variables of type III ferrules and the five principal components. Again, there may be several explanations for these subgroups and these are discussed below following a consideration of their spatial patterning. The single type III-C ferrule is not considered in this PCA plot, as it is an outlier, but its spatial location is considered in a later section.

Some of the type III ferrules were partially carved with inscriptions of '*Sigong*' (Fig. 7.16), denoting the governmental workshop, as mentioned above. Overall, 25 of the 47 type III ferrules bear this inscription (i.e. exactly half of them) in the five easternmost trenches. Both subgroups III-A and III-B include inscribed and uninscribed ferrules, and subgroup III-A has a significantly greater proportion of inscribed ferrules (Fig. 7.17). Subgroup III-C does not bear any inscription (Fig. 7.16).

Chi-square tests were employed for measuring the association of inscribed ferrules to their types. In the contingency table above (Table 7.10), the ferrules are categorised according to their values regarding the two variables concerned – type and inscription. The question is whether the two classifications of our data are independent of each other, i. e. whether membership of a particular category associated with one classification is unrelated to membership of a particular category associated with the other. We are testing for 'goodness-of-fit'. Table 7.11 shows the observed and expected values in each category.

We are now in a position to set up the significance test for the ferrule data concerning the inscriptions and the division in subgroups types III-A and III-B.

H0: types III-A and III-B have inscribed and uninscribed ferrules in similar ratios.
H1: types III-A and III-B have a different ratio of inscribed and uninscribed ferrules.
Selected significance level = 0.05

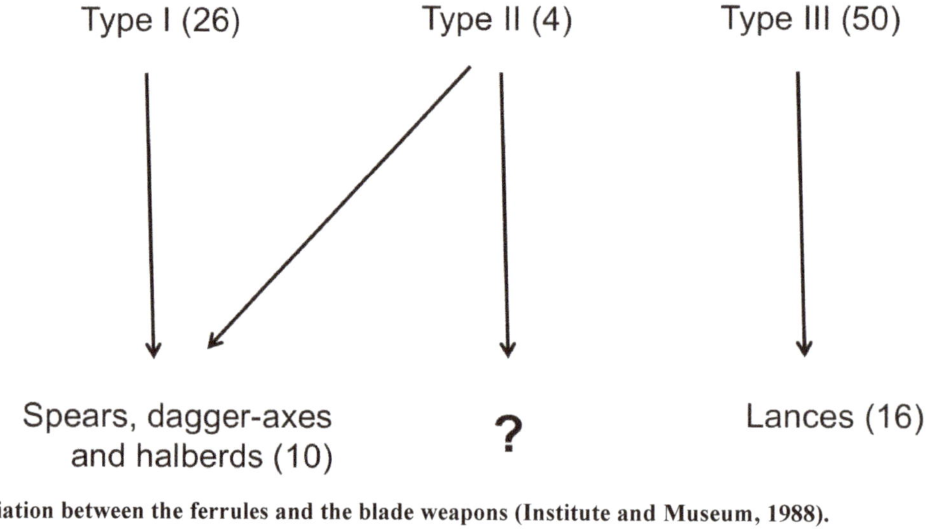

Fig. 7.13. Association between the ferrules and the blade weapons (Institute and Museum, 1988).

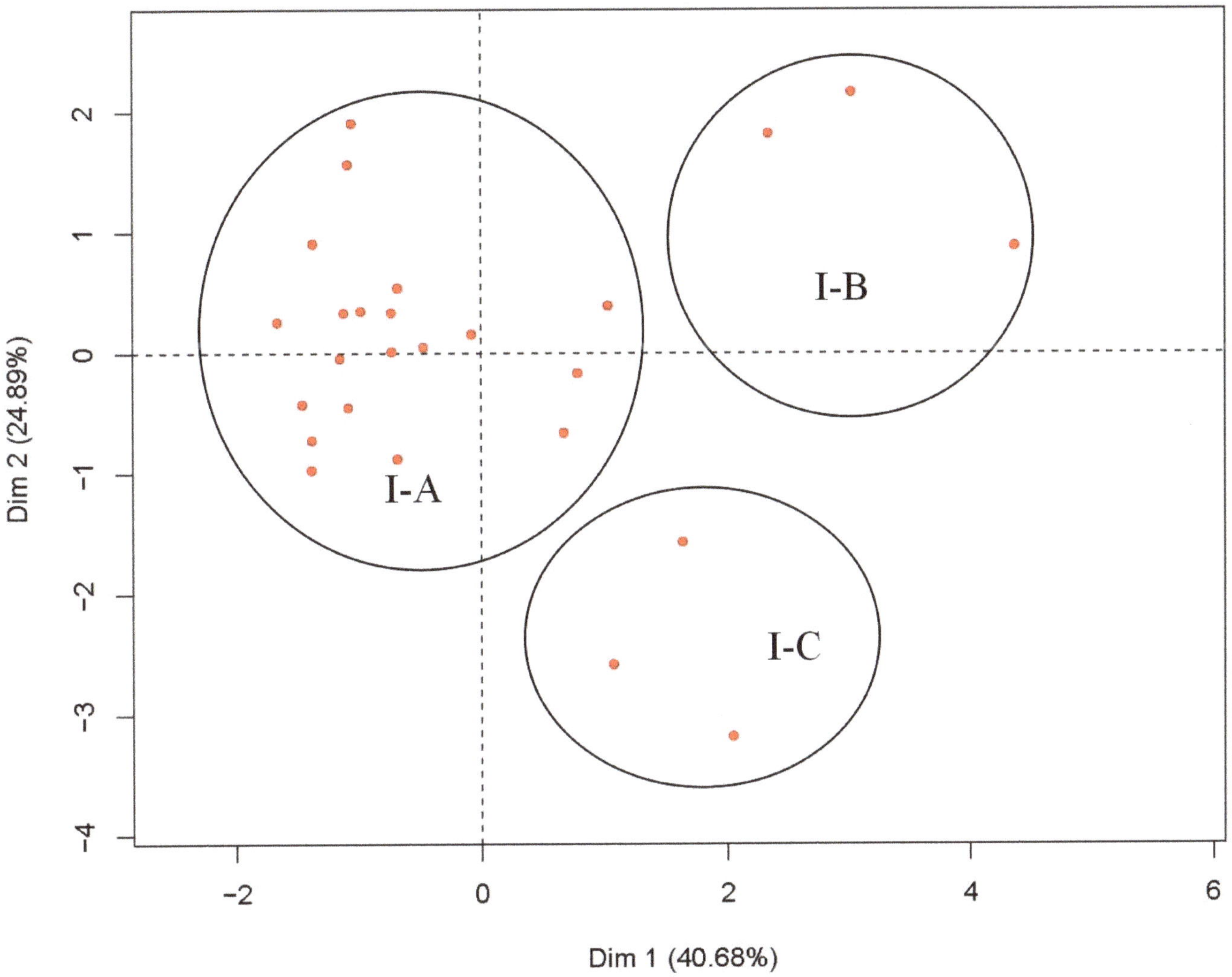

Fig. 7.14. A plot of the first two principal components of ferrule type I.

Table 7.7. The correlations between the six original variables of ferrule type I and the five principal components.

Ferrule type I	Dim.1	Dim.2	Dim.3	Dim.4	Dim.5
Length	0.899	0.029	-0.011	0.076	0.404
DiameterT1	0.639	0.525	0.171	-0.457	-0.269
DiameterT2	-0.505	0.743	-0.082	-0.260	0.325
DiameterB1	0.960	-0.015	0.074	0.091	-0.007
DiameterB2	0.146	0.684	-0.565	0.407	-0.156
Thickness	-0.160	0.443	0.811	0.347	-0.013

Table 7.8. The three subgroups within type I ferrules.

Subgroup	I-A	I-B	I-C
Number of ferrules	20	3	3
Ferrule-ID		002367 001694 001696	001711 001689 002372

In this case, V= (2-1) (2-1) = 1. The 0.05 level of significance is 3.84 in the tabulated value of chi-squared for one degree of freedom. The calculated value is larger than the tabulated value and the hypothesis of association is rejected (X2 calc = 6.00, this is > 3.84, and accordingly we reject H0). The test shows that types III-A and III-B do not have the same proportion of inscribed and uninscribed ferrules. Actually, it is worth mentioning that the chi-square test does not tell us anything about the way in which the variables are related; it simply measures departures of

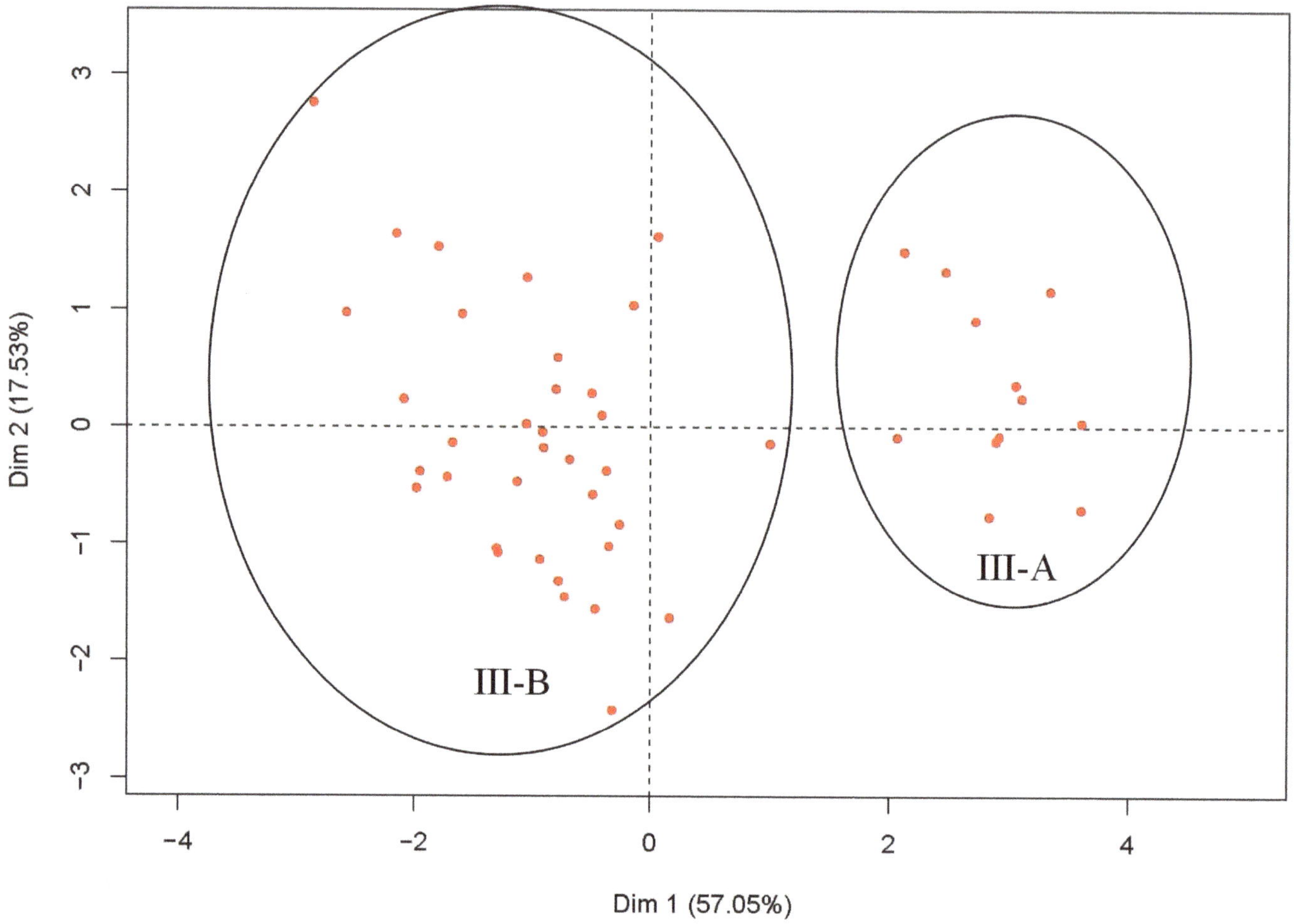

Fig. 7.15. A plot of the first two principal components of ferrule type III, with the two suggested subgroups III-A and III-B circled.

Table 7.9. The correlations between the six original variables of ferrule type III and the five principal components.

Ferrule type III	Dim.1	Dim.2	Dim.3	Dim.4	Dim.5
Length	0.583	-0.101	0.788	0.158	0.056
Diameter1	0.798	-0.154	-0.231	0.446	-0.292
Diameter2	0.908	0.295	0.001	-0.276	-0.079
Diameter1	0.844	-0.087	-0.272	0.150	0.429
Diameter2	0.922	0.234	-0.042	-0.282	-0.084
Thickness	-0.244	0.932	0.032	0.263	0.044

the observed values from expected ones. The chi-square does not indicate the strength of a relationship either; it simply tells us about the probability that a relationship exists (Shennan, 1997).

7.3.5 Coefficients of variation (CV) and degrees of standardisation

The three types of ferrules (see section 7.2.3) can be characterised both according to measurement and to typological observation, but the numerical character of the former allows us to consider standardisation on a continuous scale through a comparable measure of variation. As noted above, type II contains only 3 ferrules, and is therefore of no further relevance to this section. A coefficient of variation was calculated to assess the degree of standardisation in the subgroups of the other types, I-A, III-A, and III-B (Fig. 7.18). It also appeared important to compare the CV values on inscribed versus uninscribed type III-B ferrules; however, the results did

Fig. 7.16. '*Sigong*' inscriptions shown on types III-A and III-B, but not on the III-C ferrule.

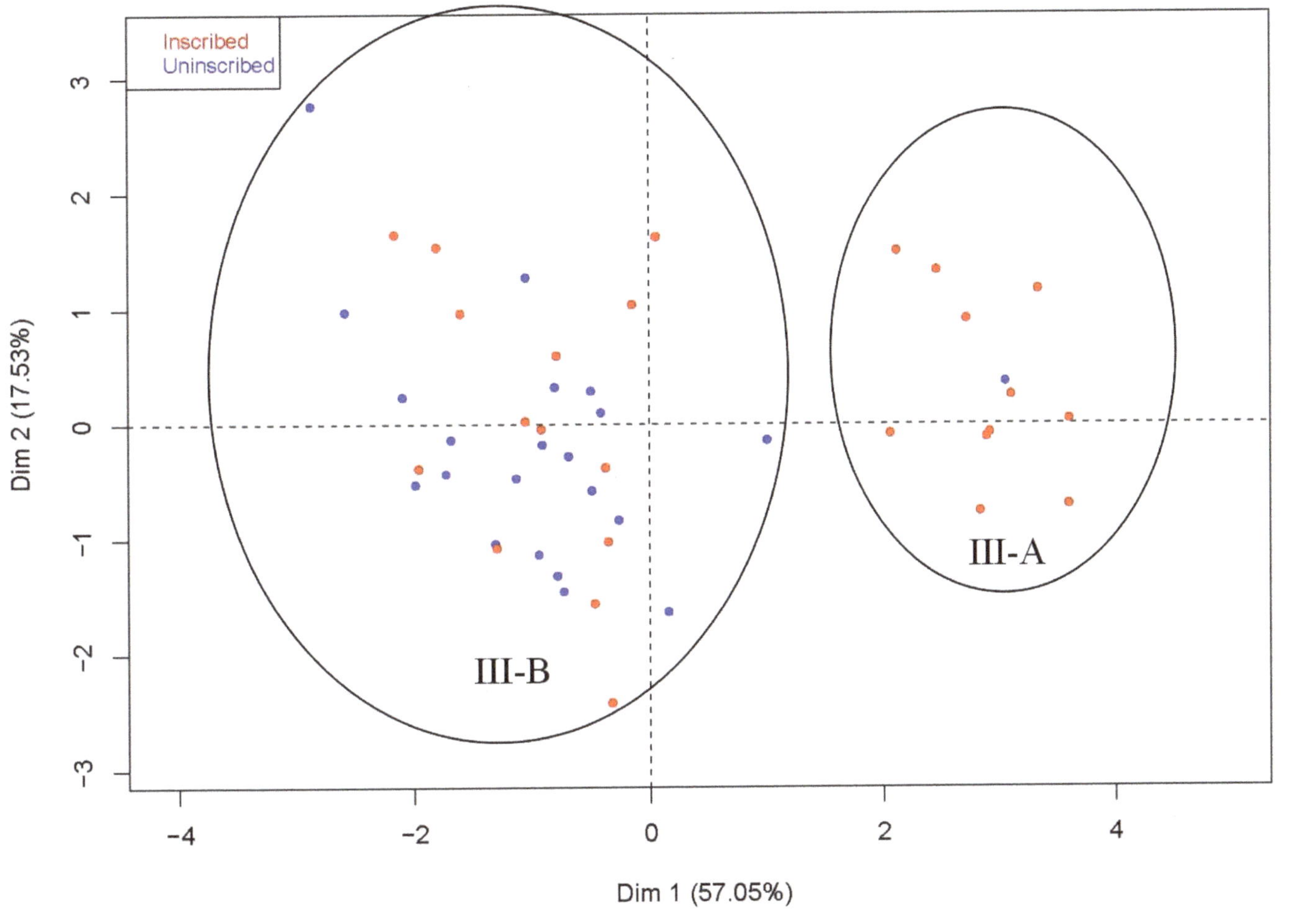

Fig. 7.17. Inscribed and uninscribed type III ferrules (blue dots refers to the uninscribed ferrules and red dots refers to the inscribed ferrules).

not show a significant difference between the two (Fig. 7.19), an aspect which perhaps provides further support to the interpretation that the governmental inscriptions were added to a random sample of otherwise identical ferrules.

CV values for the different dimensions of ferrules are generally quite low in all subgroups for five of the six parameters: length, T1-diameter, T2-diameter, B1-diameter, and B2-diameter, ranging from 0.8 to 6.5%. The CV values for the thickness of the ferrules range from 16.1 to 21.9% and therefore show comparatively greater variation.

The thickness of the ferrules within each type also seems more variable and suggests this parameter was not easily controlled from a technological perspective. Most likely, difficulties in adjusting the core and the outer mould for casting resulted in this variable thickness, with variation sometimes even noticeable within the same ferrule, ranging from 0.12 to 0.35 cm.

Table 7.10. Subgroups of type III ferrules and their inscriptions.

Subgroups	III-A	III-B	
Inscribed	11	14	25
Uninscribed	1	21	22
	12	35	47

Table 7.11. Counts of inscribed and uninscribed ferrules against their types (with the expected values for each category in brackets).

	III-A	III-B	
Inscribed	11 (6.4)	14 (18.6)	25
Uninscribed	1 (5.6)	21 (16.4)	22
	12	35	47

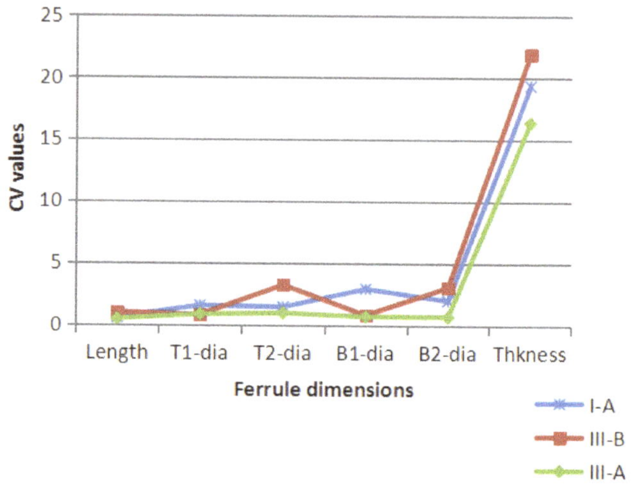

Fig. 7.18. CV values for ferrules belonging to subgroups I-A, III-A, and III-B.

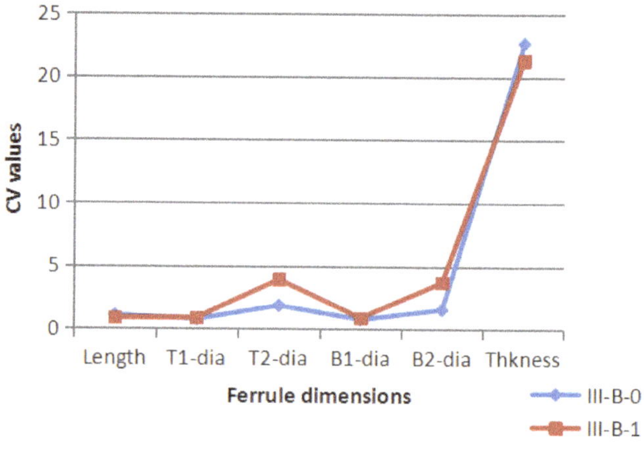

Fig. 7.19. CV values for the inscribed (III-B-1) and uninscribed (III-B-0) ferrules.

7.4 The bronze long weapons

7.4.1 Sources of data

The archaeological data regarding the long weapons were also obtain mainly from the extensive excavation report (1974–1984) and later reports of the ongoing archaeological excavations in the pits of Emperor Qin Shihuang's Terracotta Army (Fig. 7.20). During the 1974–1984 excavations, altogether 4 halberds, 1 dagger-axe, 5 bronze and 1 iron spears, and 16 lances were discovered (Institute and Team, 1988: 254). The ongoing archaeology yielded 2 more halberds, 1 dagger-axe, 4 spears, and 1 lance, which are stored in the Conservation Department of the Museum of Emperor Qin Shihuang's Terracotta Army, and were made available for this project (Jiang and Liu, 2006). The following tables (Table 7.12 and 7.13) specify the origins of the bronze long weapons.

These inscriptions on the weapons reflect not only the organisational structure of the bronze weapons production during the Qin period, but also aspects of the social status of the participants in the *Sigong*. The lance inscriptions are shorter than those found on the dagger-axes. According to these inscriptions, the lances were produced from the 15th regnal year (232 BC) to the 19th (228 BC), and the inscriptions only contain the name of '*Sigong*' and the worker, for example '*the 17th year, Sigong Wen, and worker Diao*' (Institute and Team, 1988). An interesting finding is that a worker named *Diao* also appears as the maker of a dagger-axe in the 3rd regnal year (244 BC). If both inscriptions refer to the same *Diao*, this would imply that he worked in the governmental workshop for 14 years at least, but was never promoted during this time. Obviously, a single isolated name must be treated with caution, but this might be consistent with a relatively low

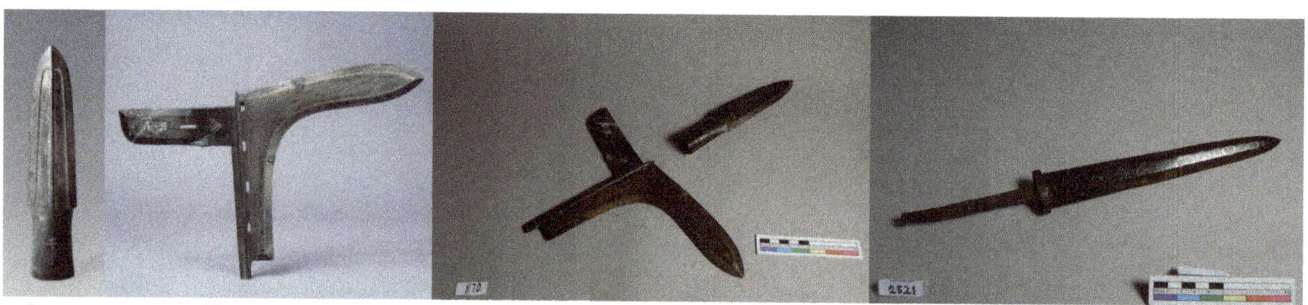

Spear Dagger-axe Halberd Lance

Fig. 7.20. Bronze spear, dagger-axe, halberd, and lance.

social status of the workers in the workshop, and indeed they might have been slaves, convicts, or conscripted soldiers (Yuan, 1990).

7.4.2 Measurements of the long weapons

The bronze halberds and dagger-axes are only a small sample with limited statistical potential, and, furthermore, some of them were not accessible because they are on display in the museum or on loan to other museums abroad. The length of the four dagger-axes provides a general idea about the degree of standardisation in their production, as they vary by only 1 or 2 mm (26.6, 26.7, 26.7, and 26.8 cm), respectively.

A total of 8 out of the 11 spears in the museum database have been observed, photographed, and measured with callipers for this particular project. The measurements include 7 variables, namely the entire length; the blade length, width, and thickness; and the handle length, width, and thickness. PCA was employed for the statistical analysis of the data. Figure 7.21 shows a plot of the first two principal components, and Table 7.14 provides the correlations between the seven variables of the spears and the five principal components. The 8 spears come from different sites: 3 from Pit 1, 2 from Pit 2, 1 from Pit 3, and 2 from other areas within the Emperor Qin Shihuang's tomb complex. Although some of them appear to cluster together, there is no obvious link between their potential group and their origins or locations in the tomb complex. It is also difficult to make inferences about the labour organisation behind their production and transportation due to the limited number of samples available.

Table 7.12. Bronze long weapons and their archaeological origins.

	Halberds	Spears	Dagger-axes	Lances
Pit 1 Excavated 1974–1984	4	5 bronze 1 iron	1	16
Pit 1 Ongoing archaeology	2	0	1	1
Pit 2	0	2	0	0
Pit 3	0	1	0	0
Other sites of the tomb complex	0	2	0	0
Total	6	11	2	17

These lances have clear inscriptions with the specific year when they were made and the specific *Sigong* or workers who were involved in the production. Thus, the investigation focused on whether there were any patterns connecting the measurements with the time period or the makers. A total of 11 out of the 17 lances have been photographed and measured in the storage of the Conservation Department in the Museum of Emperor Qin Shihuang's Terracotta Army. These lances were measured with callipers, and the measurements include 7 variables (entire length; blade length, width, and thickness; and handle length, width, and thickness) (Fig. 7.22). Plots of the first two components of the PCA are shown in Figures 7.23 and 7.24.

In Figure 7.23, four lances produced in the 17th regnal year of King Ying Zheng (later Emperor Qin Shihuang) appear to form a dispersed cluster, and two lances produced in the 15th regnal year form another cluster, while three lances produced in the 19th year are relatively dispersed in their

Table 7.13. Origins and inscriptions of the halberds or dagger-axes.

Museum-ID	Origin	Spear	Dagger-axe
00870	1974–1984 excavation	*Sigong*	The 3rd year, Lu Buwei, Sigong Long, Cheng Yi, and worker Diao
00878	1974–1984 excavation	*Sigong*	The 4th year, Lu Buwei, Sigong Long, Cheng Wo, and worker Ke
05552	1974–1984 excavation	*Sigong*	The 5th year, Lu Buwei, Sigong Long, Cheng Yi, and worker Cheng
(T19G8:0710)	1974–1984 excavation	*Sigong*	The 7th year, Lu Buwei, Sigong Zhou, Cheng Yi, and worker Jing
(T12G5)	Ongoing archaeology	*Sigong*	The 7th year, Lu Buwei, Sigong Zhou, Cheng Yi, and worker Tong
	Ongoing archaeology	*Sigong*	The 4th year, Lu Buwei, Sigong Long, Cheng Wo, and worker Ke
(T19G11:0924)	1974–1984 excavation		The 3rd year, Lu Buwei, Sigong ... (cannot be identified)
(T12G7)	Ongoing archaeology		The 10th year, Sigong Cheng Yang, and worker Zhao

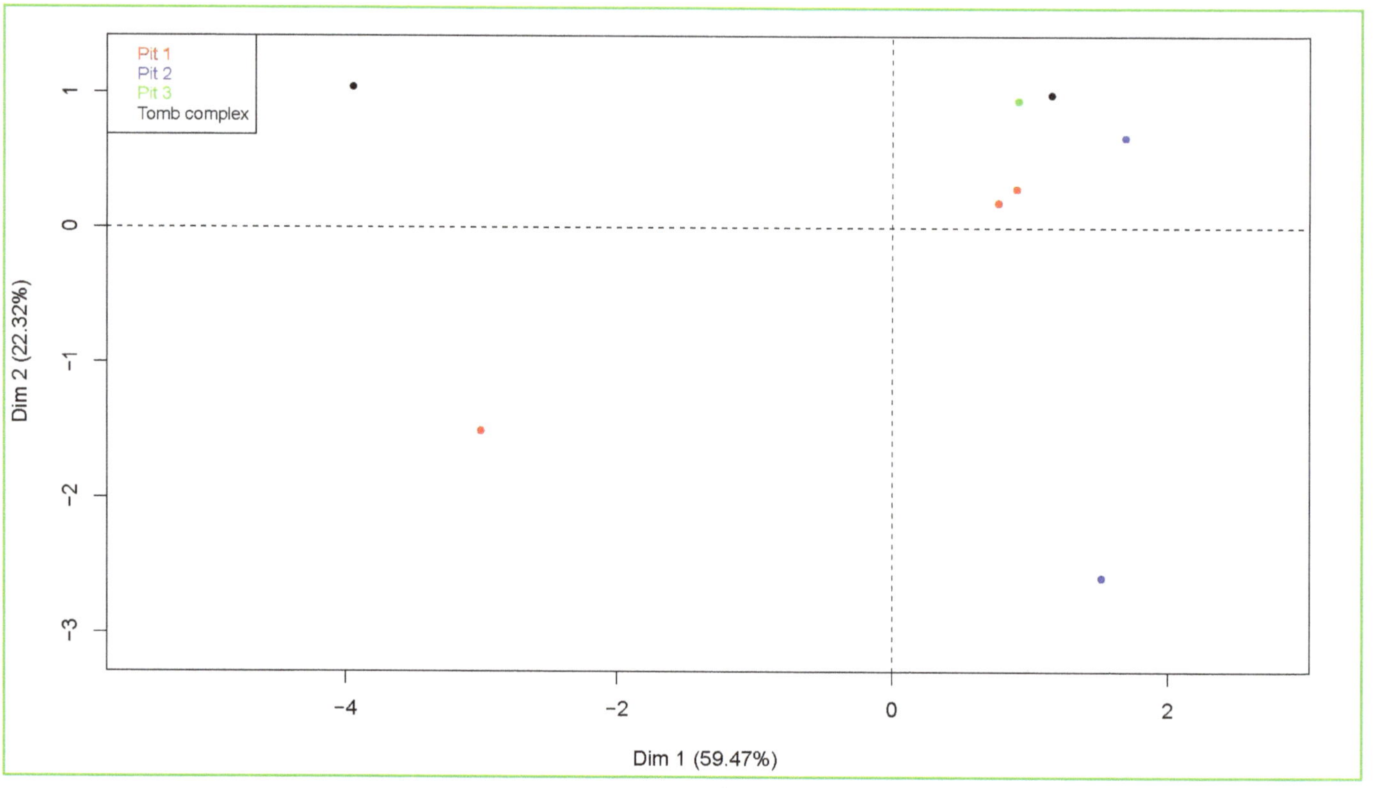

Fig. 7.21. A plot of the first two principal components of the spears.

Table 7.14. The correlations between the seven original variables of the spears and the five principal components.

Spears	Dim.1	Dim.2	Dim.3	Dim.4	Dim.5
Length	0.975	-0.018	-0.179	-0.063	-0.072
Blade length	0.986	0.061	-0.147	-0.006	0.007
Blade width	0.095	0.904	-0.170	0.375	-0.047
Blade thickness	-0.967	0.055	0.132	0.170	0.098
Handle length	0.962	0.170	0.067	0.017	0.200
Handle width	-0.315	0.840	0.119	-0.425	0.005
Handle thickness	0.517	0.048	0.848	0.093	-0.055

dimensions – one near the 18th year lance and another close to the 16th year lance. No obvious links between the dimensions of lances and *Sigong* or workers involved in the production can be discerned from Figure 7.24. Table 7.15 provides the correlations between the seven variables of the spears and the five principal components.

7.5 Spatial data

The 80 ferrules from the front trenches of Pit 1 and other bronze long weapons were mapped within a GIS environment, and these datasets were assigned point locations in this eastern portion of the pit based on a combination of museum records and excavation plans (see Chapter 1, section 1.3.2). As regards those discovered during the ongoing archaeological excavations in the other trenches of Pits 1, 2, and 3, and at other sites of the tomb complex, although they have a specific location recorded, their spatial distribution cannot be analysed at present as a coherent block and therefore they are not considered in this study.

7.5.1 Parameters affecting the spatial patterns

In addition to the two main factors that affected the location of other weapons, namely the battle formation and organisation processes, the spatial connections between the ferrules and bronze blades deserve attention for investigating the organisation of labour on the long weapons' production, assembly, delivery, and placement into the pit. Their respective spatial patterns may affect each other.

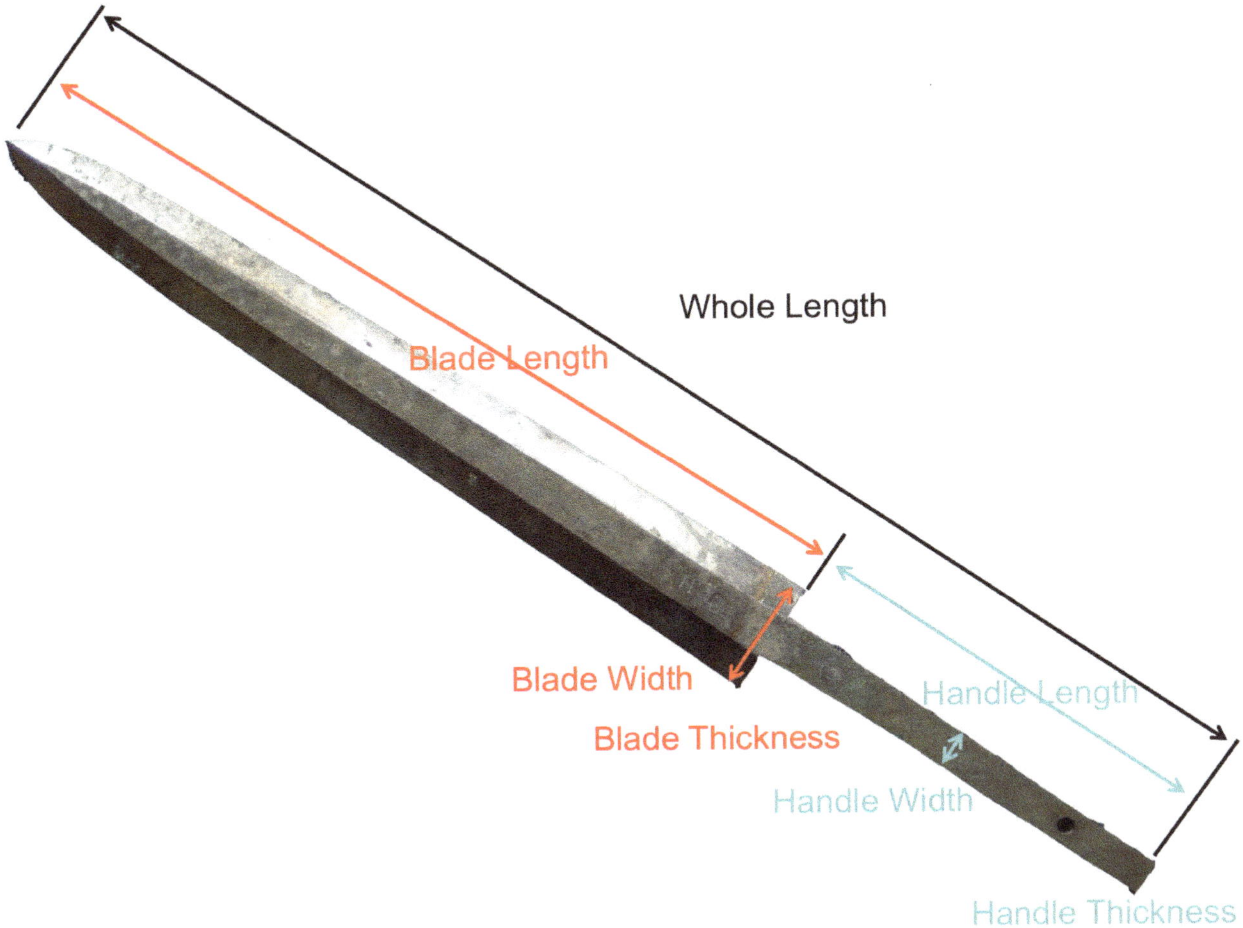

Fig. 7.22. Dimensions measured for the lances.

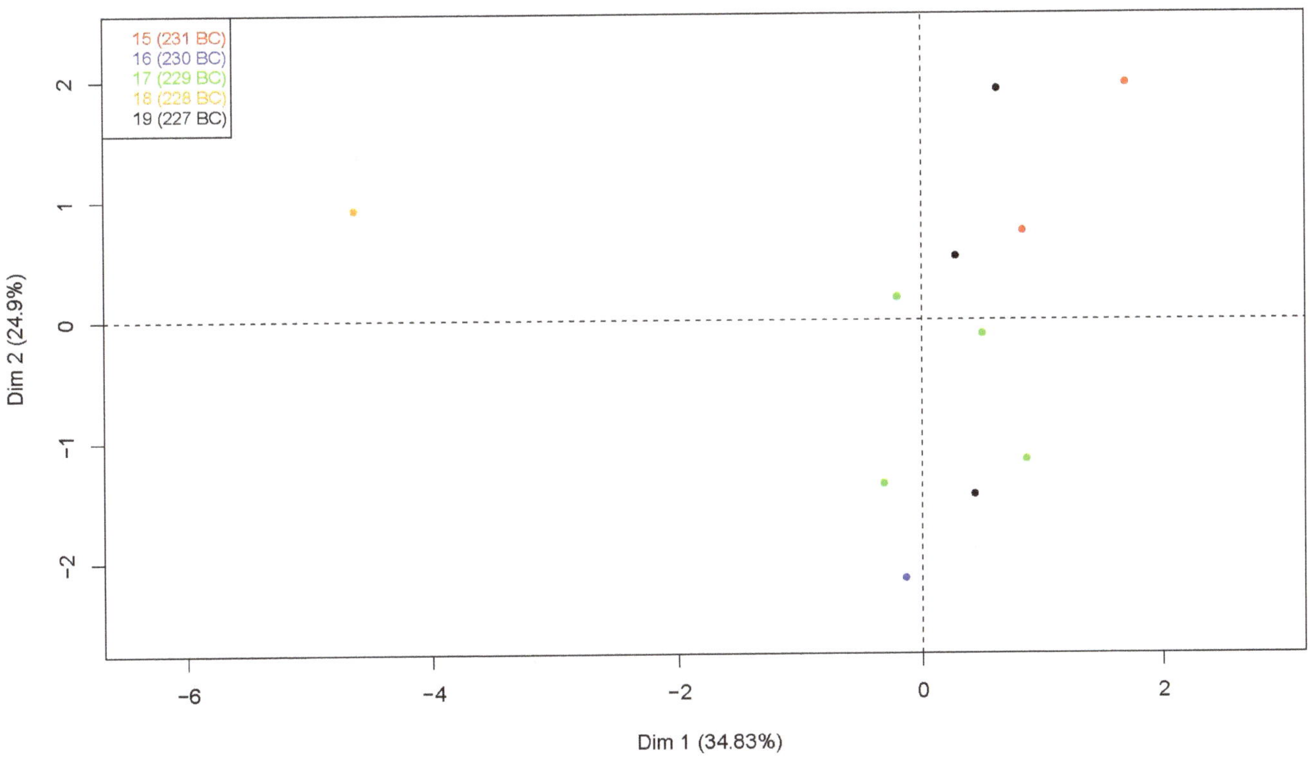

Fig. 7.23. A plot of the first two principal components of the lances coded by the dates of production.

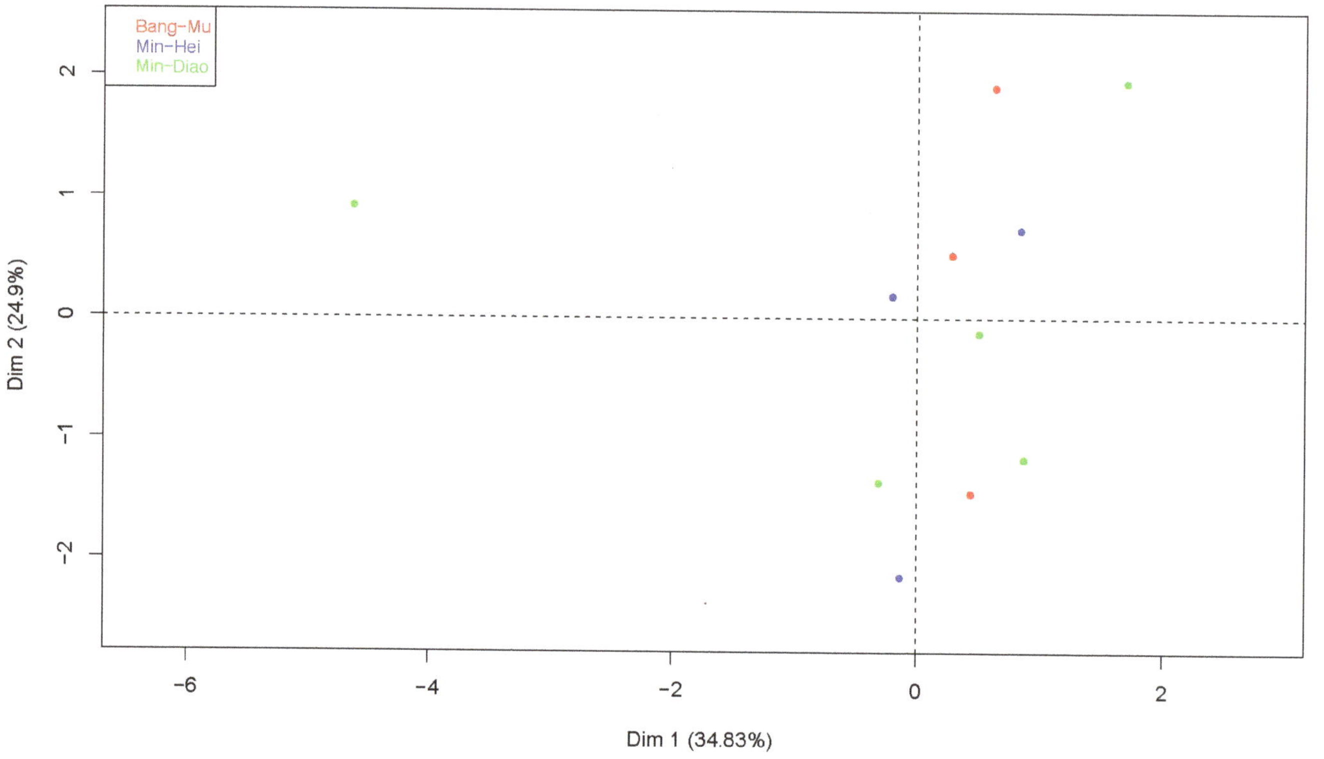

Fig. 7.24. A plot of the first two principal components of the lances coded by the craftspeople involved in their production.

Table 7.15. The correlations between the seven original variables of lances and the five principal components.

Lances	Dim.1	Dim.2	Dim.3	Dim.4	Dim.5
Length	0.964	-0.113	0.093	-0.154	0.081
Blade length	0.539	0.511	0.304	-0.222	-0.503
Blade width	-0.020	0.779	-0.464	0.040	-0.188
Blade thickness	-0.211	0.686	0.126	-0.489	0.470
Handle length	0.937	-0.224	0.0281	-0.066	0.190
Handle width	-0.110	0.356	0.806	0.414	0.083
Handle thickness	0.530	0.460	-0.327	0.517	0.225

7.5.1.1 The ferrules, long weapons, and the battle formation

The ferrules and long weapons, in contrast to the triggers and arrows found primarily in the surrounding corridors, are mostly distributed in the middle corridors of the pit (Fig. 7.25a). The long weapons with wooden or bamboo shafts were probably carried by the infantrymen or the chariot warriors who were surrounded by the archers and/or crossbowmen in the battle formation (Fig. 7.25b). This was a battle strategy employed during the Qin period. The archers or crossbowmen fired the arrows at the enemy over a long distance on the battlefield, while the infantry and the chariot warriors used the long or short weapons in close combat (Yuan, 1990).

The type I ferrules were associated with dagger-axes, spears, and halberds (Institute and Team, 1988: 270), and there is a range of archaeological evidence supporting this association. For example, four halberds unearthed during the 1974–1984 excavation in Pit 1 were all linked to wooden shafts and type I ferrules (Institute and Team, 1988: 271). More recently, a well-preserved halberd with a long wooden shaft and a ferrule was discovered intact (see above, Fig. 7.2). This was found in corridor 9 of trench 22 in Pit 1 (T22G9), lying on the paved brick floor. The entire length of the weapon is about 2.87 metres, and it has a spear and a dagger-axe fixed to one end of the shaft and a ferrule, 11.7 cm in length, attached to the other end. Even though the wooden shaft had partly decayed, traces of lacquer and red pigments can still be seen on its surface. The spear is 17.8 cm long and bears a '*Sigong*' inscription. The dagger-axe, 26.9 cm in length, is inscribed with a long sentence: '*in the 4th year (243BC), the prime minister Lu Buwei, Sigong Ji, Cheng Wo, Craftsman Ke*'. This provides specific evidence that a type I ferrule with no evidence of

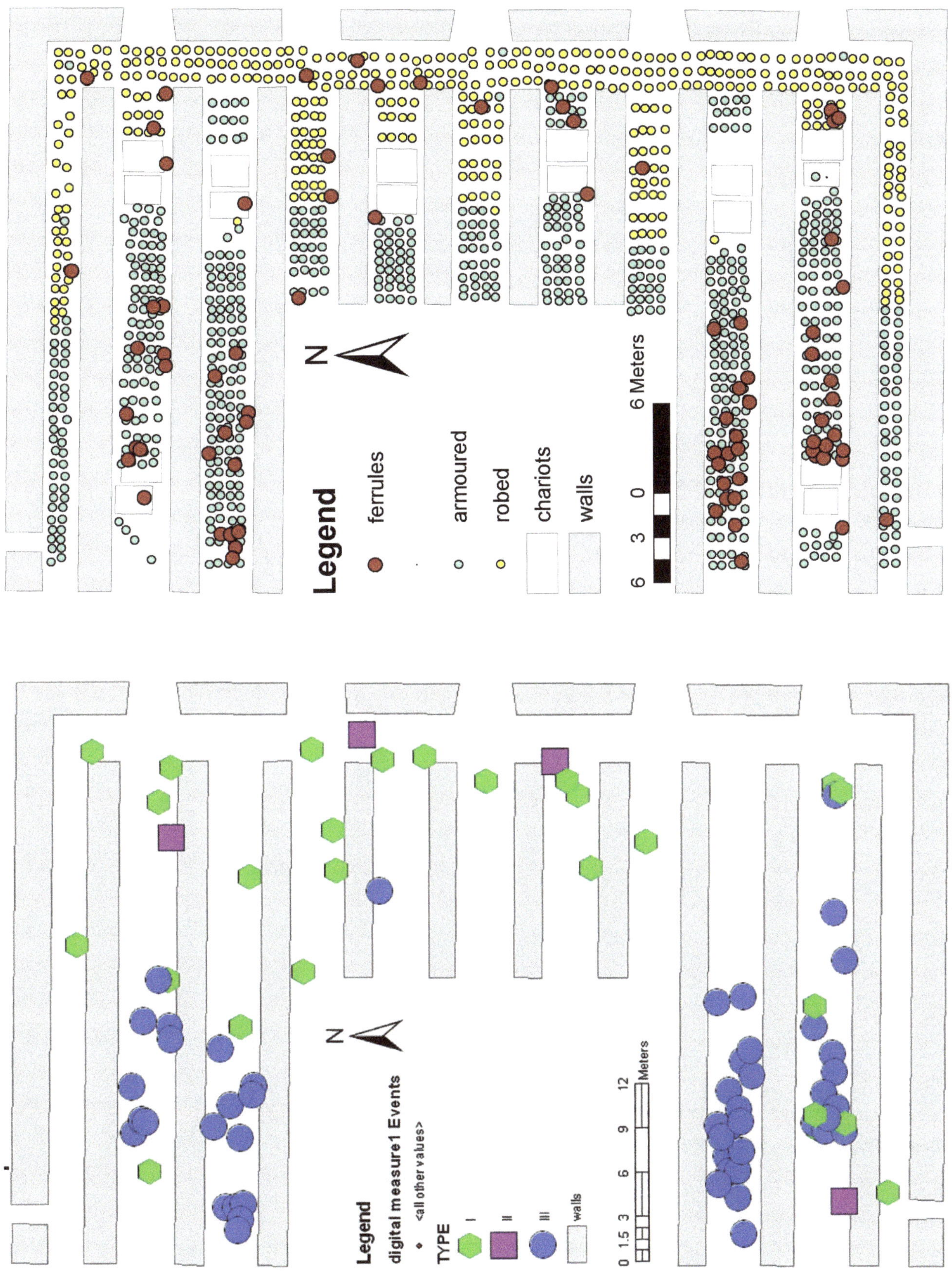

Fig. 7.25. Spatial distribution of ferrules (a) and with warriors (b) in Pit 1.

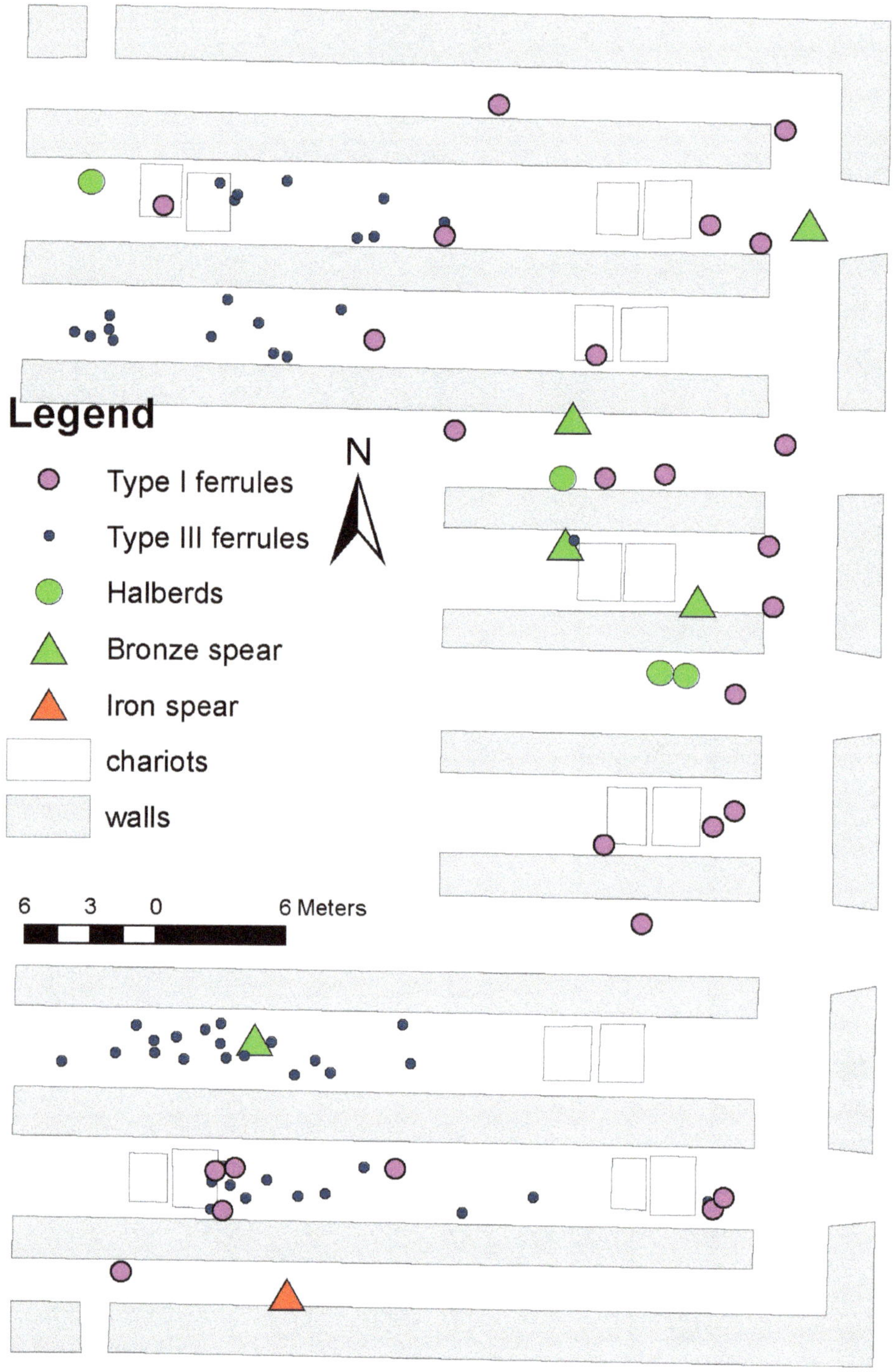

Fig. 7.26. Type I ferrules, halberds, and bronze and iron spears in Pit 1.

an inscription was assembled with the halberd (personal communications with Ma Yu and Zhao Kun). Another dagger-axe, discovered in 2005, was also fixed with a type I ferrule. It was located in corridor 7 of trench 12 in Pit 1 (T12G7). The entire length of the weapon is about 2.5 metres. The bronze dagger-axe is 27 cm in length, and the inscription on the surface reads '*in the 10th year (237BC), Sigong Cheng Yang, worker Zhao*'. This is a special case, as it was manufactured in the regnal year when chancellor Lu Buwei was dismissed. The ferrule is 11.2 cm in length (Jiang and Liu, 2006), and no inscription was identified on it.

The spatial distribution confirms the association between the type I ferrules and the halberds (or their separate components) in the five easternmost trenches in Pit 1 (Fig. 7.26). One clear example of this appears in the southernmost corridor (corridor 1), where the only iron spear discovered to date in Pit 1 was recovered, and where a single bronze type I ferrule was found as well. Even though the wooden shaft was not preserved, a possible link between this bronze ferrule and the iron spear cannot be ruled out. More confusing evidence, however, is present in corridor 3 (the third one from the south), where a bronze spear was found together with many type III ferrules, but no traces of type I ferrules were identified. Even considering this evidence, the overall pattern linking type I ferrules and long spears seems clear.

The spatial distribution shows that the type III ferrules are associated with the bronze lances in the five easternmost trenches of Pit 1 (Fig. 7.27). The type III ferrules were mainly distributed in corridors 2, 3, 9, and 10, and the lances were found in these four corridors as well. This spatial association comes to confirm earlier archaeological evidence deriving from better-preserved weapons: 7 lances in corridor 2 (second from the south) appeared assembled with the wooden shafts and type III ferrules. Again, there is also slightly confusing evidence in middle corridor 7, where a single type III ferrule seems to relate to a spear (see Fig. 7.26) rather than a lance (Fig. 7.27). Even so, the overall pattern linking type III ferrules and lances is still fairly convincing. Based on the measurements of the traces preserved, these shafts were about 218~382 cm in length and 3.5 cm in diameter (Institute and Team, 1988: 271). The variability in the lengths of the shafts may be simply due to a lack of accuracy in the measurements of partly rotten pieces, but it is also possible that shaft length varied depending on the types of associated blades and/or individual soldiers' heights.

A challenging question remains with respect to the fact that the count of ferrules is much higher than that of the long weapons, even if there is evidence for each of the ferrule types matching each type of long weapons. The type I and type II ferrules amount to a total of 30 objects on the map, while the spears, dagger-axes, and halberds only amount to 11. What was the function of the remaining 19 type I ferrules? The same question can be asked for the type III ferrules: there are 50 such ferrules in the pit, but only 17 lances to match them, leaving the other 33 without obvious links to any specific weapons. Looting by rebels shortly after the construction of the pits and damage by local farmers in later periods could be possible explanations, because it might well be that the top ends of the long weapons were more exposed to surface looking than the ferrules at the bottom of the pit.

One alternative possibility for the ferules that were not associated with long weapons is that they were perhaps attached to flags or banners flown in battle. However, no archaeological evidence has been discovered to date that would indicate that flags or banners were included in the pit. Fortunately, the terracotta warriors in the five easternmost trenches have been restored and their postures are relatively clear. The postures of these terracotta warriors in the pit may suggest the types of weapons that were held by individual warriors in the battle formation, and provide possible links between the warriors and the ferrules (Fig. 7.25b). Therefore in future it will be possible to consider if the postures of the warriors found with ferrules, but no blade ends of long weapons, are slightly different from those found with both ferrules and long weapon blades (suggesting perhaps that they were not holding the same kind of weapon), or alternatively are the same (suggesting that they were carrying similar weapons, but the blade ends have been lost).

The spatial patterning for the distribution of ferrules discussed above clearly derives from the structure of the battle formation, as there is a broad correspondence between the macroscopic types of ferrules and the different categories of warriors and weapons. However, as noted above, there are more subtle subgroups within ferrule types that are unlikely to be related to function, and may instead reflect minor variation in their manufacturing sequence or workshop. Their spatial distribution should be informative of the organisation of production. These aspects are addressed in the next section.

7.5.1.2 Workshop practice and labour organisation

7.5.1.2.1 The ferrules

In addition to the relationship between the bronze long weapons and the infantry or the warriors riding on chariots, the ferrules' spatial patterns as shown in the pit (clustered, regularly spaced, or random) were affected by the manner in which labour was organised during the production of the ferrules and their transportation to the pit. Some of the linked sequences of decisions or technical operations (*chaîne opératoire*) employed during the production and transportation of the ferrules can be traced back by interpreting the spatial patterns of the ferrules.

The spatial patterns of the subgroups in type I ferrules are demonstrated in Figure 7.28. The type I ferrules were tentatively divided into three subgroups, I-A, I-B, and I-C, based on dimensional differences arising from PCA of their measurements (see section 7.2.4.1). Each of the three subgroups I-A, I-B, and I-C appears fairly randomly dispersed across the pit (statistical assessment on subgroup I-A, see Figure 7.32).

Type III ferrules were also divided into three subgroups based on their dimensions, using PCA (see section 7.2.4.2). Subgroups III-A and III-B overlap in their spatial distribution inside the corridors (Fig. 7.29); however, exceptionally, in corridor 2 there are only III-B ferrules. The III-C subgroup includes a single ferrule with an unusual location, and it does not seem to be related to any lances (Fig. 7.27), although it is associated with a bronze spear (Fig. 7.26).

Inscriptions were carved on both type III-A and type III-B ferrules, but were not seen on the single ferrule of type III-C. Half of the type III ferrules (25 out of 50 in the five easternmost trenches) were inscribed, and the other half was left plain. Chi-square analysis has been used to test the association between the subgroups and the inscriptions mentioned above, and the result shows that there is no significant pattern to this association. It is assumed that this division is directly linked to prevailing workshop practice. Thus, the spatial patterns of the inscribed and

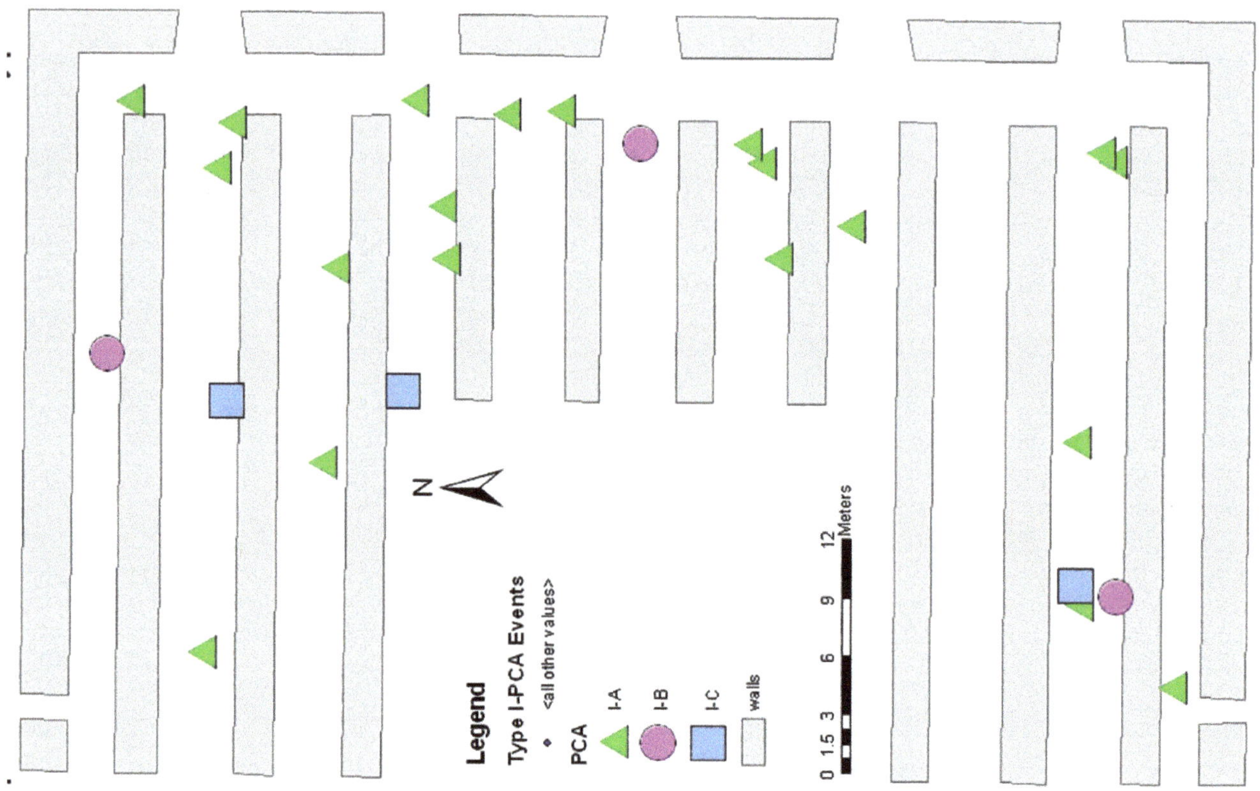

Fig. 7.28. Spatial patterns of subgroups I-A, I-B, and I-C.

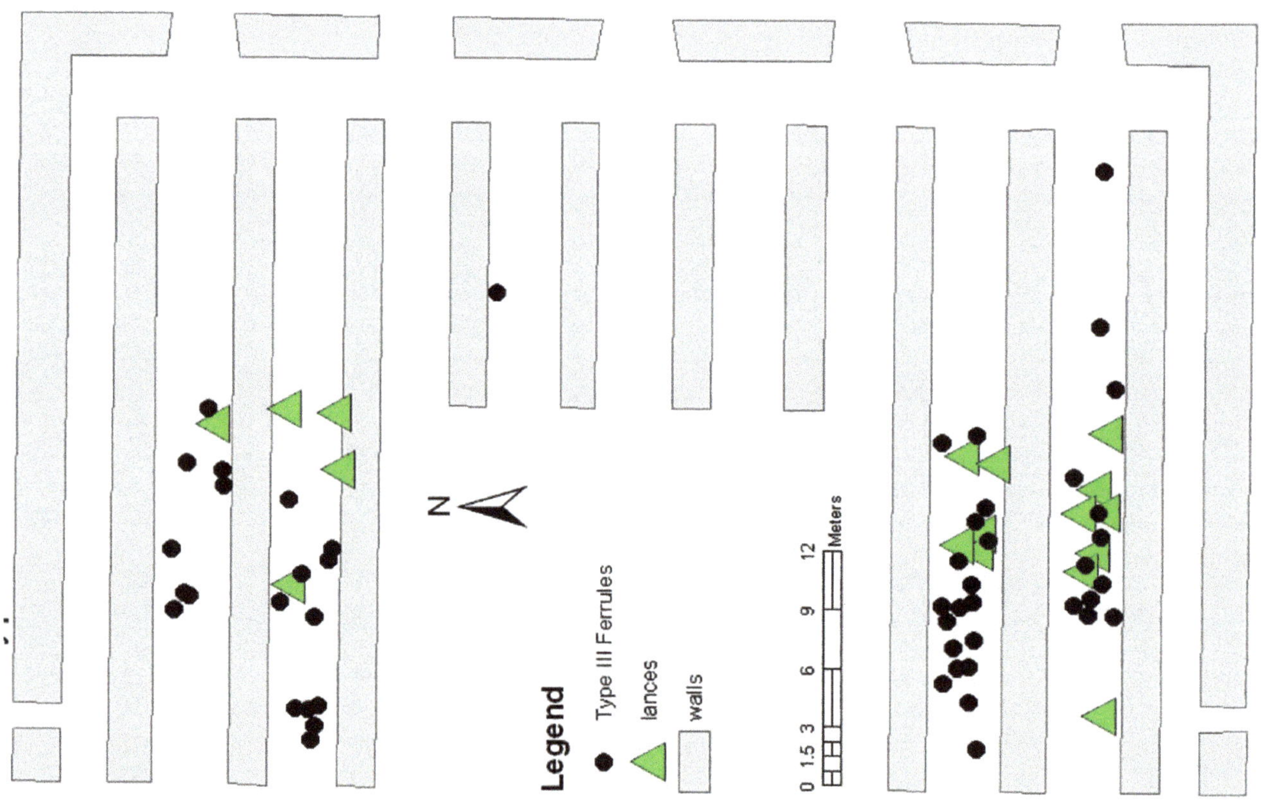

Fig. 7.27. Type III ferrules and bronze lances in Pit 1.

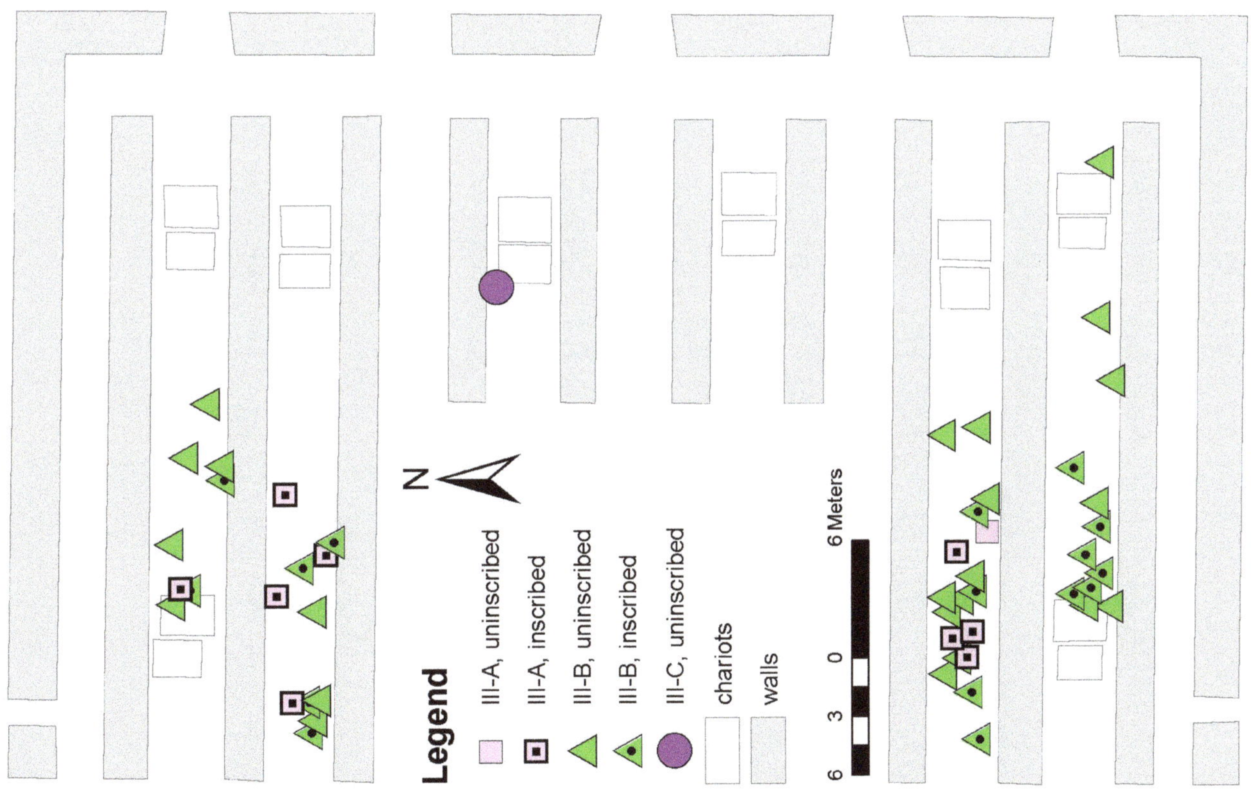

Fig. 7.29. Spatial patterns of the inscribed and uninscribed type III ferrules.

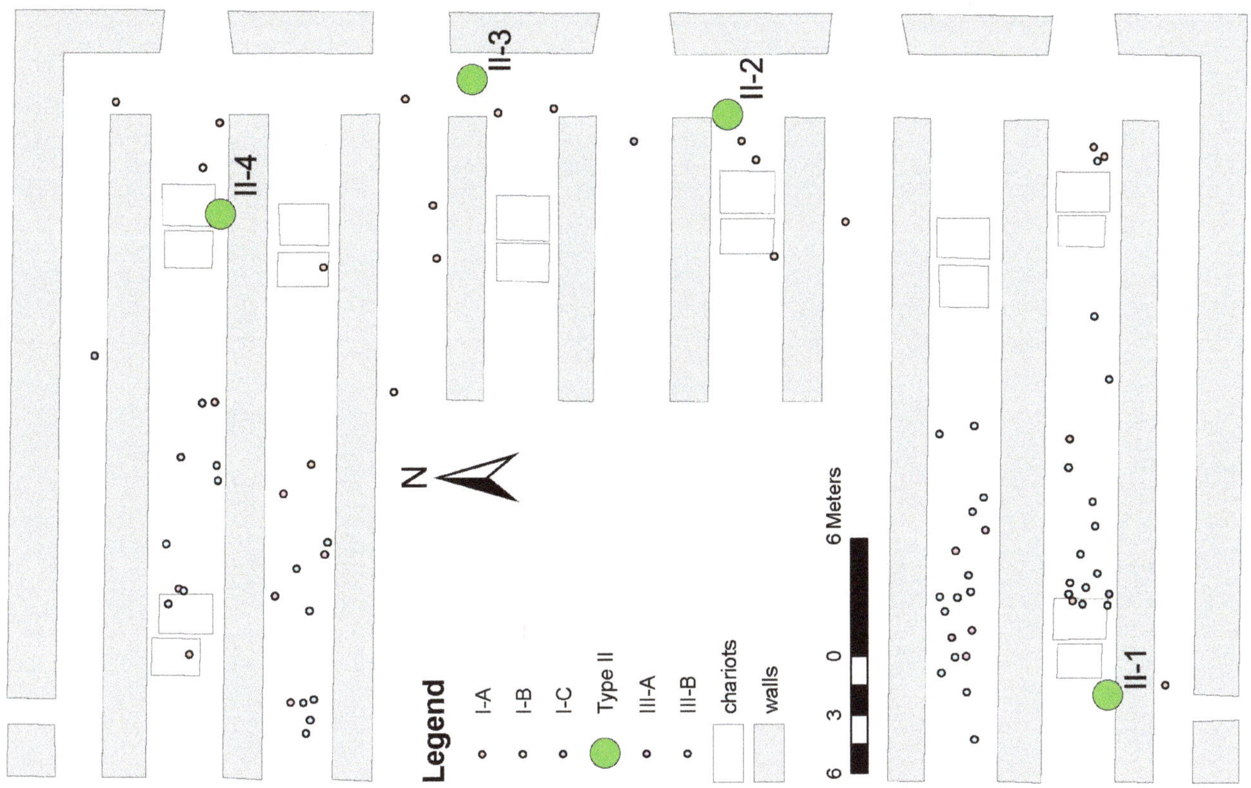

Fig. 7.30. Spatial patterns of type II ferrules.

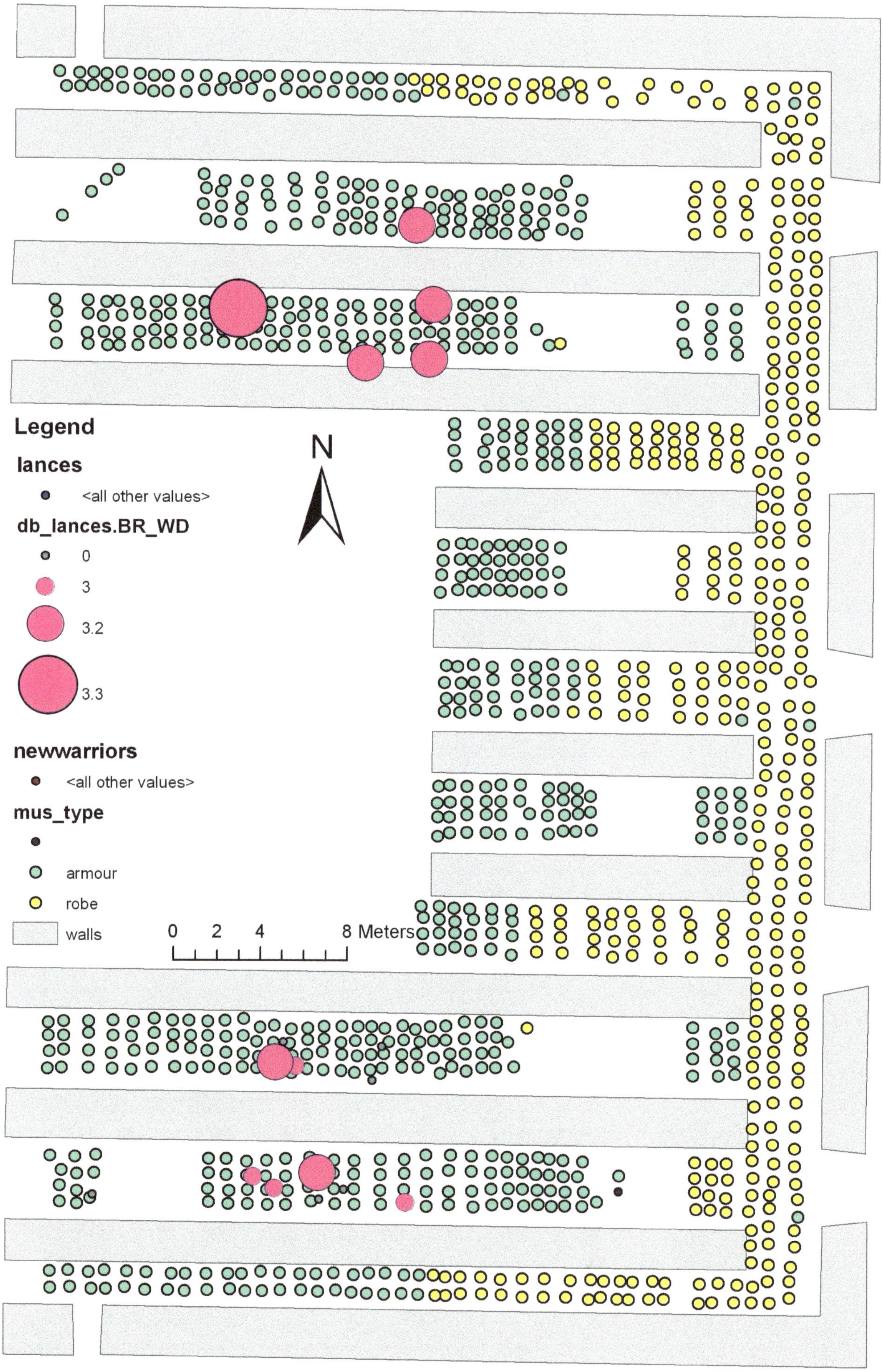

Fig. 7.31. Spatial patterns of the lance subtypes.

uninscribed ferrules (Fig. 7.29) should provide information about workshop practices and labour organisation related to their transport to the pit. The spatial patterns show a more or less overlapped distribution inside the corridors between type III-A inscribed and uninscribed ferrules, as well as between type III-B inscribed and uninscribed ferrules. Their interpretation as originating from different workshops seems improbable. Given that the inscription '*Sigong*' referred to the name of the governmental workshop in the Qin dynasty but that both inscribed and uninscribed ferrules are uniform in their sizes and distribution, one interpretation may be that some of these ferrules were sampled randomly for quality control and marked with the inscription.

Type II ferrules comprise 4 items excavated from the easternmost trenches (see Figs. 7.9 and 7.30). These four ferrules are slightly different from each other in their dimensions (9.8, 10.1, 10.3, and 11.45 cm in their length, respectively), with no obvious subgroups. They are marked in the distribution map as II-1, 2, 3, and 4. The largest one is II-2 in corridor 5, and it does not appear to be related to any blade weapons, and neither do II-3 or II-4. However, II-1 seems associated with a lance (Fig. 7.27 and 7.30).

7.5.1.2.2 The long weapons

Figure 7.31 shows the spatial distribution of the lances, discriminated by their width. It seems at first glance as if there might be a clustering of items presenting different measurements of the blade width in the northern and southern corridors, but the sample size is too small to be sure. The spatial patterns of the subtypes of spears, dagger-axe, and halberds are difficult to interpret because of the small sample size.

7.5.2 Pair correlation functions on the ferrules

Because of the small sample size of bronze spears, dagger-axes, halberds, and lances, the pair correlation function was mainly used to focus on the ferrules, in an attempt to analyse patterns of clustering, regularity, and randomness. Again, the emphasis is on controlling for the overall spatial distribution of ferrules, and looking for further spatial patterning in the ferrule subgroups within this overall pattern. Figure 7.32 shows the pair correlation result for the ferrule subgroup I-A. In this case, the result cannot be shown to depart from what we might expect if the ferrule subgroup was distributed at random. For ferrule type 3, subgroup III-A, there is a slight clustering at about 4.5 metres (Fig. 7.33). Figure 7.34 shows that subgroup III-B ferrules are in slightly regular at 2–2.5 metres, and Figures 7.35 and 7.36 present the III-B ferrules with '*Sigong*' inscription versus those without inscription, and they show a generally random pattern. However, a slight regularity is shown at 2 and 6.5 metres in Figure 7.36 (Table 7.16).

Overall, unlike bronze triggers and arrows, the bronze ferrules do not show clear spatial patterns (Table 7.16) that could be related to the workshop activity and labour organisation. A related fact to bear in mind is that, within the relatively small number of long weapons identified, the inscriptions record a relatively high number of manufacture dates and workers that would have been involved in their

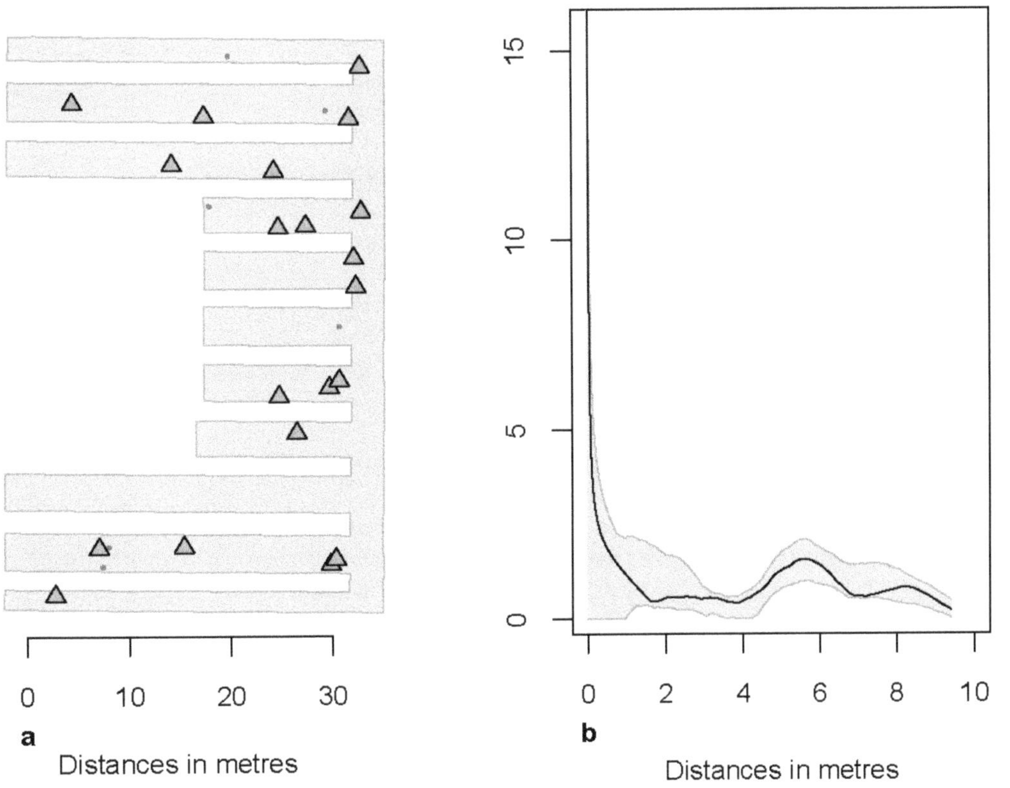

Fig. 7.32. The spatial distribution of type I-A ferrules (shown as triangles) within the five easternmost trenches of Pit 1 (a), and (b) a pair correlation function considering the spatial pattern over the first 10 metres.

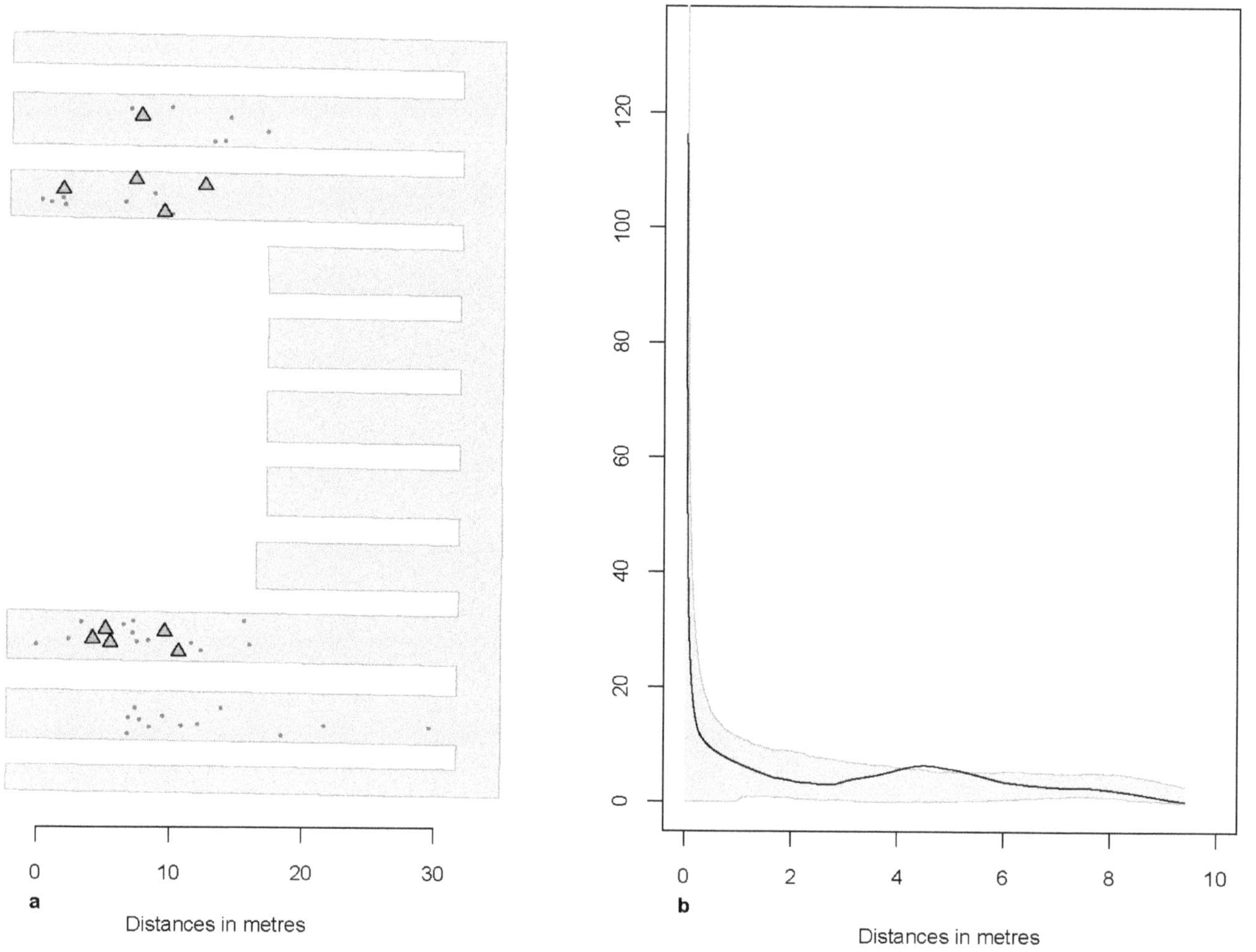

Fig. 7.33. The spatial distribution of ferrules III-A with *Sigong* inscription (shown as triangles) within the five easternmost trenches of Pit 1 (a), and (b) a pair correlation function considering the spatial pattern over the first 10 metres.

production. As the ferrules were mostly attached to these long weapons, both findings should be discussed together. Compared to the arrows and triggers, delivered directly from the workshop or arsenal, it is possible that these long weapons (produced between 244 and 228 BC), or at least some of them, were not produced specifically for the Terracotta Army, but retrieved from existing workshops or stocks where they might have been stored (and mixed) for some time. It is even possible that some of the long weapons would have been used in battle (because most of these long weapons were produced 10 or 20 years before the unification, 221 BC) before they were deposited in Pit 1, as I observed some small chips on the side of the blades when I measured them. Scientific study, including trace element analysis, may be able to confirm this hypothesis if it shows a variety of metal sources employed for the production of long weapons and ferrules. If the ferrules were attached to long weapons in different contexts and perhaps several years before they were taken to the pit, any potential spatial patterns of the ferrules distribution would have been disturbed.

However, the ferrules concentrated in the middle of the four corridors 2, 3, 9, and 10 demonstrated the association with battle formation and the relatively big activity area for workers placing ferrule-related weapons and signs into the pit. The labours were organised to arrange the warriors in these corridors and moved these weapons and signs in to equip them.

7.6 Summary and discussion

The standardisation and labour organisation behind the ferrules and the related long weapons assessed was approached in this chapter via their dimensions and spatial distributions. The dimensions of the ferrules and other bronze long weapons show a relatively high degree of standardisation. The main three types of ferrules – type I, type II, and type III – were clearly cast with different models and moulds, and are linked to different long weapons, but subgroups were also identified based on dimensional differences. Each of the subgroups shows high standardisation in their production, as indicated by the low coefficients of variation (CV). Only the thickness within each type exhibited higher levels of variability, possibly because of technical difficulties in adjusting cores and moulds or perhaps because the thickness was less important for these products and less attention was paid to it by the craftspeople. In this sense, it is worth noting that the thickness of the ferrules and the length of the arrow tangs are the dimensions showing the highest variability, and these are precisely the aspects of these

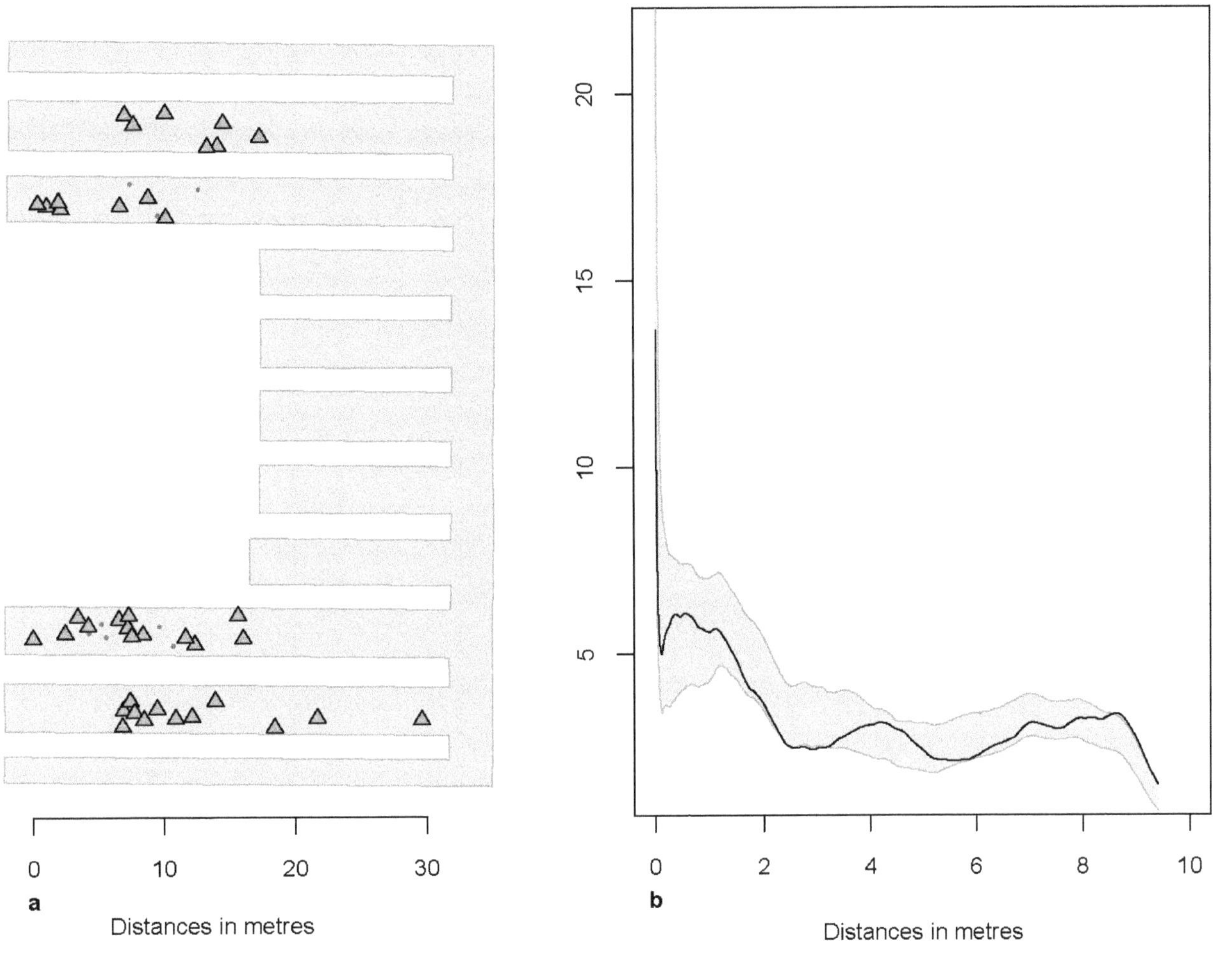

Fig. 7.34. The spatial distribution of ferrules III-B (shown as triangles) within the five easternmost trenches of Pit 1 (a), and (b) a pair correlation function considering the spatial pattern over the first 10 metres.

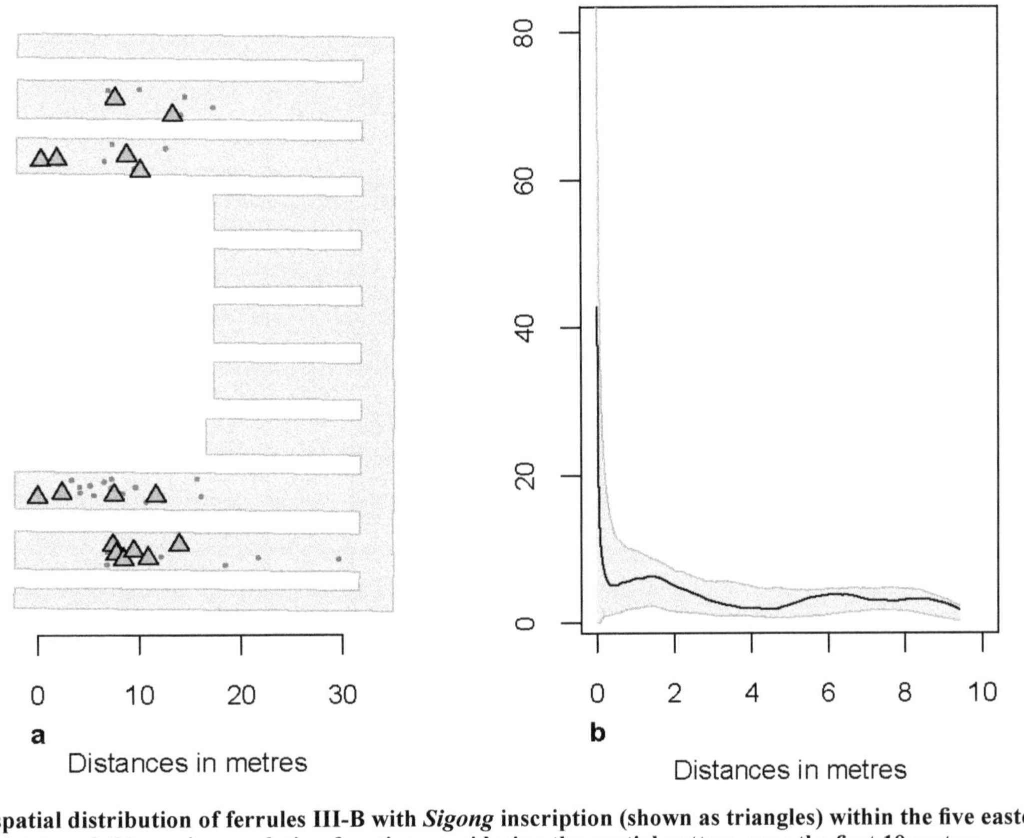

Fig. 7.35. The spatial distribution of ferrules III-B with *Sigong* inscription (shown as triangles) within the five easternmost trenches of Pit 1 (a), and (b) a pair correlation function considering the spatial pattern over the first 10 metres.

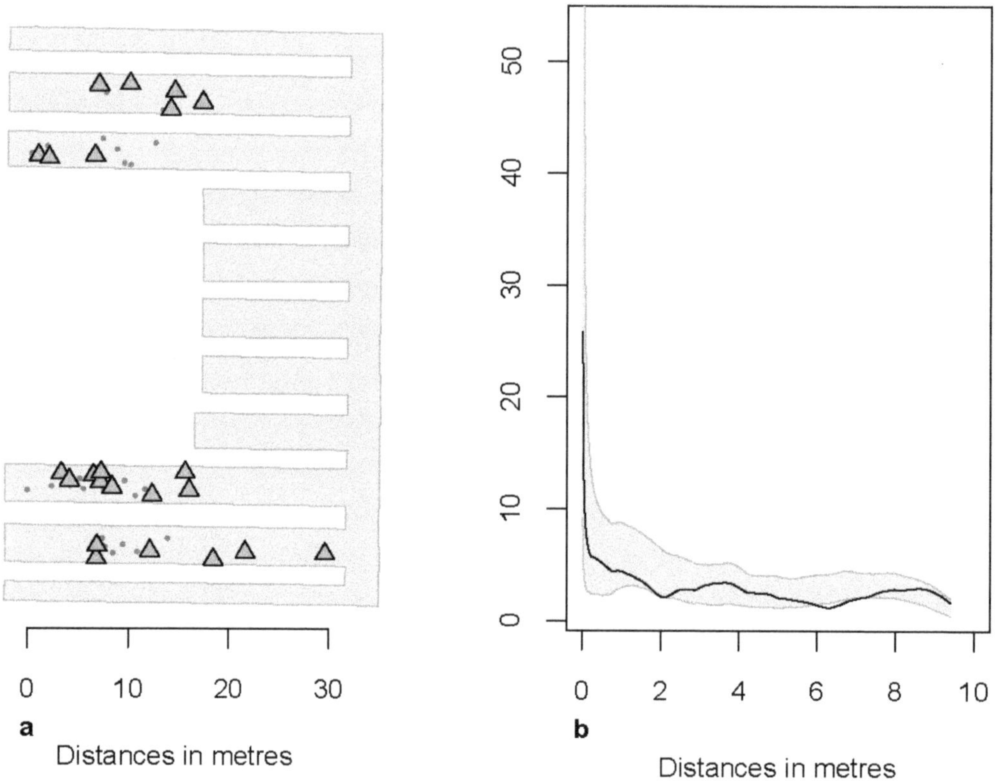

Fig. 7.36. The spatial distribution of ferrules III-B without *Sigong* inscription (shown as triangles) within the five easternmost trenches of Pit 1 (a), and (b) a pair correlation function considering the spatial pattern over the first 10 metres.

Table 7.16. Pair correlation results on the subgroups of ferrules.

Ferrules	Patterns	Distance	Figure reference
I-A	Random		Fig. 32
III-A with inscription	Clustered	4.5 m	Fig. 33
III-B	Possibly regular	2–2.5 m	Fig. 34
III-B with inscription	Random		Fig. 35
III-B without inscription	Possibly regular	2 and 6.5 m	Fig. 36

artefacts that would be less readily perceptible to the naked eye on the finished weapons. The long weapons are few in number. The 4 halberds measured show 1–2 mm differences in their length, and the 11 lances assessed show 2–3 mm differences in their width, even though they were produced in different years and 3 different craftspeople were involved in their production.

The spatial patterns in the distribution of the ferrule types were mainly affected by their association with specific long weapons and by the battle formation of the Terracotta Army of the Emperor Qin Shihuang. In contrast with the patterns shown for arrows and crossbow triggers, the spatial distribution of ferrule subgroups based on metric differences or presence/absence of inscriptions did not show any clear clustering tendencies: rather, they appeared randomly mixed, except for the interesting case of corridor 2. It was therefore suggested that some of these weapons might have not been produced specifically for the terracotta warriors, and that they might even have been used before in battle. At the same time, the long inscriptions on the bronze dagger-axes and lances indicate a structured production organisation and pyramidal supervision system that seems consistent with that suggested for the other weapons. Clearly, the spatial distribution of the bronze spears, halberds, lances, and ferrules will have to be reconsidered carefully when a larger sample is retrieved from the ongoing excavations.

Lastly, the disparity in the quantity of the ferrules and related long weapons was discussed, and several possibilities for it were considered, such as the higher incidence of looting or damage by local farmers on the top parts of these long weapons, or the potential use of some ferrules for items different from weapons. Of course, it is also conceivable that, for some reason, labourers failed to affix weapons on all the wooden shafts, but this possibility seems strongly at odds with the careful organisation and implementation of the weapons' production documented in all other aspects of this research.

8

Discussion and Conclusions

In this book, several perspectives from the bronze weapons of the Qin Terracotta Army have been considered to study the craft production, assess the standardisation, and investigate the labour organisation in the Qin Empire. This has involved a systematic and comprehensive analysis of over 40,000 bronze weapons, in a variety of types from the frozen battle formation of the Terracotta Army, recovered in 1970s to 1980s, integrating the inscriptions, dimension measures, spatial distributions, and preliminary chemical analysis. This study was preceded by a historical and cultural background of the first empire of China and an archaeological account of the underground empire for the Qin First Emperor guarded by well-equipped terracotta warriors (Chapter 1). The theoretical framework and methods were carefully tailed for tackling the specific research questions (Chapters 2 and 3), with consideration of historical documents and palaeographical evidence, in particular inscriptions on bronze weapons (Chapter 4). Three chapters (5, 6, and 7) have focused on several different types of bronze weapons respectively and investigated patterns of standardisation and possible forms of labour organisation involved in the production of each type of weapon associated with the Terracotta Army. This concluding chapter addresses the same research questions in a broader context, considering the Qin weapons as a whole, and devoting particular attention to the questions pertaining to mould casting and its limitations. A final model of the organisation of the workshop will be proposed based on the statistical results, micro-features of the artefacts, and the long inscriptions on certain weapons. The possible activity areas in Pit 1 will also be discussed further, with reference to the spatial analysis presented in preceding chapters and with a view to the likely implications for workshop practices in the Qin Dynasty. It is also worth mentioning that, given that the sample considered here represents less than one-fifth of the available weapons originally interred in Pit 1, the arguments and interpretation provided are necessarily tentative. I hope, therefore, that the interpretations offered here will suggest new research directions and strategies for the future. The archaeological theory and methods adopted for the Qin bronze weapons can also be employed to consider other objects within the tomb complex or indeed elsewhere.

The final section of this chapter will then consider the potential for future research, emphasising the need to analyse the weapons unearthed as a result of ongoing archaeological excavations within the Qin's First Emperor tomb complex and other archaeological sites; additional external evidence for workshop organisation in the production of bronze artefacts from the Shang Dynasty (1600–1050 BC) to the Qin Dynasty (221–206 BC); and the broader issue of the transition from bronze to iron weapons and from the Bronze to the Iron Age.

The discussion below begins with a summary of the main results obtained from the analyses carried out for the purpose of this book with some updated information and ongoing publications.

8.1 Standardisation

Overall, a coefficient of variation has been the summary statistical method most frequently used in the preceding chapters to assess the degree of standardisation in Qin bronze weapons. Each type of weapon and weapon part exhibits a slightly different result in this regard when compared to others. Additional comments are worth making about the weapon production process, moulds, finishing practices, and the impact of human sensory limitations.

8.1.1 Projectile Weapons

The assessment of the metric variability exhibited by the triggers and arrows provides insight into the patterns of standardisation. For this particular project, georeferencing and photographs were often used for the trigger and arrow measurements to achieve the necessary accuracy and efficiency. The trigger parts demonstrate a relatively high degree of standardisation. Measurements of trigger parts A show CV values ranging from 0.5 to 4.5%, those for parts B 1–7%, and those for parts C 1.4–6.5%. This can be understood as reflecting a high degree of standardisation with reference to the baseline suggested by Eerkens and Bettinger (2001). The varying degree of standardisation across different parts and within different trigger subgroups is likely due to several different reasons, including the use of different moulds, as well as craftspeople working in different workshop units and different phases of production over time (Chapter 5, and Li et al., 2014). It is worth remembering that, as observed in the study of the micro-features, it is obvious that different moulds were employed for each of the subtypes of each part. In contrast to the triggers, the arrowheads show an even higher degree of standardisation, within and between bundles, while the arrow tangs exhibit a low degree of standardisation (Chapter 6, and Martinón-Torres et al., 2014a).

8.1.2 Long Weapons

Archaeological remains of the long weapons mainly comprise bronze ferrules and the top blade parts, with occasionally surviving traces of the shaft between them. The measurement of these long weapons was carried out using digital callipers, especially for the ferrules. CV assessment was only carried out on the ferrules because the top blades (including dagger-axes, spears, halberds, and lances) are too small a sample to render formal measurement and statistical analysis meaningfully. CV values obtained on the subtypes of ferrules suggest a relatively high degree of standardisation, from 1 to 4.5%, of their width, height, and diameter, but relatively poor standardisation of their thickness, ranging from 17 to 22%.

8.1.3 Parameters affecting standardisation

Standardisation can be driven by several factors, which may be summarised under three overlapping categories: (a) technical procedures; (b) human behaviour; and (c) the incentives for efficiency or quality. With regards to the Qin bronze weapons, these three aspects will be addressed below and in the course of further discussion about production processes.

8.1.3.1 Models, Moulds, and Finishing Practices

The degree of standardisation exhibited by the Qin weapons is in large part the result of the models and casting moulds that were used to make them. The high degree of standardisation in some subgroups of trigger parts, such as Ag1, could be due to the same model being used throughout, with only one or a limited number of moulds being made from this, and consequently the moulded object being employed within the same workshop unit over a comparatively short period of time. In contrast, the bigger subgroups of trigger parts, such as Ag4, are less standardised and perhaps hide further subdivisions that have not been successfully discerned and/or were made using a large number of moulds. The CV values on each subtype of ferrule exhibit variation below 5% across several measurements, which is also an indication of the high degree of standardisation. It may again be that only a small number of moulds was used for casting each subtype of ferrule, but it seems there is no obvious evidence to quantify how many moulds used. However, the very high variation in thickness on the ferrules could be due to many reasons, but the use of a variable sized core during the casting process might be an explanation.

In addition to models and moulds, the finishing processes applied to the weapons are an important technical parameter affecting their degree of standardisation to some extent. The arrowheads show very high homogeneity and lower CV values in their width (0.9–1.05 cm) and length (2.6–2.9 cm). However, it is unlikely that such a large quantity of arrows was made using only a small number of moulds, and due to lack of evidence, it remains uncertain how many moulds were used specifically for arrow production. It is likely that the finishing practices, including fine grinding and polishing, were also important in achieving the very uniform final shape and size. Grinding and polishing were mainly employed to remove excess bronze and to sharpen the arrows for actual use. The marks on the arrows exhibit extreme fineness and high density on the one hand, and on the other hand, they are perfectly parallel (Li et al., 2011: 499). Thus, being different from the finishing marks on the trigger parts, a tentative interpretation is that some mechanical device such as a rotary grinding wheel was probably being used during the Qin period.

In contrast to the arrows, the finishing traces discovered on the trigger parts are coarser and not as clearly parallel (Li et al., 2011: 498). The majority of the traces are found surrounding the holes of the three mechanical parts A, B, and C, and are therefore probably the result of filing off the excess metal from the surface to make parts fit together properly. Even though the trigger parts were all produced using casting moulds, this rougher filing process could affect the degree of standardisation as assessed by metric data.

The lower and upper CV thresholds of 1.7% and 57.7% suggested by Eerkens and Bettinger (2001) were employed to assess the degree of standardisation in this project. Given that the bronze weapons were all produced using casting moulds, the CVs should theoretically be close to 1.7% with shrinkage of mould material and finishing process after casting being taken into consideration, but actually most of the CV results demonstrate comparatively high degrees of standardisation. In this sense, it should be noted that a relatively large number of craftspeople would have been involved in designing and manufacturing the models and moulds, as well as in casting and finishing the weapons. Hence, human behaviour and sensory limitations also ought to be considered when assessing the standardisation of the production. In addition, besides the limitations and inadvertent errors of the Qin mould and weapon makers, one should also bear in mind that some error would necessarily have been added during the measurement of the artefacts for the current project.

8.1.3.2 Behaviour of the craftspeople

For most ancient Chinese bronze production processes, both the models and the moulds produced from them were made of clay and involved some carving or slight alteration before casting. The involvement of different craftspeople would lead to slightly different results. Even if they probably made careful efforts to meet certain dimensional standards, their skills and learning behaviour would affect the standardisation (Eerkens and Bettinger, 2001) and the effect of these would not necessarily be identified by workshop supervisors because of their sensory limitations.

In the production of a large quantity of the weapons, 'copy errors' can also occur between the master craftspeople and their apprentices. Theoretically, copy errors accumulate from one generation to another. For example, an apprentice learning from a master artisan might not get every detail correct, and this would repeat from generation to generation. In the case of the weapons associated with the Terracotta Army, one might assume the involvement of a relatively limited number of generations in the common sense of the word, given that the entire assemblage is likely to have been produced in a short time-span. However, considering the high output, it is possible that many chains of technical transmission would have operated in practice, functioning like different generations and thus potentially adding to cumulative errors. Copy errors in the case of the bronze weapons could occur as the shape was transferred from model to mould to actual bronze weapon and back again. The slight alterations taking place during and after each casting would also result in further errors.

Some of the different subgroups identified for each trigger part are possibly related to the individual preference of craftspeople, or at least the tastes or standard procedures of a particular workshop unit. For example, notched and unnotched trigger parts B, as well as bevelled and curved parts C, seem to be technical variants or innovations adopted by some craftspeople but not others. Although it is difficult to ascertain this aspect without further knowledge of the crossbow structure and operation, it would seem that this formal variability in trigger subgroups would not affect the functioning and quality of the trigger parts.

A last important parameter for the assessment of standardisation that could be affected by individual craftspeople or workshop practice is the chemical composition of the metal used for the weapons, which would affect their hardness and quality, and where patterns of standardisation in alloy composition have been analysed and assessed in arrows (Chapter 6). Further, prompted by the results of this study, an archaeometallurgical analysis has been carried out on the subgroups of arrows by the Sino-British cooperative group (Xia et al., 2014; Martinón-Torres et al. 2014a). In the future, further archaeometallurgical analyses will be conducted on other weapons with these questions in mind.

8.1.3.3 Efficiency and quality control?

In addition to technical factors and the behaviours of craftspeople, it is worth looking deeper and more widely at issues of efficiency and quality control. Standardisation is often the result of a desire for efficiency to reduce the cost on the one hand, and, on the other hand, of the need for quality control to meet consumer demands (Hodder, 1983), political constraints (Rice, 1984), and/or functional needs. For example, the crossbow triggers were all made for the Qin First Emperor according to the supervisor's criteria and each part was standardised to enable successful assembly into a functional weapon.

It is difficult to ascertain conclusively whether all the weapons were produced specifically for the mausoleum, or whether they were taken out of existing arsenals, or if they were indeed a real army's gear. Based on the long inscriptions, some weapons were produced during the reign of King Ying Zheng (3rd to 19th regnal years, 244–227 BC), that is, before the unification, although the mausoleum was already under construction by then according to the historical documents. Aspects such as the limited batch mixing of crossbow triggers and arrows, the uniformity, and the pristine state of many weapons rather indicate that they could have been produced specifically for the mausoleum. Whatever the case, with the Emperor as the ultimate consumer, one can expect that the weapons associated with the Terracotta Army would have undergone rigorous quality control procedures.

Labour needed to be well organised and a proper supervision system had to be in place during the production processes. The following section explores this issue further.

8.2 Organisation of production

The organisational structures behind the bronze weapon production can be summarised as reflecting workshop practices, and the wider social structure of the supervision system. The following discussion is an attempt to combine the qualitative and quantitative results obtained in order to generate new observations. The discussion focuses mainly on the triggers, arrows, and long weapons.

8.2.1 Model of production organisation

Various possible organisational models for mass production in ancient China were discussed in Chapter 2. Among these, the holistic and prescriptive models were assumed to be relevant for the Shang (1600–1050 BC) bronze production (Franklin, 1983), and both were possibly employed at Anyang where the Shang bronze foundries have been discovered (Li, 2006). Bagley argued that the Houma foundry, dated to the Warring States era (476–221 BC), employed a batch production system (Bagley, 1995; 1996). According to Ledderose (2000), the Qin terracotta warriors (246–210 BC) themselves are the result of a 'modular' production system, and Han (202 BC–220 AD) lacquers were manufactured mainly according to a flow line production system (Ledderose, 2000). The terminology and defining criteria employed in the literature are rather diverse, rendering comparisons difficult.

The evidence presented here indicates that the production of weapons for the Terracotta Army would have been organised in semi-autonomous units – or a cellular production system – rather than as a flow line production. This evidence will be summarised again here, and further discussed with reference to the above-mentioned comparative models.

The investigation offered in previous chapters into the models of mass production on the finished bronze weapons has mainly focused on casting and assembly processes. Both the bronze triggers and the arrows offer clear indications of cellular production. The trigger parts A, B, and C that were assembled together to produce a working crossbow trigger show very consistent associations of part subgroups (see PCA plots in Chapter 5). If we assumed a very different production mode, such as a flow line, then one group of craftspeople would produce parts A, another group parts B, and another parts C, D, E, etc. Then, finally, all these parts would be assembled into functional triggers. This process would therefore easily lead to the mixing of the part subgroups, and hence we would see far more mixed assemblies of trigger part groups than we actually observed. The same is true of the arrows: both the visually or metrically observed features and the XRF results demonstrate similarity *within* bundles and variability *between* the bundles. As mentioned before, the arrowheads and tangs were cast separately. As such, this similarity within bundles and variability between them, visible both on the arrowheads and on the tangs, suggest that craftspeople worked in units or cells and were involved in the entire process of producing a complete bundle of one hundred working arrows, from casting through assembly to bundling. It also suggests another indication that these bundled arrows had never been used in the battlefield and were supposed to produce only for the First Emperor's afterlife (Lin and Li, 2018: 99).

By cellular production, I am referring to a situation in which a workshop might be divided into several units or cells, each of which was responsible for the production of complete objects. However, the possible size of such cells in terms of the number of craftspeople involved in each of them and their relative productivity remains open to further investigation. For example, the assembled parts of the trigger vary quite a lot in their size, although this may be partly related to the fact that only part of the pit has been excavated or that some subgroups are still unidentified in the larger ones observed to date. The same is true of arrows, in the sense that on the basis of the existing sample of arrow bundles analysed so far, it is difficult to draw a comprehensive picture of the size of any possible cells, their productivity, or the organisation of labour within cells.

Actually, the cellular form of mass production more or less approximates the working units of master and apprentices thought to have been involved in the manufacture of the terracotta warriors themselves (Yuan, 1990, 2014). Yuan (1990, 2014) argues that each master potter probably guided several apprentices to create the complete terracotta warriors, and their products would be subject to a quality control during which the master potter's name was marked (stamped or carved) on them. It is the case that 87 names of master potters have been found on the 1087 terracotta warriors (Yuan, 1990), but the updated number reaches 92 names if it includes the ongoing discoveries in the whole tomb complex (Yuan and Liu, 2009), and 68 different numerical signs (Yuan, 2014). If the weapons production was organised along similar lines, a master with a group of apprentices would form a semi-autonomous cell that cast each part of the weapons and subsequently assembled them together.

However, the question is whether the terracotta warriors were assembled together with different body parts as the bronze weapons, or whether they were individually built in the workshop unit with master and apprentices involved. On the 'modular' basis of producing Qin terracotta warriors, Ledderrose (2000) argues that the production did not operate according to a 'holistic' concept of the personality of the warriors. He supposes that the Qin artisans first modelled each part separately, fitted them together, and then reworked the facial features in a creative process to achieve the individual variety. Later, Li (2006) proposed a possible batch production based on Ledderose's modular concept, but less evidence was provided.

Separate cooperative research was conducted on the production of the terracotta warriors, particularly focusing on the marking names to investigate the creation procedures and the model of labour organisation (Li et al., 2016). The marking evidence on the terracotta warriors and its spatial clustering in the burial pit suggest that each workshop unit was responsible for turning out complete terracotta warriors (Li et al., 2016), which goes against the claim that the highly diverse, life-like appearance of the army is the result of countless combinations of a limited set of 'modules' of body parts (Ledderose, 2000: 51–73). At same time, a 3D model of the ears provides further evidence that they were individually sculptured rather than that several types of mould casting were used (Bevan et al., 2014a). The patrographic analysis showed that the clay used to build terracotta warriors was acquired locally and mixed with alluvial sand to achieve suitable paste (Quinn et al., 2017), but the slightly different chemical composition of the paste might indicate that each workshop unit prepared the clay independently (Quinn et al., in press). However, given the evidence that a master–apprentice relationship for making the terracotta warriors, or an official–worker relationship for producing the weapons, was also an important aspect of workshop organisation, the production of the terracotta warriors and their weapons, even in different procedures, is more likely to have been managed via a similar cellular system.

The cellular production behind the manufacture of the Qin bronze weapons is also deemed similar to that suggested for the Houma bronze foundry (Bagley, 1995, 1996). The remains from the latter workshop showed that batches of the different parts necessary to make a complete vessel were produced in different locations of the same workshop and then assembled together. The craftspeople at Houma thus performed separate but related tasks from mould making to casting and assembly, in the same production unit.

The peculiarity of the present case of the Qin weapons, however, is that I have arrived at this conclusion from the study of finished artefacts rather than workshop remains, and that the various assembly groups identified indicate that, most likely, more than one cell was active in the production. These are therefore methodological and interpretive developments that may be tested in future work, or applied to other case studies.

If it is the case that cellular production practices were employed in the making of the Qin bronze weapons, with several cells possibly operating at the same time, then we must consider the supervisory system and methods of control employed to organise each of the semi-autonomous cells or units to ensure overall quality and standardisation. Some clues about these aspects have been obtained from the long inscriptions engraved on the weapons, as presented in detail in Chapter 4 and summarised below.

8.2.2 Organisational structure and practices

The organisational structure and supervisory system inferred from the long inscriptions was employed to ensure the standardisation and quality of cellular production. The long inscriptions carved on the halberds provide evidence of a hierarchical supervision system in place during the Qin period. The inscriptions typically identify the Qin state chancellor, the *Sigong* workshop, the overseeing official, the individual craftsperson involved, and the regnal year when the weapon was produced. The lances bear relatively simpler long inscriptions with the regnal year, *Sigong* workshop, and producer. If the other weapons, such as triggers, arrows, and other short weapons, without such long inscriptions were produced under the same supervisory system, each cell producing triggers and arrows would have received the same standards and instructions from the supervisor or *Sigong* officials. During the production process of halberd before 237 BC, *Sigong* officials and *Cheng* would have been the monitors. Once the products were complete, they would need to pass quality control; otherwise, the craftspeople in the cell or unit would have been punished (Huang, 1990).

The supervisory system and quality control methods whose indications are visible on the weapons produced for Pit 1 had been a feature of Qin craft production since the period of chancellor Shang Yang (385–338 BC) and developed further during the Qin Kingdom and Empire. A ferrule and halberd were produced during this period and bear inscriptions of Shang Yang's name, but no names of workshop officials or craftspeople. From the reign of King Huiwen (356–311 BC) to that of King Zhao (325–251 BC), the bronze weapons started to bear the name of officials and craftspeople in addition to that of the chancellor (Yuan, 1990). The weapons for the terracotta warriors not only bear the name of chancellor Lu Buwei, but also the name of officials and the actual producers. After 237 BC, Lu Buwei is known to have been dismissed, and the long inscriptions on the lances cease to include any reference to the chancellor's name. It was only after Emperor Qin Shihuang died that the name of the new chancellor, Li Si, started to be carved on weapons, for example on a halberd produced in the 1st regnal year of the Qin Second Emperor (209 BC); moreover, the workshop is then referred to as *Yueyang* rather than *Sigong* (Jiang and Liu, 2006).

The triggers, swords, and ferrules only bear simple inscriptions for assembly or accountability, or no inscriptions at all. Why were the long inscriptions only carved on the lances and halberds, and not on the other weapons? This question will remain unanswered for the moment. It might be tempting to suggest that halberds and lances involved a more complicated manufacturing process, requiring more skilful craftspeople and stricter supervision. However, the crossbows, requiring the accurate assembly of various moving parts into a larger structure, must have also posed a complex technical challenge, which is, however, not reflected in a long inscription.

8.2.3 Craft specialisation and attached production

From the very archaeological context, as well as the patterns of standardisation and labour organisation visible in the Qin bronze weapons, it is clear that weapon makers constituted a highly specialised production group, a form of highly specialised craft industry with full-time workers working under political governmental control. The workshop reflected a political economy that showed high degrees of specialisation and standardisation in other areas as well, producing for an upper elite consumer, the Emperor.

Costin (1991: 10) argues that 'attached specialists produce several types of goods of key importance within the political economy and the status, power, or control structure of the society. These include luxury and wealth items, weaponry, and wealth-generating goods. Usually, only a limited portion of the population has access to these products.' The manufacture of these bronze weapons falls into the category of attached specialisation, and it is most likely that it involved a kind of retainer workshop (Costin 1991), as has been discussed in Chapter 2, even though the situation is slightly different in ancient China.

As outlined previously, the terms *Sigong* (寺工) or *Gong* (工) that appear in some of the inscriptions on the weapons refer to the Qin governmental workshop, which was responsible for producing the highly specialised bronze weapons, and involved skilled artisans, slaves, soldiers, and convicts (Yuan, 1990). These terms therefore denote specific forms of attached workshops. Relevant information about these terms mainly comes from Qin bamboo slips found at *Yunmeng Shuihudi* (云梦睡虎地; Shuihudi Xiaozu, 1978; Yumeng Bianxiezu, 1981). The craftspeople with *Sigong* title were normally skilled artisans in this workshop, with the responsibilities of teaching workers and controlling the quality, while a worker with the title *Gong* was the real producer, and normally of lower status. As mentioned in Chapter 4, one sentence found on a *Shuihudi* bamboo slip, *Jungong* (均工), notes that 'a slave with skills can be a *Gong*, a worker in the workshop'. Another sentence from a *Shuihudi* bamboo slip, *Junjulu* (军爵律), records the fact that if a slave or convict won military awards, he could be a *Gong* in the workshop and avoid any further punishment (Shuihudi Xiaozu, 1978; Yuan, 1990). Clearly, there is a wealth of information on workshop organisational and supervisory practices recorded in some bamboo slips recovered from the Qin tomb in China – a body of data to be explored in future work to complement and contrast the ideas put forward in this study.

The attached retainer workshops for the bronze weapons production would work under elite sponsorship but with a strong element of political coercion. The weapon producers may have held a status similar to that recorded for the Qin First Emperor's tomb-builders, who worked for the mausoleum construction project instead of being punished or paying taxes. Some tombs of the mausoleum builders were found west of the tomb complex, with many skeletons in the same pit and some pieces of pottery carved with their names and origins (Yuan, 2002). There is still little information about how long they had been involved with the tomb constructions before they died, or whether they were involved in the transportation and placement of these bronze weapons into the pits.

8.3 Spatial statistics and activity areas

As we argued in Chapter 2, the spatial patterns exhibited by technological and stylistic aspects of the weapon subgroups can be used to interpret workshop practice and labour organisation. Micro-features and metric clusters have been employed to investigate the standardisation and organization of labour in the workshop. However, these qualitative and quantitative differences among the weapons also show patterns with regard to their distribution in the pit, especially in the case of the triggers. As discussed in Chapter 2, these spatial patterns were driven by several possible factors, such as workshop practices, as well as patterns associated with the weapons' delivery and placement into the pit. In order to avoid relying solely on intuitive and unstandardised assessments of these spatial patterns, a statistical method known as a pair correlation function has been used to assess the clustering, regularity, and randomness of the patterns, above and beyond the spatial structure imposed by the corridor layout and army battle formation. Both simple distribution maps and these complex methods made it possible to explore possible activity areas in the pit that might relate to labour organization, and the processes associated with transporting weapons, storing weapons in the arsenal, and/or placing them in the pit.

One question that was raised when the above processes were discussed was whether the weapons had been used in the battlefield before they were buried in the pit. This is also a crucial factor affecting the spatial patterns. The evidence suggests that some weapons were not used on the battlefield beforehand, and hence were transported from the workshop or arsenal directly to the tomb complex. All the crossbow triggers, for example, appeared assembled without any wear marks that could have occurred during repeated use (Li et al., 2011; Li et al., 2014), and the arrows were identified as consistent within the bundles as if they were freshly made and grouped specifically for the burial pits rather than being used in the battlefield beforehand (Lin and Li, 2018). However, the long weapons – lances, halberds, and ferrules – were observed to have some small chips on their blades, and it was arguable about whether they had been used before. With this observation in mind, the spatial patterns of the weapons in the pit and their implications can be reconsidered further.

8.3.1 Spatial patterns of triggers and activity areas

The distribution maps and pair correlation functions suggest a significant clustering of several assembled crossbow trigger groups over and above the distribution of triggers as a whole, especially in the cases of assemG5, assemG8, and asssemG9 (see Chapter 5). Effectively, the same patterns also apply to the subgroups of trigger parts, given that there is a consistent correspondence among those types of parts that were assembled together. A tentative interpretation of the spatial patterns exhibited by the trigger groups is that they are most likely related to the organisation of labour in a variety of stages, from the initial production to the final arrangement in the pit. If cellular production were an appropriate model for the manufacturing and assembly of the triggers, then the assembled triggers would each have been made by specific units or cells in the workshop. Therefore, it might be assumed that batches of similar finished products were moved from the workshop to an arsenal or a simple storage, and later to the pit, without any intention to mix them, so that the spatial patterns in the pit sometimes show a clustering of triggers that were probably made by one small cell. The spatial patterns of the triggers in the pit show the 'activity areas' associated with different stages, but are possibly indicative of the entire work process from workshop practices to the tomb complex arrangement (see Chapter 5, and Li et al., 2014).

The pair correlation functions provide evidence of some trigger group clustering at small scales of up to a few metres radius, which may reflect clumps of similar products produced by the same workshop cell and preserved as a batch when they were placed in the pit. At a larger scale, however, there are also broad 'activity areas' covering parts of corridors or even entire corridors (Bevan et al., 2014b; Li et al., 2014).

The terracotta warriors were densely arranged in the pit, with each warrior about 0.67 m apart, in a fairly regular spacing (Bevan et al., 2013). Once the terracotta warriors had been arranged, it would have been very difficult for any worker to pass through the intervening gaps to equip them. It is most likely therefore that the warriors and weapons were placed into the pit simultaneously and in pre-arranged groups. Interestingly, two groups of triggers, assemG5 and assemG8, were arranged from the middle corridor to opposite directions, which may imply a pattern in the warriors' arrangement that included two groups of workers associated with triggers being responsible for the two different activity areas. Likewise, it is also interesting that the triggers with *Gong* inscriptions were mainly in the two flank corridors (see Chapter 4 and Chapter 5), along with other triggers that belong to the same subgroup. These triggers with *Gong* inscriptions also show greater consistency than other trigger subgroups, and may indicate slightly more standardised practices in the *Gong* workshop or a particularly standardised episode of production over time.

8.3.2 Placement of arrows into the pit

Despite the large quantity of arrows distributed in the pit, their spatial patterns are relatively simple, because of the consistency of metric data. However, the XRF analysis provided a promising result that could be tentatively interpreted as indicative of a model of cellular production. Spatial analysis mainly focused on the arrow bundles and on comparing those with shorter and longer tangs. The arrow bundles with different-sized tangs were mainly concentrated in the middle corridors, and a pair correlation function suggested that they were clustered up to a 7m radius, which, as with the triggers above, probably reflects the preservation of intact workshop batches as they were placed in the pit.

In addition, the six special arrows exhibited a more-or-less regular spacing in the pit, which may indicate the special function of these arrows in the battle formation.

8.3.3 Ferrules and long weapons

The ferrules and long weapons were generally found in the middle of the pit, as compared to the arrows and triggers, which were found in the surrounding corridors. Theoretically, the numbers of ferrules and long weapon blades should match (as they constituted the two ends of a long weapon). Based on the spatial distributions and one or two completely preserved examples, type I and II ferrules seem to belong to spears, dagger-axes, and halberds, while type III ferrules belonged to lances.

However, interestingly, the ferrules are much more common than the bronze long weapons in the pit. For example, in corridor 10, one lance blade was found together with eight type III ferrules. As discussed in Chapter 7, one possible explanation could be that the long weapons easily emerged out of the roof due to their long shafts and were thus more susceptible to being looted or damaged, while the other possibility could be that the ferrules were also used for flags or other ensigns.

Type I ferrules were all found in the middle corridors, but close to the east vanguard, and type II ferrules were relatively evenly spaced to the east, while type III-A and III-B ferrules were all located in the middle corridors. Type III-C is a special type found in the wooden chariot. One spear seems to be associated with it, but it is not clear what this special ferrule's function was with regard to the chariot. The spatial distribution of inscribed and uninscribed type III ferrules did not present many obvious patterns, except for a clearly identifiable group in corridor 2 that probably indicates an activity area. The ferrules with *Sigong* inscription may reflect the same type of quality control procedures involving masters and apprentices that were discussed above.

8.4 Directions for future research

Given the limitations of the available data, the arguments and interpretation offered in this study are necessarily tentative, because the three pits with the terracotta warriors represent only a corner of the entire tomb complex, and the bronze weapons considered originate only from the five easternmost trenches of Pit 1. However, it is hoped that the present research has devised a methodological model that combines typological, metric, and spatial analyses of finished artefacts for the purposes of investigating issues pertaining to craft organisation, as well as putting forth some well-grounded hypotheses. Ongoing excavations will provide an opportunity to test some of the observations made here concerning the standardisation of production and labour organisation. Furthermore, the archaeometallurgical analysis of these bronze weapons, particularly bronze arrows, has been integrated with the statistical and spatial results (see Chapter 6, Martinón-Torres et al., 2014a; Xia et al., 2014), and some other related analysis is currently underway.

In addition to these potentials, some further possibilities exploring the rich discoveries from the Qin First Emperor's tomb complex are outlined below, with a particular emphasis on broader comparisons associated with regional and temporal developments.

8.4.1 Research model for other research within the tomb complex

The statistical and spatial strategies employed above for interpreting standardisation and labour organisation have not been applied to this type of context before to any significant extent. They offer promising results and might be extended to consider other archaeological materials from the tomb complex, including the terracotta warriors themselves, the stone armours, the roof-tiles, bricks, and other bronze or iron objects. It will be fascinating to compare the production processes and labour organisation behind each type of object in order to consider the degree to which they suggest similar or different kinds of practices.

The quantity, quality, and spatial data of the bronze weapons make it well suited for this kind of approach and gain a prominent result. However, it should be possible to fruitfully apply similar approaches to other case studies, such as pottery and metal objects from other sites. Typological analysis and measurements are basic for each material discovered from a particular archaeological site, which can be used to interpret the labour practice and organization behind it, combining further with statistical and spatial data to investigate models of organization, transportation, and consumption. These strategies, benefited with the modern computing technology and proper software, may contribute to broader archaeological research projects.

8.4.2 Contemporary weapons at other sites

This study focused on the bronze weapons from the tomb complex of Qin First Emperor, and, in the future, it will be necessary, to compare these with the contemporary weapons used on the battlefield – or at least produced with this purpose in mind – to consider standardisation and mass production on a larger scale. For example, the halberds from the Qin First Emperor's tomb complex were all carved with the name of a workshop officer, *Sigong*, while a weapon found in a normal Qin tomb in Guangdong Province was produced in a workshop named *Shubang* in the 14th regnal year, 233 BC (Yuan, 1984; Jiang and Liu, 2006). This *Shubang* halberd was assumed to have been taken by a general on the battlefield and then buried locally. The comparison will be limited by the fact that there are only a small number of inscribed Qin weapons from other archaeological sites beyond the tomb complex. Other interesting comparisons may involve some Qin weapons without any inscriptions that are, however, datable based on archaeological data, or the weapons produced in other neighbouring kingdoms during the Qin period.

8.4.3 Bronze versus iron weapons

Evaluations of the standardised practices and labour organisation behind the equipment in the tomb complex have necessarily been based on the large quantity of bronze weapons recovered, but the significance of a small number of iron weapons excavated at the site could not be neglected. In addition to these, within the tomb complex, a considerable number of iron tools and implements have been discovered. These artefacts offer yet another opportunity for future research. Determining the production methods and alloy types of these different weapons and implements will be interesting in itself, but more especially as it will allow comparison between iron-based and copper-based weapons, as well as between the iron artefacts from the tomb complex and those recovered elsewhere. As such, this future study will contribute to two important topics in Chinese archaeology: firstly, the question of why most of the weapons buried in the pits were made of bronze rather than iron; secondly, the broader but related issue of the emergence and development of iron and steel in China.

On the one hand, it will therefore be useful in the future to consider Qin bronze weapons production in early Iron Age China from a wider theoretical perspective, and to dedicate further attention to the social, political, and economic changes associated with state formation during the Qin period. On the other hand, exploring weapons and other iron objects will offer a perspective on the strange overlap with the bronze production, an issue with wider implications for the transition from the Bronze Age to the Iron Age in China.

Bibliography

Arnold, D. E. and A. L. Nieves. 1992. Factors Affecting Ceramic Standardisation. In B. G. Pool, *Ceramic Production and Distribution* (pp. 93–113). Boulder, Colorado: Westview.

Baddeley, A. 2008. *Analysing Spatial Point Patterns in R.* CSIRO and University of Western Australia.

Bagley, R. W. 1980. The High Yinxu Phase (Anyang Period). In W. Fong, *The Great Bronze Age of China – An Exhibition from the People's Republic of China* (pp. 177–189). New York: The Metropolitan Museum of Art.

Bagley, R. W. 1995. What the bronzes from Hunyuan tell us about the foundry at Houma. *Orientations, January*, 46–54.

Bagley, R. W. 1996. Debris from the Houma Foundry. *Orientations, October*, 50–58.

Bagley, R. W. 1990. Shang Ritual Bronzes: Casting Techniques and Vessel Design. *Archives of Asian Art 1990 (XLIII)*, 6–20.

Bai Jiangang 白建钢 1994. The spear-warriors and military strategy in the Qin army试论军队长矛手及其战术. *Wenbo文博*, 1994:04, 61–70.

Bailey T. and T. Gatrell 1995. *Interactive Spatial Data Analysis.* Harlow: Longman.

Barnett, V. 1974. *Elements of Sampling Theory.* London: English University Press.

Benco, N. 1988. Morphological standardisation: an approach to the study of craft specialisation. In K. C. Lackey, *A pot for all reasons: ceramic ecology revisited* (pp. 57–72). Philadelphia: Temple University Press.

Bevan, A., Crema, E., Li, X. and Palmisano, A. 2013. Intensities, interactions and uncertainties: some new approaches to archaeological distributions, in Bevan, A. and Lake, M. (eds.), Computational Approaches to Archaeological Spaces. Walnut Creek: Left Coast Press.

Bevan, A. and J. Conolly 2006. Multiscalar Approaches to Settlement Pattern Analysis. In G. L. Molyneaux, *Confronting Scale in Archaeology: Issues of Theory and Practice* (pp. 217–234). New York: Springer.

Bevan, A. H., Li, X. J., Martinón-Torres, M., Green, S., Xia, Y., Zhao, K., Zhao, Z., Ma, S., Wei, C., Rehren, Th. 2014a. Computer vision, archaeological classification and China's terracotta warriors. *Journal of Archaeological Science* 49, 249–254.

Bevan, A. H., Li, Xiuzhen, Martinón-Torres, M., Rehren, Th. 2014b. Some thoughts on the quantitative spatial analysis of metallurgical and other craft activities冶金及手工业生产的空间统计学分析. In Cao, W. and Rehren, Th. (eds.). *Proceedings of International Symposium on Qin Period Metallurgy and its Social and Archaeological Context 秦时期冶金考古国际学术研讨会论文集*, 38–46. Beijing北京: Science Press 科学出版社.

Bevan, A., Li, X., Zhao, Zh., Huang, J., Laidlaw, S., Xi, n., Xia, Y., Ma Sh., and Martinón-Torres, M. 2018. Ink marks, bronze crossbows and their implications for the Qin Terracotta Army. *Heritage Science* (2018) 6:75, 2–10.

Blackman, M. J., Stein, G. J. and P. B. Vandiver 1993. The Standardisation Hypothesis and Ceramic Mass Production: Technological, Compositional and Metric Indexes of Craft Specialisation at Tell Leilan, Syria. *American Antiquity 58*, 60–68.

Blansdorf, C., Emmerling, E., and M. Petzet (eds). 2001. *The Terracotta Army of the First Chinese Emperor Qin Shihuang* 秦始皇陵兵马俑. Paris: ICOMOS.

Bleed, P. 1997. Content as Variability, Result as Selection: toward a behavior Defined of Technology. In C. B. Clark, *Rediscovering Darwin: Evolutionary Theory and Archaeological Explanation (95–107).* Arlington, Virginia: Archaeological papers of the American Anthropological Association.

Boas, C. 1940. *Race, Language and Culture.* London: The Free Press.

Chang Yong常勇 and Li Tong 李同 1983. A preliminary investigation into the concentration of mercury within the tomb complex of Qin Shihung 秦始皇陵中埋藏汞的初步研究. *Kaogu考古*, 1983:07, 659–663.

Clark, J. E. 1986. From Mountains to Molehills: A critical review of Teotihuacan's obsidian industry. In B. Isaac, *Economic aspects of prehispanic Highland Mexico* (pp. 23–74). Greenwich, Conn: JAI Press.

Clark, J. E. and W. Parry 1990. Craft Specialisation and Cultural Complexity. *Research in Economic Anthropology 12*, pp. 289–346.

Collaborative Archaeological Team of the Early Qin Culture早期秦文化联合考古队 2008. Excavation of a sacrificial site in Dabuzishan region in Lixian County, Gansu province in 2006 2006年甘肃礼县大堡子山祭祀遗迹发掘简报. *Wenwu文物*, 2008:11, 14–29.

Coren, C., L. M. Ward and J. T. Enns 1994. *Sensation and perception.* Texas: Harcourt Brace, Forth Worth.

Costin, C. 1986. *From Chiefdom to Empire State: ceramic economy among the prehispanic Wanka of Highland Peru. PhD Thesis.* Los Angeles: University of California.

Costin, C. 1991. Craft specialisation: issues in defining, documenting, and explaining the organisation of production. In M. Schiffer, *Archaeological Method and Theory Vol. 3* (pp. 1–56). Tucson and London: University of Arizona Press.

Costin, C. 2001. Craft production systems. In G. M. Price, *Archaeology at the Millennium: a sourcebook* (pp. 273–328). New York: Kluwer Academic/Plenum Publishers.

Costin, C. and M. Hagstrum 1995. Standardisation, labour investment, skill, and the organisation of ceramic production in late Prehispanic highland Peru. *American Antiquity*, 619–39.

Dang Shixue党士学 1987. A discussion about dagger-axes and halberds 戈戟小议. *Wenbo 文博*, 1987–01, 67–69.

Davis, J. L. and H. Lewis 1985. Mechanization of pottery production: a case study from the Cycladic Islands. In K. A. Stech, *Prehistoric production and exchange: the Aegean and Eastern Mediterranean* (pp. 79–92). Los Angeles: Institute of Archaeology, University of California.

Duan Qingbo段清波 2011. *A Research on the Emperor Qin Shihuang's Mausoleum Complex*秦始皇帝陵研究. Beijing北京: Peking Univeristy Press北京大学出版社.

Earle, T. 1981. Comment on P. Rice, Evolution of specialisation pottery production: a trial model. *Current Anthropology 22(3)*, 230–31.

Eerkens, J. 2000. Practice Makes Within 5% of Perfect: The Role of Visual Perception, Motor Skills, and Human Memory in Artefact Variation and Standardisation. *Current Anthropology 41(6)*, 663–668.

Eerkens J. W. and R. L. Bettinger 2001. Techniques for assessing standardisation in artefact assemblages: can we scale material variability? *American Antiquity 66*, 493–504.

Eerkens, J. W. and C. P. Lipo 2005. Cultural transmission, copy errors, and the generation of variation in material culture and the archaeological record. *Journal of Anthropological Archaeology 24*, 316–334.

Eerkens, J. W. and C. P. Lipo 2007. Cultural transmission theory and the archaeological record: providing context to understanding variation and temporal changes in material culture. *Journal of Archaeological Research 15*, 239–274.

Falkenhausen, L. von 2006. *Chinese Society in the Age of Confucius(1050–250 BC): The Archaeological Evidence.* Los Angeles: Cotsen Institute of Archaeology, UCLA.

Feinman, G., Kowalewski, S. and R. Blanton 1984. Modelling Ceramic Production and Organisational Change in the Pre-hispanic Valley of Oaxaca, Mexico. In V. D. Pritchard, *The many Dimensions of Pottery: Ceramics in Archaeology and Anthropology* (pp. 295–338). Amsterdam: Albert Egges van Giffen Institute voor Prae-en Protohistorie, University of Amsterdam.

Frankel, D. 1988. Pottery Production in Prehistoric Bronze Age Cyprus. *Journal of Mediterranean Archaeology 1(2)*, 27–55.

Franklin, U. 1983. The beginning of metallurgy in China: a comparative approach. In G. Kuwayama (ed.), *The Great Bronze Age of China: A Symposium* (pp. 94–99). Seattle: University of Washington Press.

Gansu Institute of Archaeology 甘肃省文物考古研究所, National Museum of China 中国国家博物馆, School of Archaeology and Museology, Peking University 北京大学考古文博院 et al. 2008. Archaeological investigation on the upper reaches of Xihanshui river西汉水上游考古调查报告. Beijing北京: Wenwu Press 文物出版社.

Gansu Institute of Archaeology 甘肃省文物考古研究所 2009. *The bamboo slips discovered from Fangmatan, Tianshui* 天水放马滩竹简. Beijing北京: Zhonghuashuju中华书局.

Gao Zhixi 高至喜 1964. The Warring States Period tombs found Changsha and Changde — related questions about the triggers and crossbows 记长沙、常德出土弩机的战国墓——兼谈有关弩机、弓矢的几个问题. *Wenwu文物*, 1964:06, 33–45.

Gelfand, A. E., Diggle, P. J., Fuentes, M. and P. Guttorp 2010. *Handbook of Spatial Statistics.* London: CRC/Taylor and Francis.

Gordon, A. D. (1981). *Classification: Methods for the Exploratory Analysis of Multivariate Data.* London: Chapman and Hall.

Groover, M. P. 2010. *Fundamentals of Modern Manufacturing: Materials, Processes, and Systems, 4th edition.* New Jersey: Prentice-Hall Inc.

Guo Shuzheng郭淑珍 and Wang Guancheng王关成 2000. *Military history of the Qin period 秦军事史.* Xi'an西安: Shaanxi People Education Press 陕西人民教育出版社.

Hagstrum, M. 1985. Measuring Prehistoric Ceramic Craft Specialisation: a test case in the American southwest. *Journal of Field Archaeology 12(1)*, 65–67.

Han Rubin 韩汝玢 Ma Zhaozeng马肇曾 Wang Zengjun 王曾隽and Ke Jun 柯俊 1983. A study on the corrosion surface of the bronze arrows unearthed from the pit of the Qin Shihuang's Terracotta Army 秦始皇陶俑坑出土的铜镞表面氧化层的研究. *Ziran kexueshi yanjiu 自然科学史研究 (1983)*:2, 295–302.

Hardin, M. 1979. The cognitive basis of productivity in a decorative art style: Implications of an ethnographic study for archaeologists' Taxonomies. In C. Kramer (ed.), *Ethnoarchaeology: Implications of ethnography for archaeology* (pp. 75–101). New York: Columbia University Press.

Hearn, M. 1980. The Terracotta Army of the First Emperor of Qin – An exhibition from the People's Republic of China. In W. Fong (ed.), *The Great Bronze Age of China* (pp. 353–373). New York: The Metropolitan Museum of Art.

Hebei Cultural Relics Managing Department 河北省文物管理处 1975. An Excavation Report on tomb No.44 at Yanxiadu, Yi county, Hebei province河北易县燕下都44号墓发掘报告. *Kaoku 考古*, 1975:04, 228–240.

Hebei Museum 河北省博物馆 and Taixi Archaeology Team 文管处台西考古队1977. *The Gaocheng Taixi site from the Shang Dynasty 藁城台西商代遗址*. Beijing北京: Wenwu Press文物出版社.

Hill, J. 1979. Individual Variability in Ceramics and the Study of Prehistoric Social Organisation. In J. H. Gunn (ed.), *The Individual in Prehistory* (pp. 55–108). New York: Academic Press.

He Tangkun 何堂坤. 1996. Analysis of chromium found on the surface of several bronze objects 几件表面含铬青铜器的分析. *Kaogu 考古* 1996:7, 71–76.

Hodder, I. 1983. *Symbols in action.* Cambridge: Cambridge University Press.

Hodder I. and C. Orton 1976. *Spatial Analysis in Archaeology.* Cambridge: Cambridge University Press.

Huang Shengzhang 黄盛璋 1983. A reinterpretation on Sigong inscription 寺工新考. *Kaoku 考古*, 1983:09, 829–833.

Huang Shengzhang 黄盛璋 1990. Inscriptions on the bronze weapons for the Qin Terracotta Warriors and related systems秦俑坑出土兵器铭文与相关制度发复. *Wenbo 文博*, 1990:05, 63–71.

Ilian, J., Penttinen, A., Stoyan, H. and D. Stoyan 2008. *Statistical Analysis and Modelling of Spatial Point Patterns.* New York: Wiley-Interscience.

Institute and Team 陕西省考古研究所始皇陵秦俑坑考古发掘队 1988. *Excavation Report on Pit 1 of the Qin Terra-cotta Warriors and Horses 1974–1984 秦始皇陵兵马俑坑一号坑发掘报告1974–1984.* Beijing北京: Wenwu Press文物出版社.

Institute of Archaeology in Chinese Academy of Social Science中国社会科学院考古研究所 1999. Excavation Report of *Yanshi Erlitou1959~1978*偃师二里头1959年~1978年考古发掘报告. Beijing北京: Chinese Encyclopedia Press中国大百科全书出版社.

Jiang Wenxiao 蒋文孝 and Liu Zhancheng 刘占成 2006. Research on the newly discovered bronze dagger-axe and halberd from the pit of the Terracotta Army 秦俑坑新出土的铜戈、戟研究. *Wenwu 文物, 3*, 66–71.

Ledderose, L. 2000. *Ten Thousand Things - Module and Mass Production in Chinese Art.* Princeton: Princeton University Press.

Khayutina, M. (eds). 2013. *Qin: The Eternal Empire and his Terracotta Warriors.* Zürich: Neue Züricher Zeitung Publishing.

Li Xiuzhen李秀珍 and Gao Jun高俊 2010. Standardisation and limitations in production of the Qin bronze triggers: from the inscription evidence从青铜弩机铭文年秦兵器生产标准化的局限性. *Wenbo*文博, Volume 2: 45–48.

Li, X., Martinón-Torres, M., Meeks, N. D., Xia, Y. and K. Zhao 2011. Inscriptions, filing, grinding and polishing marks on the bronze weapons from the Qin Terracotta Army in China. *Journal of Archaeological Science 38*, 492–501.

Li, X. J., Martinón-Torres, M., Meeks N. and Xia, Y. 2012. Scanning electron microscopy imaging of tool marks on Qin bronze weapons using silicone rubber impressions, in N. Meeks, C. Cartwright, A. Meek and A. Mongiatti (eds.), *Historical Technology, Materials and Conservation: SEM and Microanalysis*, 62–68. London: Archetype and The British Museum.

Li, X., Bevan, A., Martinón-Torres, M., Rehren, Th., Cao, W. Yin, X., and K. Zhao 2014. Crossbows and imperial craft organisation: the bronze triggers of China's Terracotta Army. *Antiquity*, vol.88, 126–140.

Li, X., Bevan, A., Martinón-Torres, M., Yin, X., and K. Zhao 2016. Marking practice and the making of the Qin Terracotta Army. *Journal of Anthropological Archaeology*, vol.42, 169–183.

Li Xueqin李学勤 2011. The location of the Qin's early settlement - *Zhuwu* 谈秦人初居"邾吾"的地理位置. Chutuwenxian出土文献, vol.2: 1–7.

Li, Y. 2006. *The Anyang Bronze Foundries: Archaeological Remains, Casting Technology, and Production Organisation.* ProQuest Information and Learning Company, United States.

Li, Y. 2007. Co-craft and Multicraft: section-mold casting and the organisation of craft production at the Shang capital of Anyang. In L. Shimada (ed.), *Craft production in complex society: multicraft and producer perspective* (pp. 184–223). Salt Lake City: The University of Utah Press.

Lin Jianming林剑鸣 1981. *A history of the Qin 秦史稿.* Shanghai上海: Shanghai Renmin Press上海人民出版社.

Lindesay, W. and B. Guo 1998. *The Terracotta Army of the First Emperor of China.* Hong Kong: The Guidebook Company.

Lin, J. and Li, X. 2018. *China's First Emperor and The Qin Terracotta Warriors*. Liverpool: National Museum of Liverpool Press.

Liu Zhancheng 刘占成 1982. Bronze lances discovered in the pit of Qin Terracotta Warriors秦俑坑出土的铜铍. *Wenwu 文物*, 1982: 03, 12–14.

Liu Zhancheng 刘占成 1986. A preliminary investigation on the Qin crossbows秦俑坑弓弩试探. *Wenbo 文博*, 1986:04, 66–70.

Loewe, M. 2007. The First Emperor and the Qin Empire. In J. Portal (ed.), *The First Emperor - China's Terracotta Army (pp58–79)*. London: The British Museum Press.

Loewe, M. 2010. Social distinctions, groups and privileges. In Michael Nylan and Michael Loewe (eds.), *China's Early Empires: A re-appraisal,* 296–307. Cambridge: Cambridge University Press.

Longacre, W. 1999. Standardisation and specialisation: what is the link? In S. J. Feinman (ed.), *Pottery and people* (pp. 44–58). Salt Lake City: University of Utah Press.

Longacre, W. A., Kvamme, K. L. and M. Kobayashi 1988. Southwestern Pottery Standardisation: an ethnoarchaeological view from the Philippines. *Kiva 53*, 101–112.

Ye Wansong 叶万松 1974. Excavation Report on pits of the chariots and horses from the Warring States Era at Zhongzhoulu, Luoyang洛阳中州路战国车马坑. *Kaogu 考古*, 1974:03, 171–178.

Ma, C. 1986. *Ancient Chinese Bronzes.* Oxford: Oxford University Press.

Martinón-Torres, M., Li, X., Bevan, A., Xia, Y.. Zhao, K. and Rehren Th,. 2014a. Forty thousand arms for a single emperor: from chemical data to the labour organization behind the bronze arrows of the Terracotta Army, *Journal of Archaeological Method and Theory*, Volume 21, 534–562.

Martinón-Torres, M., Li Xiuzhen, Bevan, A., Xia Yin, Zhao Kun. and Th. Rehren 2014b. Making Weapons for the First Emperor: archaeometallurgical analyses of bronze weapons from the Terracotta Army为秦始皇打造兵器：秦俑坑兵器的冶金考古学分析. In Cao, W. and Rehren, Th. (eds.). *Proceedings of International Symposium on Qin Period Metallurgy and its Social and Archaeological Context秦时期冶金考古国际学术研讨会论文集*, 38–46. Beijing北京: Science Press 科学出版社.

Martinón-Torres, M., Li, X., Xia, Y., Benzonelli, A., Bevan, A., Ma, Sh., Huang, J., Wang, L., Lan, D., Liu, J., Liu, S., Zhao, Zh., Zhao, K. and Th. Rehren. 2019. Surface chromium on Terracotta Army bronze weapons is neither an ancient anti-rust treatment nor the reason for their good preservation. *Scientific Report* (2019) 9:5289.

Matthews, R. A. 1996. *Fordism, Flexibility, and Regional Productivity Growth.* New York and London: Garland Publishing.

Miller, D. 1985. *Artifacts as Categories.* Cambridge: Cambridge University Press.

Muller, J. 1984. Mississippian specialisation and salt. *American Antiquity 49(3)*, 489–507.

Museum of Emperor Qin Shihuang's Terracotta Army 秦始皇兵马俑博物馆 and Institute of Archaeology of Shaanxi Province 陕西省考古研究所 1998. *The Bronze Chaiots and Horses Unearthed from the Qin Shihuang Mausoleum - An Excavation Report 秦始皇陵铜车马发掘报告*. Beijing北京: Wenwu Press文物出版社.

Museum of Emperor Qin Shihuang's Terracotta Army 秦始皇兵马俑博物馆 1998. *The Bronze Chaiots and Horses Unearthed from the Qin Shihuang Mausoleum – A Restoration Report.* Beijing北京: Wenwu Press文物出版社.

Museum of Emperor Qin Shihuang's Terracotta Army 秦始皇兵马俑博物馆 2009. *Excavation Report on Pit 2 of the Qin Terra-cotta Warriors and Horses*秦始皇陵二号兵马俑坑发掘报告. Beijing北京: Science Press 科学出版社.

Orton, C. 2000. *Sampling in Archaeology.* Cambridge: Cambridge University Press.

Orton, C. 2004. Point Pattern Analysis Revisited. *Archaeologia e Calcolatori 15*, 299–315.

Pines, Y. 2009. *Envisioning eternal empire: Chinse political thought of the Warring States Era*. Honolulu: University of Hawai'i Press.

Pines, Y. 2012. From historical evolution to the end of history: past, present and future from Shang Yang to the First Emperor. In Paul R. Goldin (ed.) *Dao companion to the philosophy of Han Fei*, 25–45. Berlin: Springer Verlag.

Pines, Y., von Falkenhausen, L., Shelack, G., and Yates, R. 2014. *Birth of an empire: The state of Qin revisited.* University of California Press.

Portal, J. 2007. *The First Emperor - China's Terracotta Army.* London: The British Museum Press.

Qin Bing秦兵 and Zhang Zhancheng张占民 1983. A preliminary investigation on bronze lances长铍初探. *Proceedings of the First Annual Meeting of Shaanxi Archaeological Association 陕西省考古学会第一次年会论文集*. Xi'an西安: Shaanxi Renmin Press陕西人民出版社.

Qin Ming 秦鸣 1975. A preliminary research on battle formation and weapons discovered in the Pit of the Qin Terracotta Warriors秦俑坑兵马俑军阵内容及兵器试探. *Wenwu 文物*, 1975:11, 19–23.

Quinn, P., Zhang, S., Xia, Y., and Li, X. 2017. Building the Terracotta Army: Ceramic craft technology and organisation of production at Qin Shihuang's mausoleum complex. *Antiquity,* 91 (358), 966–979.

Quinn, P., Yang, Y., Xia, Y., Ma, Sh., Wilke, D., and Li, X. (in press). Micro-geochemical Evidence for the Production and Logistics Management of Emperor Qin Shihuang's Terracotta Army, China. *Archaometry* Submission.

Rawson, J. 2007. The first emperor's tomb: the afterlife universe. In J. Portal (ed.), *The First Emperor - China's Terracotta Army* (pp. 114–151). London: The British Museum Press.

Research Committee on the Ancient Chinese Weapons 1995. *Zhongguo gudai bingqi (Ancient Chinese Weapons).* Xian: Shaanxi People Press.

Rice, P. M. 1981. Evolution of Specialisation Pottery Production: a trial model. *Current Anthropology 22(3)*, 219–240.

Rice, P. M. 1984. *Pottery analysis: a sourcebook.* Chicago: University of Chicago Press.

Rice, P. M. 1991. Specialisation, standardisation, and diversity: a retrospective. In R. B. Lange (ed.), *The ceramic legacy of Anna O. Shepard.* Colorado: University of Colorado Press.

Ripley, B. 1977. Modelling Spatial Patterns. *Journal of the Royal Statistical Society, Series B (Methodological), 39*, 172–212.

Robert, C. P. and G. Casella 2004. *Monte Carlo Statistical Methods.* New York: Springer.

Roux, V. 2003. Ceramic standardisation and intensity of production: quantitying degrees of specialisation. *American Antiquity 68*, 768–82.

Sage, S. 1992. *Ancient Sichuan and the unification of China*. Albany: State University of New York Press.

Schiffer, M. J. and J. M. Skibo 1997. Explanation of Artifact Variability. *American Antiquity 62*, 27–50.

Shaanxi Institute of Archaeology 陕西省考古研究所 2004. *Excavations of the Qin Capital Xianyang* 秦都咸阳考古报告. Beijing北京: Science Press科学出版社.

Shaanxi Institute of Archaeology 陕西省考古研究所 and Museum of Emperor Qin Shihuang's Terracotta Army秦始皇兵马俑博物馆 2000. *Excavation of the Emperor Qin Shihuang's Mausoleum Complex 1999* 秦始皇帝陵园考古报告1999. Beijing北京: Science Press科学出版社.

Shaanxi Institute of Archaeology 陕西省考古研究所 and Museum of Emperor Qin Shihuang's Terracotta Army秦始皇兵马俑博物馆 2006. *Excavation of the Emperor Qin Shihuang's Mausoleum Complex 2000* 秦始皇帝陵园考古报告2000. Beijing北京: Wenwu Press文物出版社.

Shaanxi Institute of Archaeology 陕西省考古研究所 and Museum of Emperor Qin Shihuang's Terracotta Army秦始皇兵马俑博物馆 2007. *Excavation of the Emperor Qin Shihuang's Mausoleum Complex 2001–2003*秦始皇帝陵园考古报告2001–2003. Beijing北京: Wenwu Press文物出版社.

Shelach, G. and Y. Pines 2005. Power, identity and ideology: reflections on the formation of the State of Qin (770–221 BC). In Miriam Stark (ed.) *An Archaeology of Asia,* 202–230. Oxford: Blackwell Publishers.

Shelach, G. 2014. Collapse or transformation? Anthropological and archaeological perspectives on the fall of Qin. In Pines et al., (ed.) *Birth of an empire: The state of Qin revisited.* University of California Press.

Shennan, S. 1997. *Quantifying Archaeology.* Edinburgh: Edinburgh University Press.

Shi Yan石岩 2008. *Research on Bronze Arrows of pre-Qin Dynasty in Northern China 中国北方先秦时期青铜镞研究.* Haerbin哈尔滨: Helongjiang University Press黑龙江大学出版社.

Shuihudi Qinmu Zhengli Xiaozhu 睡虎地秦墓整理小组 1978. *Shuihudi Qin bamboo slips 睡虎地秦墓竹简.* Beijing北京: Wenwu Press文物出版社.

Sichuan Committee of Cultural Relics四川省文物管理委员会 1956. An Excavation Report on tomb No.172 at Yangzishan, Chengdu成都羊子山第172号墓发掘报告. *Kaoguxuebao 考古学报*, 1956:04.

Sichuan Museum 四川省博物馆 1974. An excavation report on the Warring States Periods tombs found in Xiaotianxi, Fuling, Sichuan Province 四川涪陵地区小田溪战国土坑墓清理简报. *Wenwu 文物*, 1974:05, 13–21.

Sinopoli, C. 1988. The organisation of craft production at Vijayanagara, South India. *American Anthropologist 90(3)*, 580–597.

Sinopoli, C. 1991. *Approaches to Archaeological Ceramics.* New York: Plenum Press.

Snaft, C. 2013. Qin Government: structures, principles and practices. In M. Khayutina (ed.) *Qin: The eternal emperor and his terracotta warriors*. Zurich: Bernishes Historisches Museum.

Snaft, C. 2014. *Communication and cooperation of early imperial China: pubilizing the Qin Dynasty*. Albany: State University of New York Press.

Stark, B. 1991. *Problems in Analysis of Standardisation and Specialisation in Pottery.* Tempe: Arizona State University.

Stein, G. 1996. Producers, Patrons, and Prestige: Craft Specialists and Emergent Elites in Mesopotamia from 5500–3100BC. In B. Wailes (ed.), *Craft Specialisation and Social Evolution: In Memory of V. Gordon Childe* (pp. 26–38). Philadelphia: University of Pennsylvania.

Sun, Z. 2008. *Craft production in the Western Zhou dynasty - a case study of a jue-earrings workshop at the predynastic capital site, Zhouyuan, China.* BAR International Series 1777. Oxford: BAR Publishing.

Teghtsoonian, R. 1971. On the exponents in Stevens' law and the constant in Ekman's law. *Psychological Review 78*, 71–80.

Teng Mingyu滕铭予 2003. A*n Archaeological Examination of Qin Culture – From Vassal State to Empire 秦文化：从封国到帝国的考古学观察.* Beijing北京: Xueyuan Press学苑出版社.

Teng, M. 2014. From vassal state to empire: an archaeological examination of Qin culture. In Pines et al., (ed.) *Birth of an empire: The state of Qin revisited.* University of California Press.

Torrence, R. 1986. *Production and exchange of stone tools: prehistoric obsidian in Aegean.* Cambridge: Cambridge University Press.

Tosi, M. 1984. The notion of craft specialisation and its representations in the archaeological record of early states in the Turanian Basin. In M. Spriggs (ed.), *Marxist perspectives in archaeology* (pp. 22–52). Cambridge: Cambridge University Press.

Tsuruma, Kazuyuki 鹤间和幸(eds.) 2004. *Exhibition of the Life Size Terracotta Warriors in Janpan日本大兵马俑展.* Shenzhen深圳: Shenzhen International Printing Press深圳国际出版社.

Underhill, A. P. 2002. *Craft production and social change in northern China.* New York: Kluwer Academic/ Plenum Publishers.

Wang Xueli王学理 1980a. A research into the scientific achievement based on the Qin bronze weapons from the pits of Qin Terracotta Army秦俑坑出土青铜兵器的科技成就管窥. *Kaoguyuwenwu 考古与文物*, 1980: 03, 99–107.

Wang Xueli王学理 1980b. The Bronze Swords and Arrows from the Pit of the Qin Terracotta Warriors秦俑坑青铜剑和青铜镞. *Xibeidaxuexuebao 西北大学学报*, 1980:01, 107–115.

Wang Xueli王学理 1983. A Preliminary Discussion about the Weapons for the Qin Terracotta Warriors 秦俑兵器刍论. *Kaokuyuwenwu 考古与文物*, 1983:04, 59–80.

Wang Xueli王学理 1987. A preliminary investigation on the standardised military products in the Qin dynasty 秦代军工生产的标准化的初步考察. *Kaoguyuwenwu 考古与文物*, 1987:05, 57–67.

Wang Xueli王学理 1990. A military battle formation of the Qin dynasty一幅秦代的陈兵图：论秦俑坑的性质及其编成. *Wenbo 文博*, 1990:05, 21–35.

Wang Xueli王学理 1994. *Monographic study on the Qin Terracotta Warriors 秦俑专题研究.* Xian西安: Sanqin Press三秦出版社.

Wang Xueli王学理 1999. *Notes of the Qin Capital Xianyang 咸阳帝都记.* Xian西安: Shaanxi Renmin Press陕西人民出版社.

Wild, R. 1975. *Work Organisation: A Study of Manual Work and Mass Production.* London: John & Sons.

Wright, R. 1989. *Doing Multivariate Analysis and Prehistory: Handling Large Datasets with MV-ARCH.* Sydney: Department of Anthropology, University of Sydney.

Xianyang Cultural Relics Bureau 咸阳文物局 2002. *Treasures found in Xianyang 咸阳文物精华.* Beijing 北京: Wenwu Press文物出版社.

Xia, Yin, Li, Xiuzhen, Martinón-Torres, M., Rehren, Th., Zhao, Kun, and Zhou Tie. 2014. An initial analysis of the arrow tangs and swords from the Qin Terracotta Army秦俑坑出土铜镞铤的金相及电镜能谱初步分析. In Cao, W. and Rehren, Th. (eds.). *Proceedings of International Symposium on Qin Period Metallurgy and its Social and Archaeological Context 秦时期冶金考古国际学术研讨会论文集,* 38–46. Beijing北京: Science Press科学出版社.

Yang Hong 杨泓 1980. *Research on the Ancient Chinese Weapons 中国古代兵器论丛.* Beijing北京: Wenwu Press文物出版社.

Yang Hong. 2017. Military Armaments of the Qin and Han. In Sun Zhixin (eds.) *Age of Empires - Art of the Qin and Han Dynasties*. New York: The Metropolitan Museum of Art.

Yang, X. 2000. *Reflections of early China-decors,pictographs, and pictorial inscriptions.* Seattle and London: The Nelson-Atkins Museum of Art in association wtih the University of Washington Press.

Yates, R. 2007. The rise of Qin and the military conquest of the Warring States. In J. Portal (ed.), *The First Emperor - China's Terracotta Army* (pp. 30–57). London: The British Museum Press.

Yates, R. 2009. Law and the military in early China. In Nicola Di Cosmo (ed.) *Military Culture in Imperial China,* 23–44. Cambridge MA: Harward University Press.

Yuan Weihua袁卫华, Yuan Zhongyi袁仲一, Zhao Peizhi 赵培智, Zuo Zhenxi禚振西, Wang Xueli王学理 and Wei Yonghui韦永辉 1981a. A research into the standardisation of Qin bronze weapons found in the pits of the Terracotta Warriors秦俑坑青铜兵器工艺标准化试析. *Shaanxi Standardisation 陕西标准化*, 1981:02, 5–9.

Yuan Weihua袁卫华, Yuan Zhongyi袁仲一, Zhao Peizhi赵培智, Zuo Zhenxi禚振西, Wang Xueli王学理 and Wei Yonghui韦永辉 1981b. A research on standardisation of production during the Qin and Han period论秦汉时期标准化. *Shaanxi Standardisation 陕西标准化*, 1981:02, 10–19.

Yuan Zhongyi 袁仲一 1984. A comprehensive research on the Qin weapons' inscription and governmental control on their production秦中央督造刻辞综述. *Kaoguoyuwenwu 考古与文物*, 1984: 05, 101–112.

Yuan Zhongyi 袁仲一 1990. *Research on the Terracotta Warriors and Horses from the Mausoleum of Emperor Qin Shihuang 秦始皇陵兵马俑研究.* Beijing北京: Wenwu Press文物出版社.

Yuan Zhongyi 袁仲一 2002. *Archaeological discoveries and research on the Emperor Qin Shihuang's mausoleum 秦始皇陵考古发现与研究.* Xian西安: Shaanxi Renmin Press陕西人民出版社.

Yuan, Zhongyi 袁仲一2014. Archaeological Discoveries and Research on the Qin Terracotta Warriors秦兵马俑的考古发现与研究. Beijing北京: Wenwu Press文物出版社.

Yuan Zhongyi 袁仲一 and Cheng Xuehua 程学华 1986. Manufacturing techniques of the bronze weapons and chariots discovered from the Emperor Qin Shihuang's tomb complex秦陵出土的铜兵器和铜车马的制作工艺. *Confernce proceedings of Committee of Shaanxi Science and Technology 陕西省科学技术史学会论文集.* X'ian西安: Xian Jiaotong University Press西安交通大学出版社.

Yuang, Zhongyi 袁仲一 and Liu, Yu 刘钰 2009. New Collection and Interpretation on the Qin Pottery Inscriptions秦陶文新编. Beijing北京: Wenwu Press文物出版社.

Yunmeng Shuihudi Qinmu Bianxiezu 云梦睡虎地秦墓编写组 1981. *An excavation report on the Qin tomb found at Yunmeng Shuihudi* 云梦睡虎地秦墓. Beijing北京: Wenwu Press文物出版社.

Zhang Hai 张海 and Qin Zhen秦臻 2014. Landscape Archaeological Research on Iron-making Ruins of the Warring States period and Qin, Han Dynastiesin, in Wugang, Henan and Xiping Area河南舞钢、西平地区战国秦汉冶铁遗址的景观考古学研究. In Cao, W. and Rehren, Th. (eds.). *Proceedings of International Symposium on Qin Period Metallurgy and its Social and Archaeological Context 秦时期冶金考古国际学术研讨会论文集,* 38–46. Beijing北京: Science Press科学出版社..

Zhang, X., Yuan, S. Wu, Y., Guo, B. & Han, J. 2011. Research on the corrosion of bronze weapons from the pits of the terracotta warriors. *Mater. Res. Soc. Symp. Proc.* 1319.

Zhao Huacheng赵化成 1987. A new clue for searching the origin of the Qin culture寻找秦文化渊源的新线索. *Wenbo*文博，vol1: 7–17.

Zhao Huacheng赵化成 1989. A study on cultural relation between the Qin and Jiangrong from an archaeological perspective at the East of Gansu Province甘肃东部秦与姜戎文化的考古学探索. In Yu Weichao俞伟超, (ed.)Theory and practice of archaeological typology考古类型学的理论与实践. Beijing北京: Wenwu Press文物出版社.

Zhao, H. 2014. New explorations of early Qin culture. In Pines et al., (ed.) *Birth of an empire: The state of Qin revisited.* University of California Press.

Zhao Huangcheng 赵化成 Wang Hui 王辉 and Wei Zheng 韦正 2008. A discussion of a sacrificial site in Dabuzishan region in Lixian county, Gansu province礼县大堡子山秦子乐器坑相关问题探讨. *Wenwu 文物,* 2008:11, 54–66.

Appendix 1

Spatial and metric data on triggers: Part A

Trigger ID	Spatial ID	X	Y	Generic location	Location error	A1	A2	A3	A Type
901	148	34.105669	-26.529079	Pit 1 main	exact	7.86	2.02	1.84	2
902	77	33.159938	-25.57195	Pit 1 main	exact	7.81	2.04	1.75	2
904	91	20.860784	-25.906234	Pit 1 main	exact	7.86	2.04	1.97	2
905	86	14.533779	-26.714721	Pit 1 main	exact	7.99	2.11	1.87	2
906	99	13.630381	-26.243872	Pit 1 main	exact	7.68	1.97	1.94	4
907	102	11.591567	-25.640049	Pit 1 main	exact	7.80	2.08	1.96	2
908	58	34.201382	13.620131	Pit 1 main	exact	7.82	2.02	1.88	2
909	149	34.3053	16.990327	Pit 1 main	exact	7.83	2.07	1.85	2
910	101	10.696355	-25.812038	Pit 1 main	exact	7.65	2.04	1.78	2
911	63	33.501826	3.116224	Pit 1 main	exact	7.89	2.04	1.93	2
912	52	34.392391	21.465132	Pit 1 main	exact	7.64	1.98	1.96	4
913	59	32.710583	9.192165	Pit 1 main	poor	8.06	2.10	1.71	2
914	78	32.952122	-21.474925	Pit 1 main	poor	7.92	2.18	1.97	2
915	150	33.606594	5.511584	Pit 1 main	poor	7.65	1.47	1.49	1
918	152	31.610291	-8.585008	Pit 1 main	exact	8.24	2.04	2.03	2
919	118	18.900838	-4.311712	Pit 1 main	exact	7.94	2.02	1.89	2
920	154	19.233212	-3.294805	Pit 1 main	exact	7.87	2.04	2.04	2
921	155	18.933766	-3.095175	Pit 1 main	exact	7.99	2.05	2.05	2
922	137	18.728001	6.001707	Pit 1 main	exact	8.06	1.54	1.59	1
923	127	19.851817	7.723742	Pit 1 main	exact	8.18	1.62	1.62	1
924	131	19.831816	6.456726	Pit 1 main	exact	7.93	1.58	1.58	1
925	128	19.719986	7.316134	Pit 1 main	exact	7.88	1.54	1.57	1
926	130	19.610153	6.807441	Pit 1 main	exact	7.69	1.61	1.63	1
927	136	18.637351	6.658974	Pit 1 main	exact	7.93	1.69	1.60	1
928	138	20.810811	11.600437	Pit 1 main	exact	7.93	1.63	1.61	1
929	135	18.91818	8.419803	Pit 1 main	exact	8.11	1.61	1.59	1
930	123	21.925709	6.398791	Pit 1 main	exact	8.11	2.08	1.99	2
931	124	22.19033	5.721752	Pit 1 main	exact	8.14	2.07	2.05	2
932	126	22.331048	7.166956	Pit 1 main	exact	7.96	1.53	1.64	1
933	163	21.431484	6.557492	Pit 1 main	exact	7.58	1.48	1.54	1
934	132	20.086295	5.674169	Pit 1 main	exact	7.97	1.56	1.59	1
935	133	20.578868	5.394795	Pit 1 main	exact	7.95	1.57	1.59	1
936	25	21.448853	29.377367	Pit 1 main	poor	7.83	2.08	1.95	2

Trigger ID	Spatial ID	X	Y	Generic location	Location error	A1	A2	A3	A Type
937	44	6.475602	6.475602	Pit 1 main	poor	7.75	1.47	1.54	1
938	24	18.762581	29.917961	Pit 1 main	poor	8.12	2.09	1.74	2
939	156	28.513778	12.35631	Pit 1 main	exact	7.83	1.53	1.61	1
940	141	20.778356	13.166239	Pit 1 main	exact	8.05	1.59	1.58	1
943	35	22.496752	28.661955	Pit 1 main	exact	8.06	2.09	1.91	2
944	32	21.728948	23.548804	Pit 1 main	exact	8.08	2.09	2.01	2
946	40	26.355985	28.877476	Pit 1 main	exact	7.89	1.95	1.72	4
947	36	22.881719	28.031499	Pit 1 main	exact	8.01	2.02	1.90	2
948	39	25.235039	28.33449	Pit 1 main	exact	8.07	2.02	1.89	2
949	38	24.125749	28.355277	Pit 1 main	exact	7.92	2.01	1.87	2
950	157	23.151164	0.820272	Pit 1 main	exact	8.19	2.26	2.02	3
951	41	32.020779	29.775745	Pit 1 main	exact	7.62	2.01	1.89	2
952	43	32.420465	29.891975	Pit 1 main	exact	7.53	1.45	1.53	1
953	48	34.269167	27.43019	Pit 1 main	exact	8.15	1.59	1.62	1
954	47	33.964206	29.27371	Pit 1 main	exact	7.48	1.47	1.59	1
956	54	32.724943	19.958606	Pit 1 main	exact	8.04	2.07	2.01	2
957	169	33.918877	22.809477	Pit 1 main	poor	8.26	2.28	2.01	3
958	53	34.683898	20.732773	Pit 1 main	exact	8.02	2.07	1.92	2
959	46	33.960055	29.856813	Pit 1 main	exact	8.23	2.26	2.02	3
960	31	22.718784	24.068895	Pit 1 main	exact	7.92	2.04	1.93	2
961	34	22.765991	23.168717	Pit 1 main	exact	8.18	2.16	1.96	2
962	51	33.144621	21.865315	Pit 1 main	poor	8.03	1.99	1.87	2
963	28	20.47937	24.274925	Pit 1 main	exact	8.18	2.19	1.96	2
964	144	19.549687	19.002285	Pit 1 main	exact	8.09	2.16	1.98	2
965	50	34.121289	20.732773	Pit 1 main	poor	8.16	2.22	2.02	3
966	171	33.705328	23.250595	Pit 1 main	poor	8.05	1.52	1.51	1
967	222	29.41212	29.397728	Pit 1 main	poor	8.14	2.08	1.97	2
968	176	34.554697	16.824692	Pit 1 main	exact		2.16		2
969	37	24.039188	29.137737	Pit 1 main	exact	8.06	2.08	1.96	2
970	223	23.086751	29.521452	Pit 1 main	poor	7.89	2.09	1.75	2
971	57	32.874234	14.9208	Pit 1 main	exact	7.93	2.02	1.86	2
972	167	32.468576	25.632531	Pit 1 main	exact	8.22	2.27	1.98	3
973	158	28.596817	28.4179	Pit 1 main	exact	8.06	1.61	1.61	1
974	142	22.614815	13.541946	Pit 1 main	exact	7.91	2.11	1.98	2
975	139	21.094665	12.103157	Pit 1 main	exact	8.17	2.25	2.05	3
976	68	31.321209	-4.804873	Pit 1 main	poor	7.73	2.05	1.75	2
977	60	34.221857	9.737418	Pit 1 main	poor	7.79	2.06	1.79	2
978				Pit 1 other	T4G2	7.84	2.06	1.87	2

Trigger ID	Spatial ID	X	Y	Generic location	Location error	A1	A2	A3	A Type
979	225	7.922879	-20.702595	Pit 1 main	poor	8.12	2.15	1.91	2
1471	55	32.463186	17.22433	Pit 1 main	poor	7.83	1.98	1.85	2
1472	56	33.109812	16.732523	Pit 1 main	poor	8.22	2.19	1.98	2
1498	208	34.213518	-11.978139	Pit 1 main	exact	8.28	2.16	1.92	2
1499	65	33.149953	1.631469	Pit 1 main	exact	7.97	2.02	1.99	2
1500	64	34.821564	2.017701	Pit 1 main	exact		2.06	1.88	2
1501	61	33.756327	8.628375	Pit 1 main	exact	7.85	1.53	1.65	1
1502	49	34.069362	26.573211	Pit 1 main	exact	7.98	1.51	1.51	1
1503	71	34.052426	-5.372814	Pit 1 main	exact	8.36	2.02	1.98	2
2231	93	20.117168	-26.579856	Pit 1 main	exact		2.13	1.99	2
2232	92	20.216574	-26.208776	Pit 1 main	exact	8.02	2.05	1.86	2
2233	94	24.311596	-26.461492	Pit 1 main	exact	8.01	2.11	1.96	2
2234	95	18.83128	-25.807518	Pit 1 main	poor	7.87	2.08	1.75	2
2235	98	19.708599	-26.650064	Pit 1 main	poor	8.07	2.02	1.76	2
2236	199	14.792157	-25.388635	Pit 1 main	exact	8.19	2.13	1.91	2
2237	186	12.670258	-25.510652	Pit 1 main	exact	8.04	2.11	1.98	2
2238	96	18.26273	-25.835152	Pit 1 main	exact	8.01	2.09	1.88	2
2239	195	11.19016	28.980629	Pit 1 main	exact	7.93	2.12	1.89	2
2240	215	11.924714	-25.653903	Pit 1 main	exact	8.04	2.14	1.87	2
2241	103	11.585001	-26.397574	Pit 1 main	exact	8.07	2.11	1.82	2
2242	104	10.059564	-25.779414	Pit 1 main	exact	7.91	2.09	1.78	2
2243	105	9.189363	-25.722035	Pit 1 main	exact	7.85	2.13	1.89	2
2244	106	8.311401	-26.031949	Pit 1 main	exact	7.86	2.06	1.79	2
2245	107	7.6723	-25.728396	Pit 1 main	exact	7.96	2.09	1.81	2
2246	187	6.763962	-25.666675	Pit 1 main	exact	8.04	2.11	2.02	2
2247	109	5.711013	-25.694485	Pit 1 main	exact	7.97	2.13	1.92	2
2248	110	4.925915	-26.395367	Pit 1 main	exact	8.16	2.13	1.92	2
2249	111	4.247799	-26.376172	Pit 1 main	exact	8.01	2.13	1.77	2
2250	200	3.114904	-26.362794	Pit 1 main	exact	8.21	2.13	1.97	2
2251	90	21.451586	-25.956809	Pit 1 main	poor	7.92	2.10	1.90	2
2252	82	33.560295	-24.307966	Pit 1 main	exact	7.95	2.14	1.91	2
2253	79	32.014427	-21.256277	Pit 1 main	exact	8.04	2.10	1.75	2
2254	72	32.511583	-19.93794	Pit 1 main	exact	7.81	2.03	1.79	2
2255	75	31.593245	-16.92337	Pit 1 main	exact	7.95	2.07	1.95	2
2256	166	32.299388	-16.419768	Pit 1 main	exact	8.21	2.24	2.02	3
2257	74	34.453956	-12.542946	Pit 1 main	exact	8.24	2.14	1.94	2
2258	70	32.969332	-9.826935	Pit 1 main	exact	8.15	2.09	1.99	2
2259	66	32.629434	2.921166	Pit 1 main	exact		1.99	1.88	4

Trigger ID	Spatial ID	X	Y	Generic location	Location error	A1	A2	A3	A Type
2260	201	23.660796	-20.290116	Pit 1 main	exact	8.06	2.13	1.84	2
2261	177	22.629265	-22.361014	Pit 1 main	exact	8.02	2.02	1.91	2
2262	224	8.031137	-21.475867	Pit 1 main	exact	8.02	2.09	1.79	2
2263	147	19.333946	-14.746488	Pit 1 main	exact	8.16	2.21	2.01	3
2264	115	18.571095	-10.172154	Pit 1 main	exact	8.19	2.17	1.94	2
2265	113	17.439096	-10.70497	Pit 1 main	exact	8.13	2.22	2.02	3
2266	69	30.182927	-9.290728	Pit 1 main	exact	7.73	2.05	1.94	2
2267	114	18.542612	-10.819296	Pit 1 main	exact	8.13	2.17	1.97	2
2268	121	20.73572	-4.652206	Pit 1 main	exact	7.97	2.02	1.93	2
2269	116	17.545465	-4.1724	Pit 1 main	poor	8.04	2.05	2.01	2
2270	119	19.180578	-5.365282	Pit 1 main	exact	7.98	2.05	1.95	2
2271	172	31.308036	8.27573	Pit 1 main	exact	7.68	1.53	1.53	1
2272	62	30.871735	7.466683	Pit 1 main	exact	7.97	1.55	1.58	1
2273	134	19.708195	5.943434	Pit 1 main	exact	7.94	1.60	1.59	1
2274	145	19.960582	18.819977	Pit 1 main	exact	7.88	2.02	1.92	2
2275	146	0.147401	23.554297	Pit 1 main	exact	7.81	2.09	1.93	2
2276	42	29.989851	28.719181	Pit 1 main	exact	7.92	2.16	1.96	2
2277	220	25.05443	29.400093	Pit 1 main	exact	8.26	2.26	2.01	3
2278	67	28.19063	1.037085	Pit 1 main	poor	8.12	2.16	1.97	2
2279	26	18.373139	29.303642	Pit 1 main	exact	8.14	2.12	1.79	2
2280	16	15.438037	29.626916	Pit 1 main	exact	8.11	2.15	1.97	2
2281	30	16.593864	29.73262	Pit 1 main	exact	8.01	2.16	1.89	2
2282	19	16.070755	29.663272	Pit 1 main	exact	7.92	2.12	1.79	2
2283	20	16.750686	28.776325	Pit 1 main	exact	8.23	2.18	1.89	2
2284	217	16.27606	29.608201	Pit 1 main	exact	8.05	2.04	1.81	2
2285	21	16.387217	28.613927	Pit 1 main	exact	7.97	2.14	1.84	2
2286	17	15.651582	28.339677	Pit 1 main	exact	8.06	2.12	1.86	2
2287	13	14.079718	29.954447	Pit 1 main	exact	7.85	2.12	1.81	2
2288	12	12.281531	28.692336	Pit 1 main	exact	8.07	2.09	1.84	2
2289	14	13.437624	29.909026	Pit 1 main	exact	8.12	2.16	1.97	2
2290	196	13.563015	28.946587	Pit 1 main	exact	8.04	2.12	1.81	2
2291	15	14.510685	28.747011	Pit 1 main	exact	8.06	2.12	1.73	2
2292	188	10.134564	28.938482	Pit 1 main	exact	7.94	2.11	1.91	2
2293	11	8.691613	29.065114	Pit 1 main	exact	8.17	2.14	1.89	2
2294	9	8.131524	29.722269	Pit 1 main	exact	7.85	2.06	1.77	2
2295	10	12.627445	30.030236	Pit 1 main	exact	8.31	2.16	1.87	2
2296	7	7.598419	29.55203	Pit 1 main	exact	7.98	2.12	1.89	2
2297	5	5.704329	29.938901	Pit 1 main	exact	8.22	2.16	1.98	2

Trigger ID	Spatial ID	X	Y	Generic location	Location error	A1	A2	A3	A Type
2298	6	6.423473	29.887307	Pit 1 main	exact	7.79	2.14	1.82	2
2299	219	4.300397	29.021715	Pit 1 main	exact	8.17	2.18	1.97	2
2300	197	2.544893	29.841492	Pit 1 main	exact	8.15	2.12	1.82	2
2301	189	2.642265	29.315028	Pit 1 main	exact	8.23	2.11	1.89	2
2302	190	0.643675	29.903984	Pit 1 main	exact	8.02	2.11	1.81	2
2303	205	16.854337	-25.755526	Pit 1 main	exact	8.11	2.15	1.93	2
2304	97	14.614332	-25.879788	Pit 1 main	exact	8.08	2.08	1.89	2
2305	198	12.775463	-25.658393	Pit 1 main	exact	8.17	2.12	1.91	2
2306	85	27.579737	-26.413922	Pit 1 main	poor	8.05	2.19	1.86	2
2307	100	13.312002	-26.493856	Pit 1 main	exact	8.07	2.09	1.90	2
2308	88	22.299389	-26.486917	Pit 1 main	poor	8.11	2.19	1.78	2
2309	108	7.552892	-26.292185	Pit 1 main	exact	7.86	2.16	1.83	2
2310	191	7.746243	-25.462451	Pit 1 main	exact	8.09	2.11	1.87	2
2311	89	21.784246	-26.170666	Pit 1 main	exact	8.13	2.12	1.87	2
2312	202	5.771701	-25.375984	Pit 1 main	exact	8.23	2.14	1.93	2
2313	206	5.037919	-25.70492	Pit 1 main	exact	7.91	2.15	1.89	2
2314	207	4.000503	-25.679618	Pit 1 main	exact	8.13	2.15	1.89	2
2315	112	1.016835	-26.364434	Pit 1 main	exact	8.15	2.14	1.89	2
2316	203	-0.300977	-25.375984	Pit 1 main	exact	8.00	2.14	1.89	2
2317	204	-0.199766	-25.300075	Pit 1 main	exact	7.96	2.14	1.78	2
2318	178	33.882004	-14.574878	Pit 1 main	exact	8.01	2.05	1.87	2
2319	76	31.655381	-10.748628	Pit 1 main	exact	8.11	2.03	1.97	2
2320	73	33.907924	-12.074918	Pit 1 main	exact	8.25	2.27	2.05	3
2321	183	33.045519	11.708125	Pit 1 main	exact	7.99	2.09	1.93	2
2322	180	8.729924	-21.131185	Pit 1 main	exact	7.93	2.06	1.89	2
2323	168	17.568892	-10.531686	Pit 1 main	exact	8.14	2.27	2.02	3
2324	120	20.137346	-5.492228	Pit 1 main	exact	8.09	2.08	2.03	2
2325	117	18.912984	-5.565224	Pit 1 main	poor	8.26	2.05	2.09	2
2326	140	21.649561	12.672669	Pit 1 main	exact	7.84	1.58	1.60	1
2327	175	19.416374	11.909297	Pit 1 main	exact	7.84	1.61	1.62	1
2328	174	20.330545	11.744812	Pit 1 main	exact	7.78	1.56	1.58	1
2329	143	21.020498	16.823586	Pit 1 main	exact	7.89	2.06	1.96	2
2330	181	19.285416	17.121407	Pit 1 main	exact	8.05	2.06	1.94	2
2331	179	18.7305	16.938069	Pit 1 main	exact	8.01	2.05	1.89	2
2332	45	31.741774	28.239928	Pit 1 main	exact	7.89	1.61	1.68	1
2333	218	17.430113	29.683877	Pit 1 main	exact	7.86	2.12	1.73	2
2334	210	17.3166	29.002796	Pit 1 main	exact	7.97	2.09	1.83	2
2335	216	14.2139	29.475769	Pit 1 main	exact	7.81	2.07	1.83	2

Trigger ID	Spatial ID	X	Y	Generic location	Location error	A1	A2	A3	A Type
2336	214	11.849038	29.854147	Pit 1 main	exact	7.89	2.07	1.75	2
2337	192	10.574617	29.436132	Pit 1 main	exact	7.86	2.11	1.82	2
2338	221	9.631811	29.737968	Pit 1 main	exact	8.06	2.13	1.93	2
2339	209	11.13012	29.948741	Pit 1 main	exact	8.09	2.04	1.81	2
2340	212	9.465257	28.964958	Pit 1 main	exact	7.97	2.09	1.88	2
2341	8	7.218884	29.11886	Pit 1 main	exact	8.17	2.14	1.94	2
2342	213	8.02742	29.154147	Pit 1 main	exact	8.18	2.07	1.81	2
2343	87	24.314747	-26.094249	Pit 1 main	exact	7.92	2.13	1.97	2
2344	4	4.874686	29.922505	Pit 1 main	exact	7.88	2.03	1.75	2
2345	1	2.405659	28.996267	Pit 1 main	exact	8.00	2.09	1.79	2
2346	193	1.216068	29.248525	Pit 1 main	exact	8.09	2.11	1.84	2
2347	2	0.455338	29.241239	Pit 1 main	exact	7.98	2.09	1.76	2
2348	211	0.743644	29.267661	Pit 1 main	exact	8.06	2.13	1.86	2
2349	3	2.167348	30.072502	Pit 1 main	exact	7.94	2.09	1.79	2
2350	194	16.797904	-8.650193	Pit 1 main	exact	8.21	2.11	2.06	2
2351	184	18.719879	17.777814	Pit 1 main	exact	7.88	2.09	1.91	2
2352	162	18.074013	7.253553	Pit 1 main	exact	7.85	1.59	1.62	1
2353	129	19.610153	6.807441	Pit 1 main	exact	7.87	1.64	1.65	1
2354	165	18.608398	5.982347	Pit 1 main	exact	7.56	1.53	1.55	1
2355	159	18.360626	5.861431	Pit 1 main	exact	7.87	1.56	1.59	1
2356	160	19.220466	5.738597	Pit 1 main	exact	7.57	1.53	1.64	1
2357	173	18.200631	13.60639	Pit 1 main	exact	7.59	1.55	1.56	1
2358	170	18.183554	11.69997	Pit 1 main	exact	7.84	1.51	1.53	1
2359	22	17.925771	29.66081	Pit 1 main	exact	7.69	2.11	1.75	2
2360	23	18.113679	28.758125	Pit 1 main	poor	7.86	2.07	1.92	2
3788	83	32.811009	-20.817251	Pit 1 main	exact	7.95	2.19	1.98	2
3790	225	33.718573	-23.341457	Pit 1 main	exact	7.64	1.99	2.07	2
3793	81	31.776178	-25.53808	Pit 1 main	poor	7.77	2.04	1.94	2
3811	80	34.026357	-23.930563	Pit 1 main	exact	7.68	2.01	1.91	4
3821	151	34.005854	0.820272	Pit 1 main	poor	7.91	2.02	1.79	2
3825	84	32.315821	-26.731492	Pit 1 main	poor	7.81	2.01	1.88	2
5525	161	17.7874	6.475602	Pit 1 main	poor	8.06	1.05	1.29	5
5543				Pit 2	pit 2	7.84	2.09	1.83	2
5553	185	5.551855	24.004271	Pit 1 main	exact	8.11	2.09	2.02	2
5758				Pit 1 other	T12G6	8.12	2.01	1.98	2
5760				Pit 1 other	T12G6	8.05	1.99	1.97	2
5761				Pit 1 other	T12G6	8.01	1.99	2.02	2
5768				Pit 1 other	T12G6	7.74	2.05	1.99	2

Trigger ID	Spatial ID	X	Y	Generic location	Location error	A1	A2	A3	A Type
5771				Pit 1 other	T12G6	7.52	2.02	1.92	2
5772				Pit 1 other	T11G7	7.95	1.61	1.58	1
5774				Pit 1 other	T11G7	7.88	1.47	1.58	1
5775				Pit 1 other	T11G7	7.76	1.62	1.60	1
5819				Pit 1 other	T13G5	8.25	2.04	2.02	2
5821				Pit 1 other	T13G5	7.74	2.02	1.90	2

Appendix 2

Spatial and metric data on triggers: Part B

Trigger ID	Spatial ID	X	Y	Generic location	Location error	B1	B2	B3	B4	Notched?
901	148	34.105669	-26.529079	Pit 1 main	exact	8.14	5.21	1.91	1.57	0
902	77	33.159938	-25.57195	Pit 1 main	exact	8.56	5.20		1.44	0
904	91	20.860784	-25.906234	Pit 1 main	exact	8.56	5.42	2.11	1.55	0
905	86	14.533779	-26.714721	Pit 1 main	exact	8.60	5.13		1.44	0
906	99	13.630381	-26.243872	Pit 1 main	exact	8.61	4.99	1.93		0
907	102	11.591567	-25.640049	Pit 1 main	exact			2.02		0
908	58	34.201382	13.620131	Pit 1 main	exact	8.51	5.67	1.94	1.42	0
909	149	34.3053	16.990327	Pit 1 main	exact	8.64	5.20		1.54	0
910	101	10.696355	-25.812038	Pit 1 main	exact	8.44	5.22	1.92	1.38	0
911	63	33.501826	3.116224	Pit 1 main	exact	8.49	5.27	1.91	1.44	0
912	52	34.392391	21.465132	Pit 1 main	exact	8.81	5.62	2.03	1.36	0
913	59	32.710583	9.192165	Pit 1 main	poor	7.89	4.78	1.89	1.63	0
914	78	32.952122	-21.474925	Pit 1 main	poor	8.18	4.77	1.81		0
915	150	33.606594	5.511584	Pit 1 main	poor	8.24	5.06	1.71	1.07	1
918	152	31.610291	-8.585008	Pit 1 main	exact	8.31	5.27	2.01	1.40	1
919	118	18.900838	-4.311712	Pit 1 main	exact	8.18	5.17	1.99	1.40	1
920	154	19.233212	-3.294805	Pit 1 main	exact	8.24	5.31	2.00	1.40	1
921	155	18.933766	-3.095175	Pit 1 main	exact	8.54	5.48		1.44	0
922	137	18.728001	6.001707	Pit 1 main	exact	8.23	5.26	1.77	1.16	0
923	127	19.851817	7.723742	Pit 1 main	exact	8.23	5.26	1.48	1.16	0
924	131	19.831816	6.456726	Pit 1 main	exact	8.42	4.97	1.59		0
925	128	19.719986	7.316134	Pit 1 main	exact	8.48	4.94	1.70	1.01	0
926	130	19.610153	6.807441	Pit 1 main	exact	8.26	4.99	1.57	1.12	0
927	136	18.637351	6.658974	Pit 1 main	exact	8.27	5.20	1.65	1.05	0
928	138	20.810811	11.600437	Pit 1 main	exact	8.06	5.21	1.56	1.18	0
929	135	18.91818	8.419803	Pit 1 main	exact	8.31	4.86	1.76	1.12	0
930	123	21.925709	6.398791	Pit 1 main	exact	8.19	5.54	1.87	1.31	1
931	124	22.19033	5.721752	Pit 1 main	exact	8.24	5.33	2.01	1.39	1
932	126	22.331048	7.166956	Pit 1 main	exact	8.46	5.22	1.56	0.97	1
933	163	21.431484	6.557492	Pit 1 main	exact	8.68	5.21	1.56	1.19	0
934	132	20.086295	5.674169	Pit 1 main	exact	8.39	5.19	1.58	0.98	1
935	133	20.578868	5.394795	Pit 1 main	exact	8.38	5.16	1.56	0.98	0
936	25	21.448853	29.377367	Pit 1 main	poor	8.47	5.04	2.03	1.29	0

Trigger ID	Spatial ID	X	Y	Generic location	Location error	B1	B2	B3	B4	Notched?
937	44	6.475602	6.475602	Pit 1 main	poor	8.25	5.36	1.69	0.94	0
938	24	18.762581	29.917961	Pit 1 main	poor	8.25	4.98	1.93	1.51	0
939	156	28.513778	12.35631	Pit 1 main	exact	8.35	5.12	1.39	0.99	0
940	141	20.778356	13.166239	Pit 1 main	exact	8.32	5.15	1.84	1.24	0
943	35	22.496752	28.661955	Pit 1 main	exact	8.37	5.60	1.93	1.46	0
944	32	21.728948	23.548804	Pit 1 main	exact	8.58	4.98	1.73	1.31	0
946	40	26.355985	28.877476	Pit 1 main	exact	8.13	4.69	1.85		0
947	36	22.881719	28.031499	Pit 1 main	exact	8.48	5.62	2.09	1.49	0
948	39	25.235039	28.33449	Pit 1 main	exact	8.46	5.66	1.95	1.29	0
949	38	24.125749	28.355277	Pit 1 main	exact	8.55	5.69	2.05	1.36	0
950	157	23.151164	0.820272	Pit 1 main	exact	8.33	5.04	1.86	1.34	0
951	41	32.020779	29.775745	Pit 1 main	exact	8.57	5.46	1.98	1.44	0
952	43	32.420465	29.891975	Pit 1 main	exact	8.37	4.83	1.52	0.89	0
953	48	34.269167	27.43019	Pit 1 main	exact	8.15	5.26	1.55	0.98	0
954	47	33.964206	29.27371	Pit 1 main	exact	8.09	4.96	1.54	1.30	0
956	54	32.724943	19.958606	Pit 1 main	exact	8.25	5.53	1.93	1.65	0
957	169	33.918877	22.809477	Pit 1 main	poor	8.03	4.88	1.87		0
958	53	34.683898	20.732773	Pit 1 main	exact	8.31	5.64	2.07	1.62	0
959	46	33.960055	29.856813	Pit 1 main	exact	8.21	4.98	1.96	1.41	0
960	31	22.718784	24.068895	Pit 1 main	exact	8.36	5.50	1.84	1.47	0
961	34	22.765991	23.168717	Pit 1 main	exact	8.54	4.93	1.77	1.39	0
962	51	33.144621	21.865315	Pit 1 main	poor	8.65	5.72	1.95	1.28	0
963	28	20.47937	24.274925	Pit 1 main	exact	8.18	4.95	1.91	1.41	0
964	144	19.549687	19.002285	Pit 1 main	exact	8.54	4.89	1.77	1.38	0
965	50	34.121289	20.732773	Pit 1 main	poor	8.31	4.99	2.01	1.38	0
966	171	33.705328	23.250595	Pit 1 main	poor	8.38	5.16	1.69	1.12	1
967	222	29.41212	29.397728	Pit 1 main	poor	8.22	5.54	2.01	1.66	0
968	176	34.554697	16.824692	Pit 1 main	exact	8.31	4.81	1.59	1.47	0
969	37	24.039188	29.137737	Pit 1 main	exact	8.41	5.51	1.89	1.39	0
970	223	23.086751	29.521452	Pit 1 main	poor	8.36	4.99	2.10	1.49	0
971	57	32.874234	14.9208	Pit 1 main	exact	8.54	5.57	2.13	1.59	0
972	167	32.468576	25.632531	Pit 1 main	exact	7.97	4.94	1.82	1.52	0
973	158	28.596817	28.4179	Pit 1 main	exact	8.41	5.36	1.75	1.01	0
974	142	22.614815	13.541946	Pit 1 main	exact	8.08	5.07	1.75	1.48	0
975	139	21.094665	12.103157	Pit 1 main	exact	8.27	4.96	1.99	1.42	0
976	68	31.321209	-4.804873	Pit 1 main	poor	8.11	4.92	1.96	1.58	0
977	60	34.221857	9.737418	Pit 1 main	poor	7.99	4.91	1.77	1.52	0

Trigger ID	Spatial ID	X	Y	Generic location	Location error	B1	B2	B3	B4	Notched?
978				Pit 1 other	T4G2	8.62	4.98	1.76	1.36	0
979	225	7.922879	-20.702595	Pit 1 main	poor	8.19	5.01	1.85	1.36	0
1471	55	32.463186	17.22433	Pit 1 main	poor	8.33	5.64	2.04	1.37	0
1472	56	33.109812	16.732523	Pit 1 main	poor	8.25	4.88	1.73	1.57	0
1498	208	34.213518	-11.978139	Pit 1 main	exact	8.44	4.83	1.74	1.44	0
1499	65	33.149953	1.631469	Pit 1 main	exact	8.12	5.37	1.97	1.40	1
1500	64	34.821564	2.017701	Pit 1 main	exact	8.39	5.32	1.92	1.29	1
1501	61	33.756327	8.628375	Pit 1 main	exact	8.14	5.19	1.51	1.13	0
1502	49	34.069362	26.573211	Pit 1 main	exact	8.22	5.18	1.70	1.13	1
1503	71	34.052426	-5.372814	Pit 1 main	exact	8.35	5.45	1.95	1.42	1
2231	93	20.117168	-26.579856	Pit 1 main	exact	8.26	5.13	1.92	1.51	0
2232	92	20.216574	-26.208776	Pit 1 main	exact	7.99	4.84	1.78	1.43	0
2233	94	24.311596	-26.461492	Pit 1 main	exact	8.01	4.98	2.12	1.64	0
2234	95	18.83128	-25.807518	Pit 1 main	poor	8.05	5.01	1.91	1.41	0
2235	98	19.708599	-26.650064	Pit 1 main	poor	8.14	4.82	2.05	1.58	0
2236	199	14.792157	-25.388635	Pit 1 main	exact	8.28	4.95	1.86	1.41	0
2237	186	12.670258	-25.510652	Pit 1 main	exact	8.20	5.11	1.83	1.49	0
2238	96	18.26273	-25.835152	Pit 1 main	exact	8.09	4.94	1.98	1.53	0
2239	195	11.19016	28.980629	Pit 1 main	exact	8.16	5.17	1.88	1.49	0
2240	215	11.924714	-25.653903	Pit 1 main	exact	7.95	5.11	1.78	1.62	0
2241	103	11.585001	-26.397574	Pit 1 main	exact	8.18	4.96	1.82	1.34	0
2242	104	10.059564	-25.779414	Pit 1 main	exact	7.98	5.07	1.93	1.53	0
2243	105	9.189363	-25.722035	Pit 1 main	exact	8.21	5.23	1.79	1.39	0
2244	106	8.311401	-26.031949	Pit 1 main	exact	8.07	5.01	1.83	1.44	0
2245	107	7.6723	-25.728396	Pit 1 main	exact	8.23	5.08	1.94	1.42	0
2246	187	6.763962	-25.666675	Pit 1 main	exact	8.11	5.19	1.84	1.49	0
2247	109	5.711013	-25.694485	Pit 1 main	exact	8.08	5.14	1.63	1.44	0
2248	110	4.925915	-26.395367	Pit 1 main	exact	8.02	4.97	1.76	1.48	0
2249	111	4.247799	-26.376172	Pit 1 main	exact	8.08	5.03	1.94	1.57	0
2250	200	3.114904	-26.362794	Pit 1 main	exact	8.21	4.93	1.96	1.53	0
2251	90	21.451586	-25.956809	Pit 1 main	poor	8.51	5.18		1.42	0
2252	82	33.560295	-24.307966	Pit 1 main	exact	8.11	5.01	1.83	1.49	0
2253	79	32.014427	-21.256277	Pit 1 main	exact	8.02	4.95	1.77	1.39	0
2254	72	32.511583	-19.93794	Pit 1 main	exact	8.21	5.60	1.97	1.59	0
2255	75	31.593245	-16.92337	Pit 1 main	exact	8.33	5.71	1.96	1.29	0
2256	166	32.299388	-16.419768	Pit 1 main	exact	8.26	5.02	1.96	1.43	0
2257	74	34.453956	-12.542946	Pit 1 main	exact	8.28	4.92	1.74	1.47	0

Trigger ID	Spatial ID	X	Y	Generic location	Location error	B1	B2	B3	B4	Notched?
2258	70	32.969332	-9.826935	Pit 1 main	exact	8.49	4.83	1.71	1.34	0
2259	66	32.629434	2.921166	Pit 1 main	exact	8.12	5.42	1.99	1.39	1
2260	201	23.660796	-20.290116	Pit 1 main	exact	8.06	4.94	1.75	1.41	0
2261	177	22.629265	-22.361014	Pit 1 main	exact	8.32	5.64	1.98	1.31	0
2262	224	8.031137	-21.475867	Pit 1 main	exact	8.04	5.05	1.73	1.29	0
2263	147	19.333946	-14.746488	Pit 1 main	exact	8.14	5.02	1.73	1.34	0
2264	115	18.571095	-10.172154	Pit 1 main	exact	8.33	4.81	1.69	1.49	0
2265	113	17.439096	-10.70497	Pit 1 main	exact	8.19	4.94	1.95	1.44	0
2266	69	30.182927	-9.290728	Pit 1 main	exact	8.08	5.54	2.02	1.39	1
2267	114	18.542612	-10.819296	Pit 1 main	exact	8.01	4.73	1.61		0
2268	121	20.73572	-4.652206	Pit 1 main	exact	8.23	5.41	2.03	1.46	1
2269	116	17.545465	-4.1724	Pit 1 main	poor	8.19	5.43	1.99	1.35	1
2270	119	19.180578	-5.365282	Pit 1 main	exact	8.28	5.45	2.02	1.39	1
2271	172	31.308036	8.27573	Pit 1 main	exact	7.96	5.01	1.53	1.13	0
2272	62	30.871735	7.466683	Pit 1 main	exact	8.33	5.16	1.58	1.09	0
2273	134	19.708195	5.943434	Pit 1 main	exact	7.95	4.84	1.48	1.09	1
2274	145	19.960582	18.819977	Pit 1 main	exact	8.38	5.49	2.01	1.51	0
2275	146	0.147401	23.554297	Pit 1 main	exact	8.15	5.66	1.95	1.63	0
2276	42	29.989851	28.719181	Pit 1 main	exact	8.29	4.82	1.62	1.36	0
2277	220	25.05443	29.400093	Pit 1 main	exact	8.04	4.89	1.91	1.45	0
2278	67	28.19063	1.037085	Pit 1 main	poor	8.23	4.87	1.65	1.41	0
2279	26	18.373139	29.303642	Pit 1 main	exact	8.13	5.28	1.81	1.39	0
2280	16	15.438037	29.626916	Pit 1 main	exact	8.16	5.29	1.97	1.49	0
2281	30	16.593864	29.73262	Pit 1 main	exact	8.13	5.11	2.07	1.37	0
2282	19	16.070755	29.663272	Pit 1 main	exact	8.21	5.05	2.09	1.59	0
2283	20	16.750686	28.776325	Pit 1 main	exact	8.08	5.08	2.09	1.65	0
2284	217	16.27606	29.608201	Pit 1 main	exact	8.01	5.01	1.91	1.63	0
2285	21	16.387217	28.613927	Pit 1 main	exact	8.01	5.02	1.91	1.61	0
2286	17	15.651582	28.339677	Pit 1 main	exact	8.18	5.03	1.97	1.61	0
2287	13	14.079718	29.954447	Pit 1 main	exact	8.24	5.12	1.84	1.37	0
2288	12	12.281531	28.692336	Pit 1 main	exact	8.15	5.18	1.97	1.29	0
2289	14	13.437624	29.909026	Pit 1 main	exact	8.03	5.19	1.79	1.45	0
2290	196	13.563015	28.946587	Pit 1 main	exact	8.43	5.13		1.41	0
2291	15	14.510685	28.747011	Pit 1 main	exact	8.14	4.88	1.93	1.49	0
2292	188	10.134564	28.938482	Pit 1 main	exact	8.24	5.08	2.06	1.49	0
2293	11	8.691613	29.065114	Pit 1 main	exact	8.26	5.28	1.84	1.52	0
2294	9	8.131524	29.722269	Pit 1 main	exact	8.11	4.98	2.11	1.65	0

Trigger ID	Spatial ID	X	Y	Generic location	Location error	B1	B2	B3	B4	Notched?
2295	10	12.627445	30.030236	Pit 1 main	exact	8.12	5.26	1.86	1.33	0
2296	7	7.598419	29.55203	Pit 1 main	exact	8.19	5.02	1.94	1.37	0
2297	5	5.704329	29.938901	Pit 1 main	exact	7.82	4.83	1.84	1.61	0
2298	6	6.423473	29.887307	Pit 1 main	exact	8.25	5.03	1.89	1.35	0
2299	219	4.300397	29.021715	Pit 1 main	exact	8.12	5.13	1.85	1.27	0
2300	197	2.544893	29.841492	Pit 1 main	exact	8.14	5.04	1.97	1.65	0
2301	189	2.642265	29.315028	Pit 1 main	exact	8.27	5.07	2.04	1.45	0
2302	190	0.643675	29.903984	Pit 1 main	exact	8.05	4.89	1.91	1.46	0
2303	205	16.854337	-25.755526	Pit 1 main	exact	8.39	5.29	2.13	1.56	0
2304	97	14.614332	-25.879788	Pit 1 main	exact	8.26	4.82	1.99	1.47	0
2305	198	12.775463	-25.658393	Pit 1 main	exact	8.38	5.02	1.96	1.42	0
2306	85	27.579737	-26.413922	Pit 1 main	poor	7.74	4.95	1.82		0
2307	100	13.312002	-26.493856	Pit 1 main	exact	8.31	5.21	1.89	1.47	0
2308	88	22.299389	-26.486917	Pit 1 main	poor	8.06	4.98	1.92	1.49	0
2309	108	7.552892	-26.292185	Pit 1 main	exact	7.99	5.09	2.04	1.67	0
2310	191	7.746243	-25.462451	Pit 1 main	exact	8.17	4.91	2.02	1.63	0
2311	89	21.784246	-26.170666	Pit 1 main	exact	7.95	5.01	1.97	1.59	0
2312	202	5.771701	-25.375984	Pit 1 main	exact	8.19	5.11	1.88	1.43	0
2313	206	5.037919	-25.70492	Pit 1 main	exact	8.47	4.95	1.93	1.49	0
2314	207	4.000503	-25.679618	Pit 1 main	exact	8.05	4.95	1.93	1.57	0
2315	112	1.016835	-26.364434	Pit 1 main	exact	8.00	5.02	1.87	1.53	0
2316	203	-0.300977	-25.375984	Pit 1 main	exact	8.17	5.22	1.87	1.35	0
2317	204	-0.199766	-25.300075	Pit 1 main	exact	8.06	4.98	1.87		0
2318	178	33.882004	-14.574878	Pit 1 main	exact	8.43	5.58	1.86	1.39	0
2319	76	31.655381	-10.748628	Pit 1 main	exact	8.29	5.58	1.95	1.65	0
2320	73	33.907924	-12.074918	Pit 1 main	exact	8.21	4.93	1.96	1.47	0
2321	183	33.045519	11.708125	Pit 1 main	exact	8.54	5.65	2.14	1.56	0
2322	180	8.729924	-21.131185	Pit 1 main	exact	8.31	5.28	1.64	1.26	0
2323	168	17.568892	-10.531686	Pit 1 main	exact	8.36	5.03	2.02	1.37	0
2324	120	20.137346	-5.492228	Pit 1 main	exact	8.32	5.38	1.97	1.37	1
2325	117	18.912984	-5.565224	Pit 1 main	poor	8.24	5.26	1.93	1.32	1
2326	140	21.649561	12.672669	Pit 1 main	exact	8.34	5.19	1.55	1.06	0
2327	175	19.416374	11.909297	Pit 1 main	exact	8.61	4.84	1.68	0.98	0
2328	174	20.330545	11.744812	Pit 1 main	exact	8.35	5.29	1.57	1.26	1
2329	143	21.020498	16.823586	Pit 1 main	exact	8.52	5.65	1.87	1.49	0
2330	181	19.285416	17.121407	Pit 1 main	exact	8.41	5.61	1.98	1.54	0
2331	179	18.7305	16.938069	Pit 1 main	exact	8.55	5.65	1.93	1.39	0

Trigger ID	Spatial ID	X	Y	Generic location	Location error	B1	B2	B3	B4	Notched?
2332	45	31.741774	28.239928	Pit 1 main	exact	8.34	5.15	1.72	1.02	0
2333	218	17.430113	29.683877	Pit 1 main	exact	8.12	4.85	1.91	1.56	0
2334	210	17.3166	29.002796	Pit 1 main	exact	8.28	5.04	1.93	1.49	0
2335	216	14.2139	29.475769	Pit 1 main	exact	8.59	4.98	2.11	1.45	0
2336	214	11.849038	29.854147	Pit 1 main	exact	8.48	5.05	1.88	1.31	0
2337	192	10.574617	29.436132	Pit 1 main	exact	8.29	4.99	1.97	1.47	0
2338	221	9.631811	29.737968	Pit 1 main	exact	8.01	5.07	1.79	1.29	1
2339	209	11.13012	29.948741	Pit 1 main	exact	8.28	4.92	1.98	1.52	0
2340	212	9.465257	28.964958	Pit 1 main	exact	8.28	5.02	2.02	1.49	0
2341	8	7.218884	29.11886	Pit 1 main	exact	8.12	5.02	1.85	1.46	0
2342	213	8.02742	29.154147	Pit 1 main	exact	8.25	4.78	1.93	1.49	0
2343	87	24.314747	-26.094249	Pit 1 main	exact	8.18	5.13	1.73	1.32	0
2344	4	4.874686	29.922505	Pit 1 main	exact	8.09	4.94	1.89	1.54	0
2345	1	2.405659	28.996267	Pit 1 main	exact	8.23	4.91	1.94	1.51	0
2346	193	1.216068	29.248525	Pit 1 main	exact	8.29	4.99	2.10	1.55	0
2347	2	0.455338	29.241239	Pit 1 main	exact	8.21	5.06	2.03	1.32	0
2348	211	0.743644	29.267661	Pit 1 main	exact	8.19	5.05	1.71	1.39	0
2349	3	2.167348	30.072502	Pit 1 main	exact	8.31	5.04	1.99	1.43	0
2350	194	16.797904	-8.650193	Pit 1 main	exact	8.49	4.91	1.76	1.39	0
2351	184	18.719879	17.777814	Pit 1 main	exact	8.38	5.54	2.01	1.47	0
2352	162	18.074013	7.253553	Pit 1 main	exact	8.69	5.14	1.58	1.23	1
2353	129	19.610153	6.807441	Pit 1 main	exact	8.15	4.91	1.55	1.22	0
2354	165	18.608398	5.982347	Pit 1 main	exact	8.15	4.98	1.42	1.18	0
2355	159	18.360626	5.861431	Pit 1 main	exact	8.41	4.81	1.44	1.15	0
2356	160	19.220466	5.738597	Pit 1 main	exact	8.69	5.15	1.63	0.95	0
2357	173	18.200631	13.60639	Pit 1 main	exact	8.01	4.85	1.55	1.08	1
2358	170	18.183554	11.69997	Pit 1 main	exact	8.09	5.27	1.74	1.12	1
2359	22	17.925771	29.66081	Pit 1 main	exact	8.25	4.96	1.84	1.32	0
2360	23	18.113679	28.758125	Pit 1 main	poor	8.31	5.02	1.88	1.37	0
3788	83	32.811009	-20.817251	Pit 1 main	exact	8.17	4.74	1.93	1.35	0
3790	225	33.718573	-23.341457	Pit 1 main	exact	8.32	5.33	1.95	1.39	1
3793	81	31.776178	-25.53808	Pit 1 main	poor	7.99	5.19	1.75	1.27	0
3811	80	34.026357	-23.930563	Pit 1 main	exact	8.39	4.81	1.99	1.39	0
3821	151	34.005854	0.820272	Pit 1 main	poor	8.19	4.89	1.89	1.34	0
3825	84	32.315821	-26.731492	Pit 1 main	poor	8.25	5.31	1.99	1.45	1
5525	161	17.7874	6.475602	Pit 1 main	poor	5.03	4.62	1.51	1.22	0
5543				Pit 2	pit 2	8.25	4.98	2.02	1.49	0

Trigger ID	Spatial ID	X	Y	Generic location	Location error	B1	B2	B3	B4	Notched?
5553	185	5.551855	24.004271	Pit 1 main	exact	8.61	5.38	2.11	1.49	0
5758				Pit 1 other	T12G6	8.04	5.43	2.06	1.29	1
5760				Pit 1 other	T12G6	8.22	5.52	2.02	1.39	1
5761				Pit 1 other	T12G6	8.16	5.38	2.02	1.39	1
5768				Pit 1 other	T12G6	8.28	5.41	1.95	1.41	1
5771				Pit 1 other	T12G6	8.29	5.46	1.91	1.26	1
5772				Pit 1 other	T11G7	8.28	5.05	1.67	0.93	0
5774				Pit 1 other	T11G7	8.47	5.34	1.49	1.16	0
5775				Pit 1 other	T11G7	8.11	5.29	1.59		0
5819				Pit 1 other	T13G5	8.15	5.46	1.93	1.33	1
5821				Pit 1 other	T13G5	8.41	5.38	2.07	1.40	1

Appendix 3

Spatial and metric data on triggers: Part C

Trigger ID	Spatial ID	X	Y	Generic location	Location error	C1	C2	C3	Bevelled?
901	148	34.105669	-26.529079	Pit 1 main	exact	5.75	2.82	2.55	0
902	77	33.159938	-25.57195	Pit 1 main	exact	5.81	2.82	2.61	0
904	91	20.860784	-25.906234	Pit 1 main	exact	5.77		2.55	0
905	86	14.533779	-26.714721	Pit 1 main	exact	5.81	2.8	2.56	0
906	99	13.630381	-26.243872	Pit 1 main	exact	5.9	2.9	2.69	0
907	102	11.591567	-25.640049	Pit 1 main	exact	5.82	2.83	2.69	0
908	58	34.201382	13.620131	Pit 1 main	exact	5.94		2.61	0
909	149	34.3053	16.990327	Pit 1 main	exact	5.74	2.92	2.53	0
910	101	10.696355	-25.812038	Pit 1 main	exact	5.94	2.95	2.55	0
911	63	33.501826	3.116224	Pit 1 main	exact	5.86	2.93	2.65	0
912	52	34.392391	21.465132	Pit 1 main	exact	5.82	3.16	2.49	0
913	59	32.710583	9.192165	Pit 1 main	poor	6.03	2.89	2.67	0
914	78	32.952122	-21.474925	Pit 1 main	poor	5.87	2.73	2.39	0
915	150	33.606594	5.511584	Pit 1 main	poor	6.29	3.12	2.74	0
918	152	31.610291	-8.585008	Pit 1 main	exact	5.59	2.67	2.19	1
919	118	18.900838	-4.311712	Pit 1 main	exact	5.75		2.17	1
920	154	19.233212	-3.294805	Pit 1 main	exact	5.64	2.64	2.21	1
921	155	18.933766	-3.095175	Pit 1 main	exact	5.71		1.99	1
922	137	18.728001	6.001707	Pit 1 main	exact	6.19		2.53	0
923	127	19.851817	7.723742	Pit 1 main	exact	6.23	3.04	2.49	0
924	131	19.831816	6.456726	Pit 1 main	exact		3.15	2.53	0
925	128	19.719986	7.316134	Pit 1 main	exact	6.37	2.86	2.53	0
926	130	19.610153	6.807441	Pit 1 main	exact	6.05	3.06	2.5	0
927	136	18.637351	6.658974	Pit 1 main	exact	6.52	3.25	2.65	0
928	138	20.810811	11.600437	Pit 1 main	exact	6.51		2.62	0
929	135	18.91818	8.419803	Pit 1 main	exact	6.41	3.15	2.66	0
930	123	21.925709	6.398791	Pit 1 main	exact	5.61	2.52	2.08	1
931	124	22.19033	5.721752	Pit 1 main	exact	5.64		2.27	1
932	126	22.331048	7.166956	Pit 1 main	exact	6.58	3.18	2.63	0
933	163	21.431484	6.557492	Pit 1 main	exact	6.25	3.09	2.48	0
934	132	20.086295	5.674169	Pit 1 main	exact	6.67		2.53	0
935	133	20.578868	5.394795	Pit 1 main	exact	6.17	3.04	2.58	0
936	25	21.448853	29.377367	Pit 1 main	poor	5.88	2.86	2.61	0

Trigger ID	Spatial ID	X	Y	Generic location	Location error	C1	C2	C3	Bevelled?
937	44	6.475602	6.475602	Pit 1 main	poor	6.62	3.21	2.52	0
938	24	18.762581	29.917961	Pit 1 main	poor	6.01	2.84	2.54	0
939	156	28.513778	12.35631	Pit 1 main	exact	6.14	3.01	2.54	0
940	141	20.778356	13.166239	Pit 1 main	exact	6.72	3.37	2.38	0
943	35	22.496752	28.661955	Pit 1 main	exact	5.83	3.21	2.47	0
944	32	21.728948	23.548804	Pit 1 main	exact	5.92	2.72	2.11	0
946	40	26.355985	28.877476	Pit 1 main	exact	5.74	2.93	2.63	0
947	36	22.881719	28.031499	Pit 1 main	exact	5.82	3.33	2.58	0
948	39	25.235039	28.33449	Pit 1 main	exact	5.71	3.24	2.46	0
949	38	24.125749	28.355277	Pit 1 main	exact	5.74		2.53	0
950	157	23.151164	0.820272	Pit 1 main	exact	6.02	2.75	2.16	0
951	41	32.020779	29.775745	Pit 1 main	exact	5.61		2.53	0
952	43	32.420465	29.891975	Pit 1 main	exact	6.24	3.07	2.45	0
953	48	34.269167	27.43019	Pit 1 main	exact	6.27	3.23	2.45	0
954	47	33.964206	29.27371	Pit 1 main	exact	6.13	2.97	2.45	0
956	54	32.724943	19.958606	Pit 1 main	exact	5.69	3.39	2.54	0
957	169	33.918877	22.809477	Pit 1 main	poor	5.97	2.75	2.23	0
958	53	34.683898	20.732773	Pit 1 main	exact	5.71	3.17	2.49	0
959	46	33.960055	29.856813	Pit 1 main	exact	6.06	2.78	2.07	0
960	31	22.718784	24.068895	Pit 1 main	exact	5.83	3.17	2.48	0
961	34	22.765991	23.168717	Pit 1 main	exact	6.21	2.78	2.17	0
962	51	33.144621	21.865315	Pit 1 main	poor	5.91	3.09	2.52	0
963	28	20.47937	24.274925	Pit 1 main	exact	5.96	2.74	2.18	0
964	144	19.549687	19.002285	Pit 1 main	exact	6.22	2.89	2.35	0
965	50	34.121289	20.732773	Pit 1 main	poor	6.04	2.72	1.91	0
966	171	33.705328	23.250595	Pit 1 main	poor		3.18	2.49	0
967	222	29.41212	29.397728	Pit 1 main	poor	5.51	3.11	2.39	0
968	176	34.554697	16.824692	Pit 1 main	exact		2.82	2.33	0
969	37	24.039188	29.137737	Pit 1 main	exact	5.55	3.35	2.46	0
970	223	23.086751	29.521452	Pit 1 main	poor	6.09		2.77	0
971	57	32.874234	14.9208	Pit 1 main	exact	5.96	3.12	2.52	0
972	167	32.468576	25.632531	Pit 1 main	exact	6.15	2.77	2.13	0
973	158	28.596817	28.4179	Pit 1 main	exact	6.29		2.52	0
974	142	22.614815	13.541946	Pit 1 main	exact	5.84	2.92	2.61	0
975	139	21.094665	12.103157	Pit 1 main	exact	6.06	2.81	2.22	0
976	68	31.321209	-4.804873	Pit 1 main	poor	5.78	2.76	2.56	0
977	60	34.221857	9.737418	Pit 1 main	poor	5.65	2.73	2.55	0

Trigger ID	Spatial ID	X	Y	Generic location	Location error	C1	C2	C3	Bevelled?
978				Pit 1 other	T4G2	6.04	2.79	2.25	0
979	225	7.922879	-20.702595	Pit 1 main	poor	5.89	2.92	2.64	0
1471	55	32.463186	17.22433	Pit 1 main	poor	5.85	3.2	2.55	0
1472	56	33.109812	16.732523	Pit 1 main	poor	6.03	2.74	2.37	0
1498	208	34.213518	-11.978139	Pit 1 main	exact	5.94	2.71	2.27	0
1499	65	33.149953	1.631469	Pit 1 main	exact	5.69	2.18	2.15	1
1500	64	34.821564	2.017701	Pit 1 main	exact	5.49	2.53	2.22	1
1501	61	33.756327	8.628375	Pit 1 main	exact	6.12	3.04	2.5	0
1502	49	34.069362	26.573211	Pit 1 main	exact	6.21	3.04	2.56	0
1503	71	34.052426	-5.372814	Pit 1 main	exact	5.69		2.16	1
2231	93	20.117168	-26.579856	Pit 1 main	exact	5.77	3.06	2.61	0
2232	92	20.216574	-26.208776	Pit 1 main	exact	5.64	2.76	2.64	0
2233	94	24.311596	-26.461492	Pit 1 main	exact	5.91	2.82	2.57	0
2234	95	18.83128	-25.807518	Pit 1 main	poor	5.77	2.77	2.49	0
2235	98	19.708599	-26.650064	Pit 1 main	poor	5.96	2.97	2.63	0
2236	199	14.792157	-25.388635	Pit 1 main	exact	5.92	2.85	2.61	0
2237	186	12.670258	-25.510652	Pit 1 main	exact	5.86	2.98	2.54	0
2238	96	18.26273	-25.835152	Pit 1 main	exact	5.86	2.85	2.61	0
2239	195	11.19016	28.980629	Pit 1 main	exact	6.02	2.96	2.63	0
2240	215	11.924714	-25.653903	Pit 1 main	exact	5.85	2.86	2.57	0
2241	103	11.585001	-26.397574	Pit 1 main	exact	5.89	2.89	2.65	0
2242	104	10.059564	-25.779414	Pit 1 main	exact	5.75	2.74	2.58	0
2243	105	9.189363	-25.722035	Pit 1 main	exact	6.01	3.21	2.67	0
2244	106	8.311401	-26.031949	Pit 1 main	exact	6.05	3.07	2.67	0
2245	107	7.6723	-25.728396	Pit 1 main	exact	5.91	2.83	2.59	0
2246	187	6.763962	-25.666675	Pit 1 main	exact	5.89	2.93	2.58	0
2247	109	5.711013	-25.694485	Pit 1 main	exact	5.68	2.92	2.63	0
2248	110	4.925915	-26.395367	Pit 1 main	exact	5.89	2.85	2.58	0
2249	111	4.247799	-26.376172	Pit 1 main	exact	5.82	2.87	2.59	0
2250	200	3.114904	-26.362794	Pit 1 main	exact	5.92	3.04	2.78	0
2251	90	21.451586	-25.956809	Pit 1 main	poor	5.72	2.86	2.59	0
2252	82	33.560295	-24.307966	Pit 1 main	exact	5.85	2.94	2.63	0
2253	79	32.014427	-21.256277	Pit 1 main	exact	5.85	2.75	2.59	0
2254	72	32.511583	-19.93794	Pit 1 main	exact	5.89	3.29	2.34	0
2255	75	31.593245	-16.92337	Pit 1 main	exact	5.76		2.45	0
2256	166	32.299388	-16.419768	Pit 1 main	exact	6.04	2.81	2.16	0
2257	74	34.453956	-12.542946	Pit 1 main	exact	5.94	2.74	2.16	0

Trigger ID	Spatial ID	X	Y	Generic location	Location error	C1	C2	C3	Bevelled?
2258	70	32.969332	-9.826935	Pit 1 main	exact	6.08	2.69	2.34	0
2259	66	32.629434	2.921166	Pit 1 main	exact	5.71	2.73	2.14	1
2260	201	23.660796	-20.290116	Pit 1 main	exact	5.97	2.91	2.61	0
2261	177	22.629265	-22.361014	Pit 1 main	exact	5.63	3.12	2.42	0
2262	224	8.031137	-21.475867	Pit 1 main	exact	5.82	2.86	2.65	0
2263	147	19.333946	-14.746488	Pit 1 main	exact	5.99	2.75	2.24	0
2264	115	18.571095	-10.172154	Pit 1 main	exact	5.97	2.67	2.29	0
2265	113	17.439096	-10.70497	Pit 1 main	exact	6.02	2.85	2.14	0
2266	69	30.182927	-9.290728	Pit 1 main	exact	5.69		2.16	1
2267	114	18.542612	-10.819296	Pit 1 main	exact	5.95	2.73	2.13	0
2268	121	20.73572	-4.652206	Pit 1 main	exact	5.73	2.65	2.14	1
2269	116	17.545465	-4.1724	Pit 1 main	poor	5.74		2.14	1
2270	119	19.180578	-5.365282	Pit 1 main	exact	5.85		2.29	1
2271	172	31.308036	8.27573	Pit 1 main	exact	6.35	3.27	2.48	0
2272	62	30.871735	7.466683	Pit 1 main	exact	6.41	3.09	2.51	0
2273	134	19.708195	5.943434	Pit 1 main	exact	6.62	3.16	2.62	0
2274	145	19.960582	18.819977	Pit 1 main	exact	5.88		2.52	0
2275	146	0.147401	23.554297	Pit 1 main	exact	5.79	3.27	2.45	0
2276	42	29.989851	28.719181	Pit 1 main	exact	5.99	2.92	2.25	0
2277	220	25.05443	29.400093	Pit 1 main	exact	6.03	2.78	2.11	0
2278	67	28.19063	1.037085	Pit 1 main	poor	6.07	2.87	2.25	0
2279	26	18.373139	29.303642	Pit 1 main	exact		2.89	2.52	0
2280	16	15.438037	29.626916	Pit 1 main	exact	6.06	2.92	2.58	0
2281	30	16.593864	29.73262	Pit 1 main	exact	5.96	2.93	2.71	0
2282	19	16.070755	29.663272	Pit 1 main	exact	5.92	2.92	2.52	0
2283	20	16.750686	28.776325	Pit 1 main	exact	5.86	2.99	2.71	0
2284	217	16.27606	29.608201	Pit 1 main	exact	5.84	2.77	2.69	0
2285	21	16.387217	28.613927	Pit 1 main	exact	5.68	2.87	2.65	0
2286	17	15.651582	28.339677	Pit 1 main	exact	5.89	2.88	2.67	0
2287	13	14.079718	29.954447	Pit 1 main	exact	5.89	2.92	2.65	0
2288	12	12.281531	28.692336	Pit 1 main	exact	6.03	3.06	2.71	0
2289	14	13.437624	29.909026	Pit 1 main	exact	5.75	2.84	2.58	0
2290	196	13.563015	28.946587	Pit 1 main	exact	5.96	2.87	2.65	0
2291	15	14.510685	28.747011	Pit 1 main	exact	5.97	2.86	2.65	0
2292	188	10.134564	28.938482	Pit 1 main	exact	5.85	2.82	2.57	0
2293	11	8.691613	29.065114	Pit 1 main	exact	5.89		2.52	0
2294	9	8.131524	29.722269	Pit 1 main	exact	5.86	2.89	2.65	0

Trigger ID	Spatial ID	X	Y	Generic location	Location error	C1	C2	C3	Bevelled?
2295	10	12.627445	30.030236	Pit 1 main	exact	5.92	2.78	2.65	0
2296	7	7.598419	29.55203	Pit 1 main	exact	5.79		2.64	0
2297	5	5.704329	29.938901	Pit 1 main	exact	5.84	2.94	2.65	0
2298	6	6.423473	29.887307	Pit 1 main	exact	5.99	2.89	2.54	0
2299	219	4.300397	29.021715	Pit 1 main	exact	5.95	2.89	2.54	0
2300	197	2.544893	29.841492	Pit 1 main	exact	5.94	2.81	2.65	0
2301	189	2.642265	29.315028	Pit 1 main	exact	5.98	2.84	2.57	0
2302	190	0.643675	29.903984	Pit 1 main	exact	5.91		2.64	0
2303	205	16.854337	-25.755526	Pit 1 main	exact	6.26		2.86	0
2304	97	14.614332	-25.879788	Pit 1 main	exact	5.81	2.77	2.64	0
2305	198	12.775463	-25.658393	Pit 1 main	exact	6.03	3.01	2.59	0
2306	85	27.579737	-26.413922	Pit 1 main	poor	6.03	2.86	2.65	0
2307	100	13.312002	-26.493856	Pit 1 main	exact	5.79	2.76	2.55	0
2308	88	22.299389	-26.486917	Pit 1 main	poor	5.98	2.87	2.78	0
2309	108	7.552892	-26.292185	Pit 1 main	exact	5.89	3.04	2.71	0
2310	191	7.746243	-25.462451	Pit 1 main	exact	5.91	2.79	2.59	0
2311	89	21.784246	-26.170666	Pit 1 main	exact	5.86	2.77	2.57	0
2312	202	5.771701	-25.375984	Pit 1 main	exact	5.51	2.84	2.68	0
2313	206	5.037919	-25.70492	Pit 1 main	exact	5.94	2.84	2.67	0
2314	207	4.000503	-25.679618	Pit 1 main	exact	5.79	2.85	2.57	0
2315	112	1.016835	-26.364434	Pit 1 main	exact	5.84	2.85	2.63	0
2316	203	-0.300977	-25.375984	Pit 1 main	exact	5.69	2.84	2.63	0
2317	204	-0.199766	-25.300075	Pit 1 main	exact	5.94	2.84	2.63	0
2318	178	33.882004	-14.574878	Pit 1 main	exact	5.82		2.53	0
2319	76	31.655381	-10.748628	Pit 1 main	exact	6.11		2.61	0
2320	73	33.907924	-12.074918	Pit 1 main	exact	6.05	2.87	2.23	0
2321	183	33.045519	11.708125	Pit 1 main	exact	5.87	3.23	2.47	0
2322	180	8.729924	-21.131185	Pit 1 main	exact	5.62	2.83	2.5	0
2323	168	17.568892	-10.531686	Pit 1 main	exact	6.05		2.24	0
2324	120	20.137346	-5.492228	Pit 1 main	exact	5.76		2.24	1
2325	117	18.912984	-5.565224	Pit 1 main	poor	5.74	2.62	2.29	1
2326	140	21.649561	12.672669	Pit 1 main	exact	6.32	3.11	2.55	0
2327	175	19.416374	11.909297	Pit 1 main	exact	6.49	3.26	2.77	0
2328	174	20.330545	11.744812	Pit 1 main	exact	6.14	2.71	2.63	0
2329	143	21.020498	16.823586	Pit 1 main	exact	5.87	3.26	2.45	0
2330	181	19.285416	17.121407	Pit 1 main	exact	5.79	2.38	2.48	0
2331	179	18.7305	16.938069	Pit 1 main	exact	5.86	2.79	2.52	0

Trigger ID	Spatial ID	X	Y	Generic location	Location error	C1	C2	C3	Bevelled?
2332	45	31.741774	28.239928	Pit 1 main	exact	6.37	3.14	2.62	0
2333	218	17.430113	29.683877	Pit 1 main	exact	6.01	3.04	2.69	0
2334	210	17.3166	29.002796	Pit 1 main	exact	5.91	2.96	2.69	0
2335	216	14.2139	29.475769	Pit 1 main	exact	5.82	2.91	2.67	0
2336	214	11.849038	29.854147	Pit 1 main	exact	5.99	3.01	2.68	0
2337	192	10.574617	29.436132	Pit 1 main	exact	5.85	2.82	2.51	0
2338	221	9.631811	29.737968	Pit 1 main	exact	6.03	2.89	2.52	0
2339	209	11.13012	29.948741	Pit 1 main	exact		2.75	2.75	0
2340	212	9.465257	28.964958	Pit 1 main	exact	5.81	2.79	2.69	0
2341	8	7.218884	29.11886	Pit 1 main	exact	5.93	2.94	2.69	0
2342	213	8.02742	29.154147	Pit 1 main	exact	5.97	2.94	2.74	0
2343	87	24.314747	-26.094249	Pit 1 main	exact	5.73	2.95	2.68	0
2344	4	4.874686	29.922505	Pit 1 main	exact	5.92	2.89	2.65	0
2345	1	2.405659	28.996267	Pit 1 main	exact	5.94	2.83	2.65	0
2346	193	1.216068	29.248525	Pit 1 main	exact	5.93	2.85	2.59	0
2347	2	0.455338	29.241239	Pit 1 main	exact	5.71	2.89	2.54	0
2348	211	0.743644	29.267661	Pit 1 main	exact	5.85	2.79	2.58	0
2349	3	2.167348	30.072502	Pit 1 main	exact	5.87	2.85	2.67	0
2350	194	16.797904	-8.650193	Pit 1 main	exact	5.93	2.63	2.24	0
2351	184	18.719879	17.777814	Pit 1 main	exact	5.91	3.04	2.53	0
2352	162	18.074013	7.253553	Pit 1 main	exact	5.91	2.83	2.23	0
2353	129	19.610153	6.807441	Pit 1 main	exact	6.23	3.04	2.53	0
2354	165	18.608398	5.982347	Pit 1 main	exact	6.25	3.18	2.45	0
2355	159	18.360626	5.861431	Pit 1 main	exact	6.75	3.36	2.31	0
2356	160	19.220466	5.738597	Pit 1 main	exact	6.45	2.92	2.52	0
2357	173	18.200631	13.60639	Pit 1 main	exact	6.25	3.03	2.53	0
2358	170	18.183554	11.69997	Pit 1 main	exact	6.47	3.24	2.53	0
2359	22	17.925771	29.66081	Pit 1 main	exact	5.89	2.85	2.58	0
2360	23	18.113679	28.758125	Pit 1 main	poor	5.89	2.92	2.58	0
3788	83	32.811009	-20.817251	Pit 1 main	exact	5.85	2.71	2.12	0
3790	225	33.718573	-23.341457	Pit 1 main	exact	5.65	2.49	2.19	1
3793	81	31.776178	-25.53808	Pit 1 main	poor	5.86	2.97	2.61	0
3811	80	34.026357	-23.930563	Pit 1 main	exact	5.79	2.78	2.53	0
3821	151	34.005854	0.820272	Pit 1 main	poor	5.82	2.82	2.53	0
3825	84	32.315821	-26.731492	Pit 1 main	poor		2.64	2.17	1
5525	161	17.7874	6.475602	Pit 1 main	poor	4.96	2.41	2.08	0
5543				Pit 2	pit 2	5.72	2.69	2.53	0

Trigger ID	Spatial ID	X	Y	Generic location	Location error	C1	C2	C3	Bevelled?
5553	185	5.551855	24.004271	Pit 1 main	exact	5.73	2.95	2.53	0
5758				Pit 1 other	T12G6	5.65		2.17	1
5760				Pit 1 other	T12G6	5.73		2.19	1
5761				Pit 1 other	T12G6	5.58	2.52	2.18	1
5768				Pit 1 other	T12G6	5.73	2.59	2.16	1
5771				Pit 1 other	T12G6	5.57		2.18	1
5772				Pit 1 other	T11G7	6.26	3.01	2.53	0
5774				Pit 1 other	T11G7	6.55	3.24	2.29	0
5775				Pit 1 other	T11G7		3.04	2.63	0
5819				Pit 1 other	T13G5	5.72		2.32	1
5821				Pit 1 other	T13G5	5.82		2.01	1

Appendix 4

Spatial and metric data on triggers: Part D and E

Trigger ID	Spatial ID	X	Y	Generic location	Location error	D1R	D1L	D1W	D2	E1R	E1L	E1W	E2
901	148	34.105669	-26.529079	Pit 1 main	exact		1.00		3.19		1.04		3.27
902	77	33.159938	-25.57195	Pit 1 main	exact		1.06		3.24		1.08		3.27
904	91	20.860784	-25.906234	Pit 1 main	exact		0.98				0.94		3.16
905	86	14.533779	-26.714721	Pit 1 main	exact		1.02				0.92		3.25
906	99	13.630381	-26.243872	Pit 1 main	exact		1.22				1.22		
907	102	11.591567	-25.640049	Pit 1 main	exact		1.17				1.18		
908	58	34.201382	13.620131	Pit 1 main	exact		1.06	1.06					
909	149	34.3053	16.990327	Pit 1 main	exact	1.08				1.05			
910	101	10.696355	-25.812038	Pit 1 main	exact		1.14				1.12		
911	63	33.501826	3.116224	Pit 1 main	exact	1.11					0.99		3.29
912	52	34.392391	21.465132	Pit 1 main	exact		1.05	1.05		0.96			3.02
913	59	32.710583	9.192165	Pit 1 main	poor		1.02		3.27		1.07		
914	78	32.952122	-21.474925	Pit 1 main	poor		0.95				0.95		
915	150	33.606594	5.511584	Pit 1 main	poor		1.19			1.13			
918	152	31.610291	-8.585008	Pit 1 main	exact	0.94					1.11		
919	118	18.900838	-4.311712	Pit 1 main	exact	0.95				0.95			
920	154	19.233212	-3.294805	Pit 1 main	exact	0.93				1.02			
921	155	18.933766	-3.095175	Pit 1 main	exact		1.18				1.18		
922	137	18.728001	6.001707	Pit 1 main	exact		1.08			1.05			
923	127	19.851817	7.723742	Pit 1 main	exact		1.15			1.05			
924	131	19.831816	6.456726	Pit 1 main	exact	1.05					1.07		
925	128	19.719986	7.316134	Pit 1 main	exact	1.05					1.05		3.31
926	130	19.610153	6.807441	Pit 1 main	exact	0.98				0.95			
927	136	18.637351	6.658974	Pit 1 main	exact		1.08				1.17		
928	138	20.810811	11.600437	Pit 1 main	exact		1.15			1.05			
929	135	18.91818	8.419803	Pit 1 main	exact		1.09	1.15		1.02			
930	123	21.925709	6.398791	Pit 1 main	exact	0.90				0.88			
931	124	22.19033	5.721752	Pit 1 main	exact	0.98					1.07	1.13	
932	126	22.331048	7.166956	Pit 1 main	exact	1.05					1.16	1.16	
933	163	21.431484	6.557492	Pit 1 main	exact		1.07	1.13			1.06	1.06	
934	132	20.086295	5.674169	Pit 1 main	exact		1.21	1.25		1.09			

Trigger ID	Spatial ID	X	Y	Generic location	Location error	D1R	D1L	D1W	D2	E1R	E1L	E1W	E2
935	133	20.578868	5.394795	Pit 1 main	exact	1.04					1.09		3.61
936	25	21.448853	29.377367	Pit 1 main	poor		1.03	1.13			1.05	1.18	
937	44	6.475602	6.475602	Pit 1 main	poor		1.08	1.08			0.98	0.98	
938	24	18.762581	29.917961	Pit 1 main	poor		1.09	1.16			1.08	1.18	
939	156	28.513778	12.35631	Pit 1 main	exact		1.16	1.18			1.15	1.16	
940	141	20.778356	13.166239	Pit 1 main	exact		1.15	1.16		1.05			
943	35	22.496752	28.661955	Pit 1 main	exact		0.95	1.00			0.95	0.98	
944	32	21.728948	23.548804	Pit 1 main	exact	0.88				0.88			
946	40	26.355985	28.877476	Pit 1 main	exact		1.05	1.06			1.09	1.10	
947	36	22.881719	28.031499	Pit 1 main	exact		0.87	0.91		0.95			
948	39	25.235039	28.33449	Pit 1 main	exact		0.94	1.07		0.94			
949	38	24.125749	28.355277	Pit 1 main	exact		1.05	1.07			0.97	0.99	
950	157	23.151164	0.820272	Pit 1 main	exact	0.94				0.94			
951	41	32.020779	29.775745	Pit 1 main	exact	1.00				1.00			
952	43	32.420465	29.891975	Pit 1 main	exact		1.06	1.15		1.01			
953	48	34.269167	27.43019	Pit 1 main	exact	1.03					1.02	1.02	
954	47	33.964206	29.27371	Pit 1 main	exact		1.01	1.02		1.04			
956	54	32.724943	19.958606	Pit 1 main	exact		0.98	1.05			0.97	1.05	
957	169	33.918877	22.809477	Pit 1 main	poor	0.95				0.95			
958	53	34.683898	20.732773	Pit 1 main	exact		0.95	0.99			0.98	1.06	
959	46	33.960055	29.856813	Pit 1 main	exact	0.90				0.91			
960	31	22.718784	24.068895	Pit 1 main	exact		0.98	1.08			0.95	0.99	
961	34	22.765991	23.168717	Pit 1 main	exact	0.88				0.85			
962	51	33.144621	21.865315	Pit 1 main	poor		0.9	0.93			0.96	0.99	
963	28	20.47937	24.274925	Pit 1 main	exact	0.90				0.89			
964	144	19.549687	19.002285	Pit 1 main	exact	0.87				0.87			
965	50	34.121289	20.732773	Pit 1 main	poor	0.93				0.92			
966	171	33.705328	23.250595	Pit 1 main	poor		1.16	1.23		1.05			
967	222	29.41212	29.397728	Pit 1 main	poor		1.01	1.06			0.97	0.98	
968	176	34.554697	16.824692	Pit 1 main	exact	0.90				0.88			
969	37	24.039188	29.137737	Pit 1 main	exact		1.00	1.05			0.98	0.99	
970	223	23.086751	29.521452	Pit 1 main	poor		1.16	1.18			1.06	1.06	
971	57	32.874234	14.9208	Pit 1 main	exact	0.98					0.96	1.05	
972	167	32.468576	25.632531	Pit 1 main	exact	0.97				0.91			
973	158	28.596817	28.4179	Pit 1 main	exact		1.09	1.11			1.00	1.1	
974	142	22.614815	13.541946	Pit 1 main	exact		1.09	1.25			1.19	1.29	

Trigger ID	Spatial ID	X	Y	Generic location	Location error	D1R	D1L	D1W	D2	E1R	E1L	E1W	E2
975	139	21.094665	12.103157	Pit 1 main	exact	0.93				0.93			
976	68	31.321209	-4.804873	Pit 1 main	poor		0.89		3.33		1.01		3.22
977	60	34.221857	9.737418	Pit 1 main	poor		1.15		3.23		0.97		3.16
978				Pit 1 other	T4G2	0.87				0.85			
979	225	7.922879	-20.702595	Pit 1 main	poor		1.12	1.18		1.02			
1471	55	32.463186	17.22433	Pit 1 main	poor		1.04	1.06		0.94			
1472	56	33.109812	16.732523	Pit 1 main	poor	0.91				0.91			
1498	208	34.213518	-11.978139	Pit 1 main	exact	0.91				0.88			
1499	65	33.149953	1.631469	Pit 1 main	exact	0.92				0.92			
1500	64	34.821564	2.017701	Pit 1 main	exact	0.95					1.1	1.11	
1501	61	33.756327	8.628375	Pit 1 main	exact		1.12	1.17			1.09	1.15	
1502	49	34.069362	26.573211	Pit 1 main	exact		1.06	1.06		0.99			
1503	71	34.052426	-5.372814	Pit 1 main	exact	0.92				0.92			
2231	93	20.117168	-26.579856	Pit 1 main	exact		1.2	1.22			1.17	1.17	
2232	92	20.216574	-26.208776	Pit 1 main	exact		1.00	1.09			1.05	1.08	
2233	94	24.311596	-26.461492	Pit 1 main	exact		0.98	1.04			1.04	1.09	
2234	95	18.83128	-25.807518	Pit 1 main	poor		0.95		3.34		1.01		3.34
2235	98	19.708599	-26.650064	Pit 1 main	poor		1.01	1.08		1.07			
2236	199	14.792157	-25.388635	Pit 1 main	exact		1.04	1.11			1.04	1.09	
2237	186	12.670258	-25.510652	Pit 1 main	exact		1.11	1.23			1.14	1.18	
2238	96	18.26273	-25.835152	Pit 1 main	exact		1.09	1.09			1.04	1.18	
2239	195	11.19016	28.980629	Pit 1 main	exact		1.14	1.19			1.21	1.3	
2240	215	11.924714	-25.653903	Pit 1 main	exact		1.3	1.35			1.21	1.24	
2241	103	11.585001	-26.397574	Pit 1 main	exact		1.03	1.13			1.07	1.13	
2242	104	10.059564	-25.779414	Pit 1 main	exact		1.04	1.13			0.86		3.25
2243	105	9.189363	-25.722035	Pit 1 main	exact		1.12	1.33			1.18	1.21	
2244	106	8.311401	-26.031949	Pit 1 main	exact		1.13	1.17			1.07	1.13	
2245	107	7.6723	-25.728396	Pit 1 main	exact		1.12	1.12			1.06	1.13	
2246	187	6.763962	-25.666675	Pit 1 main	exact		1.25	1.29			1.23	1.29	
2247	109	5.711013	-25.694485	Pit 1 main	exact		1.19	1.26			1.18	1.23	
2248	110	4.925915	-26.395367	Pit 1 main	exact		1.27	1.34			1.22	1.29	
2249	111	4.247799	-26.376172	Pit 1 main	exact		1.06	1.12		1.03			
2250	200	3.114904	-26.362794	Pit 1 main	exact		1.01	1.05			1.01	1.14	
2251	90	21.451586	-25.956809	Pit 1 main	poor	1.05					1.07	1.15	
2252	82	33.560295	-24.307966	Pit 1 main	exact		1.18	1.20			1.26	1.26	

Trigger ID	Spatial ID	X	Y	Generic location	Location error	D1R	D1L	D1W	D2	E1R	E1L	E1W	E2
2253	79	32.014427	-21.256277	Pit 1 main	exact		1.17	1.17		1.05			
2254	72	32.511583	-19.93794	Pit 1 main	exact		1.04	1.06			0.95	0.95	
2255	75	31.593245	-16.92337	Pit 1 main	exact		0.97	1.03		0.88			
2256	166	32.299388	-16.419768	Pit 1 main	exact	0.95				0.95			
2257	74	34.453956	-12.542946	Pit 1 main	exact	0.86					1.01	1.07	
2258	70	32.969332	-9.826935	Pit 1 main	exact	0.83					1.13	1.15	
2259	66	32.629434	2.921166	Pit 1 main	exact	0.90				0.88			
2260	201	23.660796	-20.290116	Pit 1 main	exact		1.03	1.05			1.03	1.12	
2261	177	22.629265	-22.361014	Pit 1 main	exact		1.00	1.00		0.97			
2262	224	8.031137	-21.475867	Pit 1 main	exact		1.13	1.13			1.05	1.14	
2263	147	19.333946	-14.746488	Pit 1 main	exact	0.94					0.91	0.97	
2264	115	18.571095	-10.172154	Pit 1 main	exact	0.86				0.86			
2265	113	17.439096	-10.70497	Pit 1 main	exact	0.91				0.93			
2266	69	30.182927	-9.290728	Pit 1 main	exact	0.93				0.92			3.29
2267	114	18.542612	-10.819296	Pit 1 main	exact	0.88				0.82			
2268	121	20.73572	-4.652206	Pit 1 main	exact	0.92				0.92			
2269	116	17.545465	-4.1724	Pit 1 main	poor	0.90				0.91			
2270	119	19.180578	-5.365282	Pit 1 main	exact	0.91				0.95			
2271	172	31.308036	8.27573	Pit 1 main	exact		1.00		3.41		1.00		3.36
2272	62	30.871735	7.466683	Pit 1 main	exact		1.10	1.10			1.12	1.12	
2273	134	19.708195	5.943434	Pit 1 main	exact		1.09	1.09		1.05			
2274	145	19.960582	18.819977	Pit 1 main	exact		0.90	0.95			0.89	0.98	
2275	146	0.147401	23.554297	Pit 1 main	exact		0.98	1.02			0.97	0.97	
2276	42	29.989851	28.719181	Pit 1 main	exact	0.89				0.91			
2277	220	25.05443	29.400093	Pit 1 main	exact	0.93				0.94			
2278	67	28.19063	1.037085	Pit 1 main	poor	0.85				0.89			
2279	26	18.373139	29.303642	Pit 1 main	exact		1.06	1.08			1.16	1.24	
2280	16	15.438037	29.626916	Pit 1 main	exact		1.17	1.29			1.21	1.23	
2281	30	16.593864	29.73262	Pit 1 main	exact		1.10	1.13			1.07	1.13	
2282	19	16.070755	29.663272	Pit 1 main	exact		1.12	1.21		1.02			
2283	20	16.750686	28.776325	Pit 1 main	exact		1.05	1.11			1.03	1.16	
2284	217	16.27606	29.608201	Pit 1 main	exact		1.05	1.18			1.04	1.12	
2285	21	16.387217	28.613927	Pit 1 main	exact		1.08	1.15			1.06	1.09	
2286	17	15.651582	28.339677	Pit 1 main	exact		1.02	1.13			1.06	1.12	
2287	13	14.079718	29.954447	Pit 1 main	exact		1.05	1.13			1.05	1.17	
2288	12	12.281531	28.692336	Pit 1 main	exact		1.11	1.11			1.07	1.21	

Trigger ID	Spatial ID	X	Y	Generic location	Location error	D1R	D1L	D1W	D2	E1R	E1L	E1W	E2
2289	14	13.437624	29.909026	Pit 1 main	exact		1.18	1.22			1.09	1.23	
2290	196	13.563015	28.946587	Pit 1 main	exact	1.06					1.1	1.13	
2291	15	14.510685	28.747011	Pit 1 main	exact		1.08	1.13			1.06	1.17	
2292	188	10.134564	28.938482	Pit 1 main	exact		1.09	1.11			1.11	1.15	
2293	11	8.691613	29.065114	Pit 1 main	exact		1.09	1.22			1.04	1.22	
2294	9	8.131524	29.722269	Pit 1 main	exact		1.1	1.15			1.02	1.1	
2295	10	12.627445	30.030236	Pit 1 main	exact	1.08					1.01	1.14	
2296	7	7.598419	29.55203	Pit 1 main	exact		1.11	1.15			1.02	1.11	
2297	5	5.704329	29.938901	Pit 1 main	exact		1.11	1.16		1.03			
2298	6	6.423473	29.887307	Pit 1 main	exact	1.09				1.04			
2299	219	4.300397	29.021715	Pit 1 main	exact		1.08	1.19			1.09	1.18	
2300	197	2.544893	29.841492	Pit 1 main	exact		1.08	1.16			1.03	1.14	
2301	189	2.642265	29.315028	Pit 1 main	exact		1.11	1.12			0.95	1.08	
2302	190	0.643675	29.903984	Pit 1 main	exact		1.01	1.14			1.09	1.14	
2303	205	16.854337	-25.755526	Pit 1 main	exact		1.26	1.31			1.24	1.29	
2304	97	14.614332	-25.879788	Pit 1 main	exact		1.03	1.13			1.02	1.11	
2305	198	12.775463	-25.658393	Pit 1 main	exact		1.08	1.15			0.96	1.09	
2306	85	27.579737	-26.413922	Pit 1 main	poor		1.12	1.15			1.08	1.08	
2307	100	13.312002	-26.493856	Pit 1 main	exact		1.17	1.34			1.17		3.21
2308	88	22.299389	-26.486917	Pit 1 main	poor		1.05	1.06			1.09	1.16	
2309	108	7.552892	-26.292185	Pit 1 main	exact		1.05	1.17			0.99	1.08	
2310	191	7.746243	-25.462451	Pit 1 main	exact		1.09	1.16			1.09	1.14	
2311	89	21.784246	-26.170666	Pit 1 main	exact	0.98					1.08	1.11	
2312	202	5.771701	-25.375984	Pit 1 main	exact		1.09	1.15			1.02	1.06	
2313	206	5.037919	-25.70492	Pit 1 main	exact		1.12	1.19			1.19	1.21	
2314	207	4.000503	-25.679618	Pit 1 main	exact		1.03	1.21			1.2	1.2	
2315	112	1.016835	-26.364434	Pit 1 main	exact		1.11	1.14			1.09	1.12	
2316	203	-0.300977	-25.375984	Pit 1 main	exact		1.12	1.15		1.00			
2317	204	-0.199766	-25.300075	Pit 1 main	exact		1.07	1.19			1.12	1.12	
2318	178	33.882004	-14.574878	Pit 1 main	exact		1.01	1.06			0.95	0.99	
2319	76	31.655381	-10.748628	Pit 1 main	exact		1.01	1.01			0.89		2.96
2320	73	33.907924	-12.074918	Pit 1 main	exact	0.92					0.89		
2321	183	33.045519	11.708125	Pit 1 main	exact		1.02	1.02		0.95			
2322	180	8.729924	-21.131185	Pit 1 main	exact		1.17	1.22			1.07		3.31
2323	168	17.568892	-10.531686	Pit 1 main	exact	0.97				0.96			
2324	120	20.137346	-5.492228	Pit 1 main	exact	0.91				0.92			

Trigger ID	Spatial ID	X	Y	Generic location	Location error	D1R	D1L	D1W	D2	E1R	E1L	E1W	E2
2325	117	18.912984	-5.565224	Pit 1 main	poor	0.94				0.94			
2326	140	21.649561	12.672669	Pit 1 main	exact		1.13	1.13		1.03			
2327	175	19.416374	11.909297	Pit 1 main	exact		1.09	1.09			1.17	1.18	
2328	174	20.330545	11.744812	Pit 1 main	exact	1.15					1.19	1.23	
2329	143	21.020498	16.823586	Pit 1 main	exact		0.96	1.02			0.94	0.97	
2330	181	19.285416	17.121407	Pit 1 main	exact		0.94	0.98			0.96	1.09	
2331	179	18.7305	16.938069	Pit 1 main	exact		1.02	1.05			0.95	0.98	
2332	45	31.741774	28.239928	Pit 1 main	exact		1.07	1.08		1.02			
2333	218	17.430113	29.683877	Pit 1 main	exact		1.04	1.17			1.06	1.15	
2334	210	17.3166	29.002796	Pit 1 main	exact		1.05	1.16			1.01	1.09	
2335	216	14.2139	29.475769	Pit 1 main	exact		1.09	1.09		1.08			
2336	214	11.849038	29.854147	Pit 1 main	exact		1.01	1.14			1.11	1.19	
2337	192	10.574617	29.436132	Pit 1 main	exact		1.01	1.17			1.05	1.09	
2338	221	9.631811	29.737968	Pit 1 main	exact		1.05	1.15			1.15	1.19	
2339	209	11.13012	29.948741	Pit 1 main	exact		1.09	1.16			1.13	1.14	
2340	212	9.465257	28.964958	Pit 1 main	exact		1.02	1.07			1.08	1.09	
2341	8	7.218884	29.11886	Pit 1 main	exact		1.11	1.13			1.03	1.14	
2342	213	8.02742	29.154147	Pit 1 main	exact		1.04	1.13		1.01			
2343	87	24.314747	-26.094249	Pit 1 main	exact		1.24	1.26			1.22	1.24	
2344	4	4.874686	29.922505	Pit 1 main	exact		1.09	1.09			1.05	1.15	
2345	1	2.405659	28.996267	Pit 1 main	exact		1.04	1.13			1.06	1.13	
2346	193	1.216068	29.248525	Pit 1 main	exact		1.11	1.16			1.11	1.16	
2347	2	0.455338	29.241239	Pit 1 main	exact		1.07	1.11			1.04	1.09	
2348	211	0.743644	29.267661	Pit 1 main	exact		1.04	1.12			1.09	1.12	
2349	3	2.167348	30.072502	Pit 1 main	exact		1.08	1.09			1.31	1.33	
2350	194	16.797904	-8.650193	Pit 1 main	exact	0.86				0.88			
2351	184	18.719879	17.777814	Pit 1 main	exact		0.97	0.99			0.91	1.02	
2352	162	18.074013	7.253553	Pit 1 main	exact	0.95				0.95			
2353	129	19.610153	6.807441	Pit 1 main	exact		1.15	1.15			1.05		3.32
2354	165	18.608398	5.982347	Pit 1 main	exact		1.03		3.38		1.03		3.63
2355	159	18.360626	5.861431	Pit 1 main	exact	1.06				1.07			
2356	160	19.220466	5.738597	Pit 1 main	exact		1.05	1.08			1		3.37
2357	173	18.200631	13.60639	Pit 1 main	exact		1.19	1.19			1.09		3.54
2358	170	18.183554	11.69997	Pit 1 main	exact	0.95				1.03			
2359	22	17.925771	29.66081	Pit 1 main	exact		1.04	1.13			1.12	1.35	
2360	23	18.113679	28.758125	Pit 1 main	poor		1.07	1.2			1.11	1.18	

Trigger ID	Spatial ID	X	Y	Generic location	Location error	D1R	D1L	D1W	D2	E1R	E1L	E1W	E2
3788	83	32.811009	-20.817251	Pit 1 main	exact		0.95	0.96			0.86		3.28
3790	225	33.718573	-23.341457	Pit 1 main	exact	0.92					1.06	1.13	
3793	81	31.776178	-25.53808	Pit 1 main	poor		1.21	1.26		1.06			
3811	80	34.026357	-23.930563	Pit 1 main	exact		1.05	1.16			1.05	1.12	
3821	151	34.005854	0.820272	Pit 1 main	poor		1.01		3.29	1			3.21
3825	84	32.315821	-26.731492	Pit 1 main	poor		1.08	1.1			1.07		3.28
5525	161	17.7874	6.475602	Pit 1 main	poor	1.41			3.31	1.29			3.19
5543				Pit 2	pit 2		0.93		3.17		0.96		3.26
5553	185	5.551855	24.004271	Pit 1 main	exact		1.02	1.07			0.99	1.03	
5758				Pit 1 other	T12G6	0.94				0.93			
5760				Pit 1 other	T12G6		1.03	1.03			1.1	1.1	
5761				Pit 1 other	T12G6	0.95				0.95			
5768				Pit 1 other	T12G6	0.95				0.95			
5771				Pit 1 other	T12G6	0.93				0.89			
5772				Pit 1 other	T11G7		1.11	1.11			1.05	1.06	
5774				Pit 1 other	T11G7	0.99					1.01	1.06	
5775				Pit 1 other	T11G7	1.01					1.02		3.46
5819				Pit 1 other	T13G5	0.93					1.04		3.11
5821				Pit 1 other	T13G5	0.87				0.88			

Appendix 5

Correspondence between trigger part groups

Trigger ID	A Group	B Group	B notes	C Groups	C Notes	AssemG	Assemblage	Assem Group
915	Ag1	Bg1n	notched	Cg2	curved	2	Ag1-Bg1n-Cg2	G-1
932	Ag1	Bg1n	notched	Cg2	curved	2	Ag1-Bg1n-Cg2	G-1
934	Ag1	Bg1n	notched	Cg2	curved	2	Ag1-Bg1n-Cg2	G-1
966	Ag1	Bg1n	notched	Cg2	curved	2	Ag1-Bg1n-Cg2	G-1
1502	Ag1	Bg1n	notched	Cg2	curved	2	Ag1-Bg1n-Cg2	G-1
2273	Ag1	Bg1n	notched	Cg2	curved	2	Ag1-Bg1n-Cg2	G-1
2352	Ag1	Bg1n	notched	Cg3c	curved	3	Ag1-Bg1n-Cg3c	G-1
2357	Ag1	Bg1n	notched	Cg2	curved	2	Ag1-Bg1n-Cg2	G-1
2358	Ag1	Bg1n	notched	Cg2	curved	2	Ag1-Bg1n-Cg2	G-1
922	Ag1	Bg1u	unnotched	Cg2	curved	5	Ag1-Bg1u-Cg2	G-2
923	Ag1	Bg1u	unnotched	Cg2	curved	5	Ag1-Bg1u-Cg2	G-2
924	Ag1	Bg1u	unnotched	Cg1	curved	4	Ag1-Bg1u-Cg1	G-2
925	Ag1	Bg1u	unnotched	Cg2	curved	5	Ag1-Bg1u-Cg2	G-2
926	Ag1	Bg1u	unnotched	Cg2	curved	5	Ag1-Bg1u-Cg2	G-2
927	Ag1	Bg1u	unnotched	Cg2	curved	5	Ag1-Bg1u-Cg2	G-2
928	Ag1	Bg1u	unnotched	Cg2	curved	5	Ag1-Bg1u-Cg2	G-2
933	Ag1	Bg1u	unnotched	Cg2	curved	5	Ag1-Bg1u-Cg2	G-2
935	Ag1	Bg1u	unnotched	Cg2	curved	5	Ag1-Bg1u-Cg2	G-2
937	Ag1	Bg1u	unnotched	Cg2	curved	5	Ag1-Bg1u-Cg2	G-2
939	Ag1	Bg1u	unnotched	Cg2	curved	5	Ag1-Bg1u-Cg2	G-2
940	Ag1	Bg1u	unnotched	Cg2	curved	5	Ag1-Bg1u-Cg2	G-2
952	Ag1	Bg1u	unnotched	Cg2	curved	5	Ag1-Bg1u-Cg2	G-2
953	Ag1	Bg1u	unnotched	Cg2	curved	5	Ag1-Bg1u-Cg2	G-2
954	Ag1	Bg1u	unnotched	Cg2	curved	5	Ag1-Bg1u-Cg2	G-2
973	Ag1	Bg1u	unnotched	Cg2	curved	5	Ag1-Bg1u-Cg2	G-2
1501	Ag1	Bg1u	unnotched	Cg2	curved	5	Ag1-Bg1u-Cg2	G-2
2271	Ag1	Bg1u	unnotched	Cg2	curved	5	Ag1-Bg1u-Cg2	G-2
2272	Ag1	Bg1u	unnotched	Cg2	curved	5	Ag1-Bg1u-Cg2	G-2
2326	Ag1	Bg1u	unnotched	Cg2	curved	5	Ag1-Bg1u-Cg2	G-2
2327	Ag1	Bg1u	unnotched	Cg2	curved	5	Ag1-Bg1u-Cg2	G-2
2328	Ag1	Bg1n	notched	Cg1	curved	1	Ag1-Bg1n-Cg1	G-1
2332	Ag1	Bg1u	unnotched	Cg2	curved	5	Ag1-Bg1u-Cg2	G-2
2353	Ag1	Bg1u	unnotched	Cg2	curved	5	Ag1-Bg1u-Cg2	G-2

Trigger ID	A Group	B Group	B notes	C Groups	C Notes	AssemG	Assemblage	Assem Group
2354	Ag1	Bg1u	unnotched	Cg2	curved	5	Ag1-Bg1u-Cg2	G-2
2355	Ag1	Bg1u	unnotched	Cg1	curved	4	Ag1-Bg1u-Cg1	G-2
2356	Ag1	Bg1u	unnotched	Cg2	curved	5	Ag1-Bg1u-Cg2	G-2
929	Ag1	Bg2u	unnotched	Cg2	curved	13	Ag1-Bg2u-Cg2	G-2
944	Ag2	Bg1u	unnotched	Cg3c	curved	7	Ag2-Bg1u-Cg3c	G-4
961	Ag2	Bg1u	unnotched	Cg3c	curved	7	Ag2-Bg1u-Cg3c	G-4
964	Ag2	Bg1u	unnotched	Cg3c	curved	7	Ag2-Bg1u-Cg3c	G-4
968	Ag2	Bg1u	unnotched	Cg3c	curved	7	Ag2-Bg1u-Cg3c	G-4
1472	Ag2	Bg1u	unnotched	Cg3c	curved	7	Ag2-Bg1u-Cg3c	G-4
1498	Ag2	Bg1u	unnotched	Cg3c	curved	7	Ag2-Bg1u-Cg3c	G-4
2247	Ag2	Bg1u	unnotched	Cg1	curved	6	Ag2-Bg1u-Cg1	G-4
2257	Ag2	Bg1u	unnotched	Cg3c	curved	7	Ag2-Bg1u-Cg3c	G-4
2258	Ag2	Bg1u	unnotched	Cg3c	curved	7	Ag2-Bg1u-Cg3c	G-4
2264	Ag2	Bg1u	unnotched	Cg3c	curved	7	Ag2-Bg1u-Cg3c	G-4
2267	Ag2	Bg1u	unnotched	Cg3c	curved	7	Ag2-Bg1u-Cg3c	G-4
2276	Ag2	Bg1u	unnotched	Cg3c	curved	7	Ag2-Bg1u-Cg3c	G-4
2278	Ag2	Bg1u	unnotched	Cg3c	curved	7	Ag2-Bg1u-Cg3c	G-4
2322	Ag2	Bg1u	unnotched	Cg1	curved	6	Ag2-Bg1u-Cg1	G-4
2343	Ag2	Bg1u	unnotched	Cg1	curved	6	Ag2-Bg1u-Cg1	G-4
2347	Ag2	Bg1u	unnotched	Cg1	curved	6	Ag2-Bg1u-Cg1	G-4
2348	Ag2	Bg1u	unnotched	Cg1	curved	6	Ag2-Bg1u-Cg1	G-4
2350	Ag2	Bg1u	unnotched	Cg3c	curved	7	Ag2-Bg1u-Cg3c	G-4
918	Ag2	Bg2n	notched	Cg3b	bevelled	8	Ag2-Bg2n-Cg3b	G-3
919	Ag2	Bg2n	notched	Cg3b	bevelled	8	Ag2-Bg2n-Cg3b	G-3
920	Ag2	Bg2n	notched	Cg3b	bevelled	8	Ag2-Bg2n-Cg3b	G-3
921	Ag2	Bg2n	notched	Cg3b	bevelled	8	Ag2-Bg2n-Cg3b	G-3
930	Ag2	Bg2n	notched	Cg3b	bevelled	8	Ag2-Bg2n-Cg3b	G-3
931	Ag2	Bg2n	notched	Cg3b	bevelled	8	Ag2-Bg2n-Cg3b	G-3
1499	Ag2	Bg2n	notched	Cg3b	bevelled	8	Ag2-Bg2n-Cg3b	G-3
1500	Ag2	Bg2n	notched	Cg3b	bevelled	8	Ag2-Bg2n-Cg3b	G-3
1503	Ag2	Bg2n	notched	Cg3b	bevelled	8	Ag2-Bg2n-Cg3b	G-3
2259	Ag2	Bg2n	notched	Cg3b	bevelled	8	Ag2-Bg2n-Cg3b	G6
2266	Ag2	Bg2n	notched	Cg3b	bevelled	8	Ag2-Bg2n-Cg3b	G-3
2268	Ag2	Bg2n	notched	Cg3b	bevelled	8	Ag2-Bg2n-Cg3b	G-3
2269	Ag2	Bg2n	notched	Cg3b	bevelled	8	Ag2-Bg2n-Cg3b	G-3
2270	Ag2	Bg2n	notched	Cg3b	bevelled	8	Ag2-Bg2n-Cg3b	G-3
2324	Ag2	Bg2n	notched	Cg3b	bevelled	8	Ag2-Bg2n-Cg3b	G-3
2325	Ag2	Bg2n	notched	Cg3b	bevelled	8	Ag2-Bg2n-Cg3b	G-3

Trigger ID	A Group	B Group	B notes	C Groups	C Notes	AssemG	Assemblage	Assem Group
3790	Ag2	Bg2n	notched	Cg3b	bevelled	8	Ag2-Bg2n-Cg3b	G-3
3825	Ag2	Bg2n	notched	Cg3b	bevelled	8	Ag2-Bg2n-Cg3b	G-3
904	Ag2	Bg2u	unnotched	Cg1	curved	9	Ag2-Bg2u-Cg1	G-4
905	Ag2	Bg2u	unnotched	Cg1	curved	9	Ag2-Bg2u-Cg1	G-4
906	Ag2	Bg2u	unnotched	Cg1	curved	9	Ag2-Bg2u-Cg1	G-7
907	Ag2	Bg2u	unnotched	Cg1	curved	9	Ag2-Bg2u-Cg1	G-4
908	Ag2	Bg2u	unnotched	Cg1	curved	9	Ag2-Bg2u-Cg1	G-4
909	Ag2	Bg2u	unnotched	Cg1	curved	9	Ag2-Bg2u-Cg1	G-4
910	Ag2	Bg2u	unnotched	Cg1	curved	9	Ag2-Bg2u-Cg1	G-4
911	Ag2	Bg2u	unnotched	Cg1	curved	9	Ag2-Bg2u-Cg1	G-4
912	Ag2	Bg2u	unnotched	Cg1	curved	9	Ag2-Bg2u-Cg1	G-7
913	Ag2	Bg2u	unnotched	Cg1	curved	9	Ag2-Bg2u-Cg1	G-4
914	Ag2	Bg2u	unnotched	Cg3c	curved	11	Ag2-Bg2u-Cg3c	G-4
936	Ag2	Bg2u	unnotched	Cg1	curved	9	Ag2-Bg2u-Cg1	G-4
938	Ag2	Bg2u	unnotched	Cg1	curved	9	Ag2-Bg2u-Cg1	G-4
943	Ag2	Bg2u	unnotched	Cg1	curved	9	Ag2-Bg2u-Cg1	G-4
946	Ag2	Bg2u	unnotched	Cg1	curved	9	Ag2-Bg2u-Cg1	G-7
947	Ag2	Bg2u	unnotched	Cg1	curved	9	Ag2-Bg2u-Cg1	G-4
948	Ag2	Bg2u	unnotched	Cg1	curved	9	Ag2-Bg2u-Cg1	G-4
949	Ag2	Bg2u	unnotched	Cg1	curved	9	Ag2-Bg2u-Cg1	G-4
951	Ag2	Bg2u	unnotched	Cg1	curved	9	Ag2-Bg2u-Cg1	G-4
956	Ag2	Bg2u	unnotched	Cg1	curved	9	Ag2-Bg2u-Cg1	G-4
958	Ag2	Bg2u	unnotched	Cg1	curved	9	Ag2-Bg2u-Cg1	G-4
960	Ag2	Bg2u	unnotched	Cg1	curved	9	Ag2-Bg2u-Cg1	G-4
967	Ag2	Bg2u	unnotched	Cg1	curved	9	Ag2-Bg2u-Cg1	G-4
969	Ag2	Bg2u	unnotched	Cg1	curved	9	Ag2-Bg2u-Cg1	G-4
970	Ag2	Bg2u	unnotched	Cg1	curved	9	Ag2-Bg2u-Cg1	G-4
971	Ag2	Bg2u	unnotched	Cg1	curved	9	Ag2-Bg2u-Cg1	G-4
974	Ag2	Bg2u	unnotched	Cg1	curved	9	Ag2-Bg2u-Cg1	G-4
976	Ag2	Bg2u	unnotched	Cg1	curved	9	Ag2-Bg2u-Cg1	G-4
977	Ag2	Bg2u	unnotched	Cg1	curved	9	Ag2-Bg2u-Cg1	G-4
979	Ag2	Bg2u	unnotched	Cg1	curved	9	Ag2-Bg2u-Cg1	G-4
1471	Ag2	Bg2u	unnotched	Cg1	curved	9	Ag2-Bg2u-Cg1	G-4
2231	Ag2	Bg2u	unnotched	Cg1	curved	9	Ag2-Bg2u-Cg1	G-4
2232	Ag2	Bg2u	unnotched	Cg1	curved	9	Ag2-Bg2u-Cg1	G-4
2233	Ag2	Bg2u	unnotched	Cg1	curved	9	Ag2-Bg2u-Cg1	G-4
2234	Ag2	Bg2u	unnotched	Cg1	curved	9	Ag2-Bg2u-Cg1	G-4
2235	Ag2	Bg2u	unnotched	Cg1	curved	9	Ag2-Bg2u-Cg1	G-4

Trigger ID	A Group	B Group	B notes	C Groups	C Notes	AssemG	Assemblage	Assem Group
2236	Ag2	Bg2u	unnotched	Cg1	curved	9	Ag2-Bg2u-Cg1	G-4
2237	Ag2	Bg2u	unnotched	Cg1	curved	9	Ag2-Bg2u-Cg1	G-4
2238	Ag2	Bg2u	unnotched	Cg1	curved	9	Ag2-Bg2u-Cg1	G-4
2239	Ag2	Bg2u	unnotched	Cg1	curved	9	Ag2-Bg2u-Cg1	G-4
2240	Ag2	Bg2u	unnotched	Cg1	curved	9	Ag2-Bg2u-Cg1	G-4
2241	Ag2	Bg2u	unnotched	Cg1	curved	9	Ag2-Bg2u-Cg1	G-4
2242	Ag2	Bg2u	unnotched	Cg1	curved	9	Ag2-Bg2u-Cg1	G-4
2243	Ag2	Bg2u	unnotched	Cg1	curved	9	Ag2-Bg2u-Cg1	G-4
2244	Ag2	Bg2u	unnotched	Cg1	curved	9	Ag2-Bg2u-Cg1	G-4
2245	Ag2	Bg2u	unnotched	Cg1	curved	9	Ag2-Bg2u-Cg1	G-4
2246	Ag2	Bg2u	unnotched	Cg1	curved	9	Ag2-Bg2u-Cg1	G-4
2248	Ag2	Bg2u	unnotched	Cg1	curved	9	Ag2-Bg2u-Cg1	G-4
2249	Ag2	Bg2u	unnotched	Cg1	curved	9	Ag2-Bg2u-Cg1	G-4
2250	Ag2	Bg2u	unnotched	Cg1	curved	9	Ag2-Bg2u-Cg1	G-4
2251	Ag2	Bg2u	unnotched	Cg1	curved	9	Ag2-Bg2u-Cg1	G-4
2252	Ag2	Bg2u	unnotched	Cg1	curved	9	Ag2-Bg2u-Cg1	G-4
2253	Ag2	Bg2u	unnotched	Cg1	curved	9	Ag2-Bg2u-Cg1	G-4
2254	Ag2	Bg2u	unnotched	Cg2	curved	10	Ag2-Bg2u-Cg2	G-4
2255	Ag2	Bg2u	unnotched	Cg2	curved	10	Ag2-Bg2u-Cg2	G-4
2260	Ag2	Bg2u	unnotched	Cg1	curved	9	Ag2-Bg2u-Cg1	G-4
2261	Ag2	Bg2u	unnotched	Cg1	curved	9	Ag2-Bg2u-Cg1	G-4
2262	Ag2	Bg2u	unnotched	Cg1	curved	9	Ag2-Bg2u-Cg1	G-4
2274	Ag2	Bg2u	unnotched	Cg1	curved	9	Ag2-Bg2u-Cg1	G-4
2275	Ag2	Bg2u	unnotched	Cg1	curved	9	Ag2-Bg2u-Cg1	G-4
2279	Ag2	Bg2u	unnotched	Cg1	curved	9	Ag2-Bg2u-Cg1	G-4
2280	Ag2	Bg2u	unnotched	Cg1	curved	9	Ag2-Bg2u-Cg1	G-4
2281	Ag2	Bg2u	unnotched	Cg1	curved	9	Ag2-Bg2u-Cg1	G-4
2282	Ag2	Bg2u	unnotched	Cg1	curved	9	Ag2-Bg2u-Cg1	G-4
2283	Ag2	Bg2u	unnotched	Cg1	curved	9	Ag2-Bg2u-Cg1	G-4
2284	Ag2	Bg2u	unnotched	Cg1	curved	9	Ag2-Bg2u-Cg1	G-4
2285	Ag2	Bg2u	unnotched	Cg1	curved	9	Ag2-Bg2u-Cg1	G-4
2286	Ag2	Bg2u	unnotched	Cg1	curved	9	Ag2-Bg2u-Cg1	G-4
2287	Ag2	Bg2u	unnotched	Cg1	curved	9	Ag2-Bg2u-Cg1	G-4
2288	Ag2	Bg2u	unnotched	Cg1	curved	9	Ag2-Bg2u-Cg1	G-4
2289	Ag2	Bg2u	unnotched	Cg1	curved	9	Ag2-Bg2u-Cg1	G-4
2290	Ag2	Bg2u	unnotched	Cg1	curved	9	Ag2-Bg2u-Cg1	G-4
2291	Ag2	Bg2u	unnotched	Cg1	curved	9	Ag2-Bg2u-Cg1	G-4
2292	Ag2	Bg2u	unnotched	Cg1	curved	9	Ag2-Bg2u-Cg1	G-4

Trigger ID	A Group	B Group	B notes	C Groups	C Notes	AssemG	Assemblage	Assem Group
2293	Ag2	Bg2u	unnotched	Cg1	curved	9	Ag2-Bg2u-Cg1	G-4
2294	Ag2	Bg2u	unnotched	Cg1	curved	9	Ag2-Bg2u-Cg1	G-4
2295	Ag2	Bg2u	unnotched	Cg1	curved	9	Ag2-Bg2u-Cg1	G-4
2296	Ag2	Bg2u	unnotched	Cg1	curved	9	Ag2-Bg2u-Cg1	G-4
2297	Ag2	Bg2u	unnotched	Cg1	curved	9	Ag2-Bg2u-Cg1	G-4
2298	Ag2	Bg2u	unnotched	Cg1	curved	9	Ag2-Bg2u-Cg1	G-4
2299	Ag2	Bg2u	unnotched	Cg1	curved	9	Ag2-Bg2u-Cg1	G-4
2300	Ag2	Bg2u	unnotched	Cg1	curved	9	Ag2-Bg2u-Cg1	G-4
2301	Ag2	Bg2u	unnotched	Cg1	curved	9	Ag2-Bg2u-Cg1	G-4
2302	Ag2	Bg2u	unnotched	Cg1	curved	9	Ag2-Bg2u-Cg1	G-4
2303	Ag2	Bg2u	unnotched	Cg1	curved	9	Ag2-Bg2u-Cg1	G-4
2304	Ag2	Bg2u	unnotched	Cg1	curved	9	Ag2-Bg2u-Cg1	G-4
2305	Ag2	Bg2u	unnotched	Cg1	curved	9	Ag2-Bg2u-Cg1	G-4
2306	Ag2	Bg2u	unnotched	Cg1	curved	9	Ag2-Bg2u-Cg1	G-4
2307	Ag2	Bg2u	unnotched	Cg1	curved	9	Ag2-Bg2u-Cg1	G-4
2308	Ag2	Bg2u	unnotched	Cg1	curved	9	Ag2-Bg2u-Cg1	G-4
2309	Ag2	Bg2u	unnotched	Cg1	curved	9	Ag2-Bg2u-Cg1	G-4
2310	Ag2	Bg2u	unnotched	Cg1	curved	9	Ag2-Bg2u-Cg1	G-4
2311	Ag2	Bg2u	unnotched	Cg1	curved	9	Ag2-Bg2u-Cg1	G-4
2312	Ag2	Bg2u	unnotched	Cg1	curved	9	Ag2-Bg2u-Cg1	G-4
2313	Ag2	Bg2u	unnotched	Cg1	curved	9	Ag2-Bg2u-Cg1	G-4
2314	Ag2	Bg2u	unnotched	Cg1	curved	9	Ag2-Bg2u-Cg1	G-4
2315	Ag2	Bg2u	unnotched	Cg1	curved	9	Ag2-Bg2u-Cg1	G-4
2316	Ag2	Bg2u	unnotched	Cg1	curved	9	Ag2-Bg2u-Cg1	G-4
2317	Ag2	Bg2u	unnotched	Cg1	curved	9	Ag2-Bg2u-Cg1	G-4
2318	Ag2	Bg2u	unnotched	Cg1	curved	9	Ag2-Bg2u-Cg1	G-4
2319	Ag2	Bg2u	unnotched	Cg1	curved	9	Ag2-Bg2u-Cg1	G-4
2321	Ag2	Bg2u	unnotched	Cg1	curved	9	Ag2-Bg2u-Cg1	G-4
2329	Ag2	Bg2u	unnotched	Cg1	curved	9	Ag2-Bg2u-Cg1	G-4
2330	Ag2	Bg2u	unnotched	Cg3c	curved	11	Ag2-Bg2u-Cg3c	G-4
2331	Ag2	Bg2u	unnotched	Cg1	curved	9	Ag2-Bg2u-Cg1	G-4
2333	Ag2	Bg2u	unnotched	Cg1	curved	9	Ag2-Bg2u-Cg1	G-4
2334	Ag2	Bg2u	unnotched	Cg1	curved	9	Ag2-Bg2u-Cg1	G-4
2335	Ag2	Bg2u	unnotched	Cg1	curved	9	Ag2-Bg2u-Cg1	G-4
2336	Ag2	Bg2u	unnotched	Cg1	curved	9	Ag2-Bg2u-Cg1	G-4
2337	Ag2	Bg2u	unnotched	Cg1	curved	9	Ag2-Bg2u-Cg1	G-4
2338	Ag2	Bg2u	unnotched	Cg1	curved	9	Ag2-Bg2u-Cg1	G-4
2339	Ag2	Bg2u	unnotched	Cg1	curved	9	Ag2-Bg2u-Cg1	G-4

Trigger ID	A Group	B Group	B notes	C Groups	C Notes	AssemG	Assemblage	Assem Group
2340	Ag2	Bg2u	unnotched	Cg1	curved	9	Ag2-Bg2u-Cg1	G-4
2341	Ag2	Bg2u	unnotched	Cg1	curved	9	Ag2-Bg2u-Cg1	G-4
2342	Ag2	Bg2u	unnotched	Cg1	curved	9	Ag2-Bg2u-Cg1	G-4
2344	Ag2	Bg2u	unnotched	Cg1	curved	9	Ag2-Bg2u-Cg1	G-4
2345	Ag2	Bg2u	unnotched	Cg1	curved	9	Ag2-Bg2u-Cg1	G-4
2346	Ag2	Bg2u	unnotched	Cg1	curved	9	Ag2-Bg2u-Cg1	G-4
2349	Ag2	Bg2u	unnotched	Cg1	curved	9	Ag2-Bg2u-Cg1	G-4
2351	Ag2	Bg2u	unnotched	Cg1	curved	9	Ag2-Bg2u-Cg1	G-4
2359	Ag2	Bg2u	unnotched	Cg1	curved	9	Ag2-Bg2u-Cg1	G-4
2360	Ag2	Bg2u	unnotched	Cg1	curved	9	Ag2-Bg2u-Cg1	G-4
3788	Ag2	Bg2u	unnotched	Cg3c	curved	11	Ag2-Bg2u-Cg3c	G-4
3793	Ag2	Bg2u	unnotched	Cg1	curved	9	Ag2-Bg2u-Cg1	G-4
3811	Ag2	Bg2u	unnotched	Cg1	curved	9	Ag2-Bg2u-Cg1	G-7
3821	Ag2	Bg2u	unnotched	Cg1	curved	9	Ag2-Bg2u-Cg1	G-4
5553	Ag2	Bg2u	unnotched	Cg1	curved	9	Ag2-Bg2u-Cg1	G-4
901	Ag2	Bg2u	unnotched	Cg1	curved	9	Ag2-Bg2u-Cg1	G-4
902	Ag2	Bg2u	unnotched	Cg1	curved	9	Ag2-Bg2u-Cg1	G-4
962	Ag2	Bg2u	unnotched	Cg1	curved	9	Ag2-Bg2u-Cg1	G-4
963	Ag2	Bg2u	unnotched	Cg3c	curved	11	Ag2-Bg2u-Cg3c	G-4
950	Ag3	Bg2u	unnotched	Cg3c	curved	12	Ag3-Bg2u-Cg3c	G-5
957	Ag3	Bg2u	unnotched	Cg3c	curved	12	Ag3-Bg2u-Cg3c	G-5
959	Ag3	Bg2u	unnotched	Cg3c	curved	12	Ag3-Bg2u-Cg3c	G-5
965	Ag3	Bg2u	unnotched	Cg3c	curved	12	Ag3-Bg2u-Cg3c	G-5
972	Ag3	Bg2u	unnotched	Cg3c	curved	12	Ag3-Bg2u-Cg3c	G-5
975	Ag3	Bg2u	unnotched	Cg3c	curved	12	Ag3-Bg2u-Cg3c	G-5
2256	Ag3	Bg2u	unnotched	Cg3c	curved	12	Ag3-Bg2u-Cg3c	G-5
2263	Ag3	Bg2u	unnotched	Cg3c	curved	12	Ag3-Bg2u-Cg3c	G-5
2265	Ag3	Bg2u	unnotched	Cg3c	curved	12	Ag3-Bg2u-Cg3c	G-5
2277	Ag3	Bg2u	unnotched	Cg3c	curved	12	Ag3-Bg2u-Cg3c	G-5
2320	Ag3	Bg2u	unnotched	Cg3c	curved	12	Ag3-Bg2u-Cg3c	G-5
2323	Ag3	Bg2u	unnotched	Cg3c	curved	12	Ag3-Bg2u-Cg3c	G-5
5525	Ag4	Bspecial	Bspecial	special	Cspecial	14	Ag4-Bg3- Cg4	G-8

Appendix 6

Inscriptions on triggers

Trigger ID	Inscription code	Location	Symbol
901	22	Aa	
901	22	Bc	
901	22	Ca	
902	63	Aa	
902	63	Be	
902	63	Ca	
904	11_9	Bb	
905	56_62	Bd	
906	12	Aa	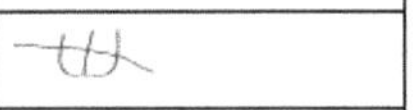
906	12	Ca	
907			
908	7_4	Bb	
909	4_2	Ab	
909	4_2	Bd	
910			
911			
912	4_9	Bb	
913	36	Ca	
913	37	Aa	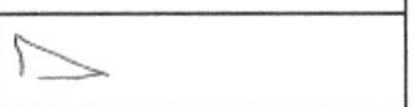
913	40	Bc	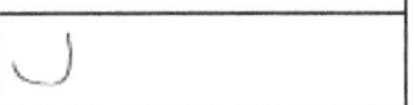
914	2_4	Ac	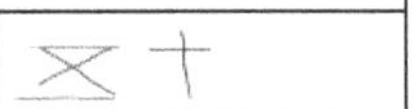
915			
918	32	Bc	
918	33	Ca	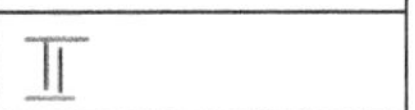
919			
920	2_4_2	Ac	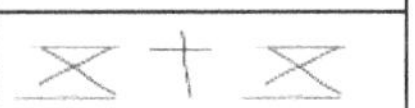
920	2_4_2	Bc	
920	2_4_2	Ca	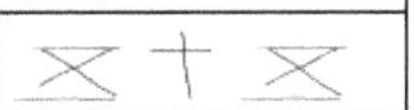
921	11_9	E	
922			
923			
924			

Trigger ID	Inscription code	Location	Symbol
925			
926			
927			
928			
929			
930			
931	57	Bc	
932			
933			
934			
935			
936			
937	1	Bc	
938			
939			
940			
943	39	Bb	
944	14	Ba	
946	3	Aa	
946	3	Bc	
946	3	Ca	
947	73_3	Bb	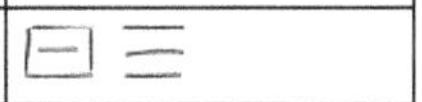
947	73_3	Cb	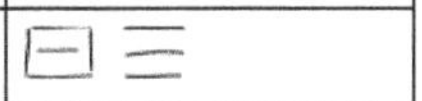
948	9_41	Bb	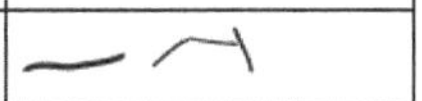
949	41	Bb	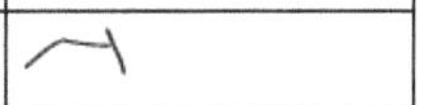
950			
951	12_4		
952			
953	4_4	Bc	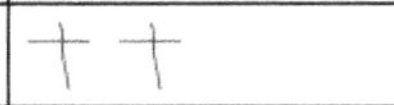
954	74	E	
956	4_9	Bb	
957	62	Ba	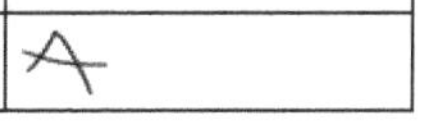

Trigger ID	Inscription code	Location	Symbol
958	75_12	Bb	
959			
960	7_11	Bb	
961	15	Ba	
961	15	D	
962	7	Bb	
963			
964	16	Ba	
965	42_43	Ba	
966			
967	5	Bb	
967	5	Cb	
968	1	Ab	
968	7	D	
969	76	Bb	
970	12_5	Aa	
970	12_5	Bc	
970	12_5	Ca	
971			
972	4	Ab	
973			
974	4_3	Aa	
974	4_7	Bc	
974	4_7	Ca	
975	21	Ac	
975	21	Bd	
975	21	Cb	
976	17	Bc	
976	35	Ab	
977	77	Bc	
977	77	Ca	
978			
979	2	Ba	
1471	7_2	B	
1472			
1498	2	B	

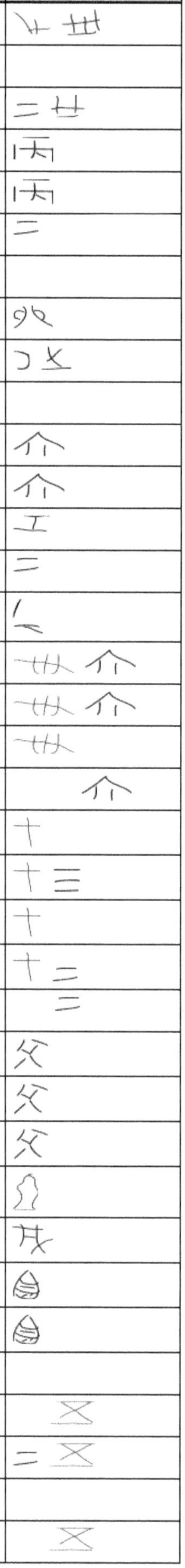

Trigger ID	Inscription code	Location	Symbol
1499	4_10	Aa	
1499	4_10	Bc	
1499	4_10	C	
1500			
1501			
1502			
1503			
2231	6_7	Aa	
2231	6_7	Bc	
2231	6_7	Ca	
2232	23	Ab	
2232	23	Bd	
2232	23	Cb	
2233	13	Bc	
2234			
2235			
2236	18	Bc	
2237	3	Aa	
2237	3	Bc	
2237	3	Ca	
2237	3	D	
2237	3	E	
2238	59	C	
2238	19_58	Bc	
2239	20	Aa	
2239	20	Bc	
2239	20	Ca	
2240	7	D	
2240	9	Bc	
2241	50	C	
2241	60	B	
2242	4_9	B	
2243	11_3	Aa	
2243	11_3	Bc	
2243	11_3	Ca	
2244	7	Aa	

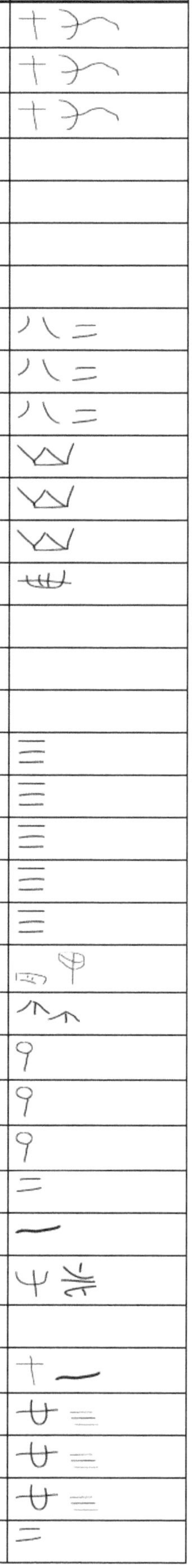

Trigger ID	Inscription code	Location	Symbol
2245	2	Ba	
2245	4_7	Aa	
2246	44	Aa	
2246	44	Bc	
2246	44	Cc	
2247	20	Bc	
2247	20	Ca	
2248	50	Ca	
2249			
2250	34	Ca	
2251	24	Ca	
2251	61	Bc	
2252	4_9	Aa	
2252	4_9	Bc	
2252	4_9	Ca	
2253			
2254	2_53	Bb	
2255			
2256	4_3	B	
2257	25	E	
2258	3	E	
2259			
2260	54	Bc	
2261	2_78	Ab	
2261	2_78	Bb	
2261	2_78	Cb	
2262	30	Ac	
2262	30	Bc	
2263	5_62_38	B	
2264	6	Aa	
2264	6	B	
2264	6	D	
2264	6	E	
2265	4	B	
2266	11_6	Aa	
2266	11_6	Bc	

Trigger ID	Inscription code	Location	Symbol
2267	7_2	E	
2267	9_2	D	
2268			
2269	13_9	A	
2270	4	A	
2271			
2272	7	Aa2	
2273	7	D	
2274	64	Bb	
2275	3	D	
2275	4_64	Bb	
2276			
2277	1_4	Ba	
2278	3	C	
2278	3	D	
2279	1	Ba	
2279	8	Ca	
2280	8_4	Aa	
2280	8_4	Bc	
2280	8_4	Ca	
2281	13_9	Bc	
2282	13	Bc	
2283			
2284	1	Ba	
2285	26	Aa	
2285	26	Bc	
2286	1	Ba	
2286	2	Aa	
2286	2	Bc	
2286	52	Ca	
2287	8	Ca	
2287	55	Bc	
2288	1	Ba	
2288	44	Ca	
2289	4_2	Aa	

Trigger ID	Inscription code	Location	Symbol
2289	4_2	Bc	
2289	4_2	Ca	
2290	1	Ba	
2290	20	Aa	
2290	20	Bc	
2291	51	Bc	
2292	79	Bc	
2293	1	Ba	
2293	8	Ac	
2293	8	Bc	
2293	8	Ca	
2294	17_9	Aa	
2294	17_9	Bc	
2294	17_9	Ca	
2295	50	Ab	
2296			
2297	1	Ba	
2297	27	Aa	
2297	27	Bc	
2298	13	Ab	
2298	13	Bd	
2298	13	Cb	
2299	6	Aa	
2299	6	Bc	
2299	6	Ca	
2299	6	D	
2299	6	E	
2300			
2301	62	Aa	
2301	62	Bc	
2301	62	Ca	
2302	65	Ca	
2302	2_9	Cb	
2303			
2304	2	Aa	
2304	4_10	Bc	
2304	4_5	Ca	
2305	28	Aa	
2305	28	Bc	
2306	80	A	
2306	80	Bc	
2307	3	Aa	
2307	3	Bc	
2307	3	Ca	
2308	1_50	Ba	
2309	1	Ba	
2309	5	Ca	
2309	24	Aa	
2309	24	Bc	
2310	12_5	Aa	
2310	12_5	Bc	
2310	12_7	Ca	
2311	4_2	Bc	
2311	9_29	Aa	
2312	4	Aa	
2312	4	Bc	
2312	45	Ca	
2313	49	Bc	
2314	3	Bc	
2315	1	Bc	
2315	6	Aa	
2316	1	Ba	
2316	19	Bc	
2317	13_9	Ab	
2317	13_9	Bd	
2318			
2319	66	Bb	
2320	43	B	
2321	18	Bb	
2322	2_4	Bc	
2322	2_4	Ca	
2323	11_30	A	

Trigger ID	Inscription code	Location	Symbol
2323	11_30	Bd	
2323	11_30	Cb	
2324	13_9	Bc	
2325	81	Bc	
2326			
2327			
2328			
2329	4_2	Bb	
2330	2	Bb	
2331			
2332			
2333	82	Aa	
2334	4	Ab	
2335	1	Ba	
2336	1	Ba	
2336	67	Bc	
2337	4	Cb	
2337	12_2_5	Bd	
2338	9_48	Bc	
2339	25	Bc	
2339	25	Ca	
2340	1	Ba	
2341	4_10	Ab	
2341	4_10	Bd	
2342	1	Ba	
2342	3	Cb	
2343	1	Ba	
2343	11_2	Aa	
2343	11_2	Bc	
2344	4_7	Aa	
2344	4_7	Bc	
2345	31	Bc	
2345	31	Ca	
2346	1	Ba	
2346	47	Ca	
2347	3	Ab	
2347	3	Bd	
2348	1	Ba	
2348	4	Aa	
2348	46	Bc	
2349	11	Cb	
2349	68	Ab	
2350			
2351			
2352			
2353			
2354			
2355			
2356	69	B	
2357			
2358			
2359	1	Ca	
2359	70	Bc	
2359	71	Ca	
2359	2_4_7	Ab	
2360	11_6	Bc	
2360	11_6	Ca	
3788	4_5	Ab	
3790	13_7	Aa	
3790	13_7	Bc	
3790	13_7	Ca	
3793	4	Ab	
3811			
3821	12_9	Bd	
3821	12_9	Ca	
3825	7	Ab	
3825	7	Cb	
5525			
5543	4_4	Ab	
5553			
5758	10	Aa	
5758	12_10	Bc	

Trigger ID	Inscription code	Location	Symbol
5758	12_10	Ca	
5760	5	Ab	
5760	5	Bd	
5760	5	Cb	
5761			
5768	12_6	Bd	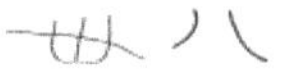
5768	12_6	Cb	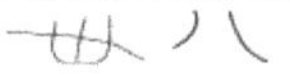
5771	11_4	Bd	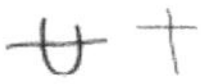
5771	12_4	Ab	
5771	12_4	Cb	
5772			
5774			
5775			
5819	2	B	
5819	2	C	
5821	1	Ba	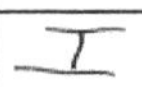
5821	19	C	
5821	72	C	

Appendix 7

Count and metric data on arrow bundles

7a

Bundle ID	Count	Head W1	Head W2	Head W3	Head W4	Head W5	Head W6	Mean	Std Dev	CV
001002	120	0.90	0.90	0.90	0.90	0.90	0.90	0.90	0.00	0.00
001003	101	0.95	0.95	0.95	0.95	0.95	0.95	0.95	0.00	0.00
001004	103	0.95	0.90	0.95	0.90	0.90	0.95	0.93	0.03	2.96
001005	100	0.95	1.00	0.95	0.95	0.95	0.90	0.95	0.03	3.33
001009	160	0.95	0.95	0.90	0.95	0.95	1.00	0.95	0.03	3.33
001017	93	0.95	0.90	0.95	0.95	0.95	0.90	0.93	0.03	2.77
001018	96	0.95	1.00	0.90	1.00	1.00	0.95	0.97	0.04	4.22
001020	106	0.95	0.95	0.95	0.95	0.95	0.95	0.95	0.00	0.00
001021	99	0.90	0.90	0.95	0.90	0.90	0.90	0.91	0.02	2.25
001022	102	1.05	1.05	0.95	1.05	0.95	1.00	1.01	0.05	4.88
001026	96	1.00	1.00	0.95	0.95	1.00	1.00	0.98	0.03	2.63
001027	102	0.90	0.90	0.90	0.90	0.95	0.95	0.92	0.03	2.82
001028	94	0.90	0.95	0.90	0.95	0.90	0.90	0.92	0.03	2.82
001030	98	1.00	0.95	0.95	0.90	0.90	0.90	0.93	0.04	4.37
001042	109	0.95	0.95	0.90	0.90	0.90	0.95	0.93	0.03	2.96
001043	105	1.00	1.00	0.95	0.95	0.95	0.95	0.97	0.03	2.67
001050	95	0.95	0.95	0.95	1.00	0.95	0.95	0.96	0.02	2.13
001055	90	1.00	0.85	0.95	0.95	1.00	1.00	0.96	0.06	6.10
001056	116	1.00	1.00	0.95	0.95	1.00	0.95	0.98	0.03	2.81
001062	92	0.95	0.95	0.95	0.95	0.95	0.95	0.95	0.00	0.00
001106	143	0.95	0.95	0.95	0.90	0.95	0.95	0.94	0.02	2.17
001107	101	0.90	0.95	0.95	0.90	0.95		0.93	0.03	2.94
001110	119	1.05	0.90	0.90	0.95	0.90	0.95	0.94	0.06	6.21
001112	99	0.95	0.95	1.00	0.90	0.90	1.00	0.95	0.04	4.71
001113	99	0.90	0.90	0.95	0.90	1.00	0.95	0.93	0.04	4.37
001114	122	0.95	1.00	0.90	0.90	0.95	0.95	0.94	0.04	4.00
001116	100	0.95	0.95	0.95	0.95	0.95	0.95	0.95	0.00	0.00
001117	101	1.00	0.90	0.95	1.00	1.00	0.95	0.97	0.04	4.22
001118	100	0.95	1.00	0.95	1.00	1.00	0.95	0.98	0.03	2.81
001119	146	1.00	0.95	1.00	1.00	0.95	1.00	0.98	0.03	2.63
001120	160	0.95	0.95	1.05	0.95	1.00	0.95	0.98	0.04	4.29
001136	154	0.95	0.95	1.00	1.05	0.95	0.95	0.98	0.04	4.29

Bundle ID	Count	Head W1	Head W2	Head W3	Head W4	Head W5	Head W6	Mean	Std Dev	CV
001139	150	0.95	0.95	0.95	0.95	0.95	0.95	0.95	0.00	0.00
001140	150	1.05	1.00	1.00	0.95	0.90	0.95	0.98	0.05	5.38
001149	125	0.95	0.95	1.00	0.95	0.95		0.96	0.02	2.33
001151	100	1.00	0.95	1.00	0.95	0.95	0.95	0.97	0.03	2.67
001152	92	1.00	1.00	0.95	1.00	1.00	0.95	0.98	0.03	2.63
001155	200	1.00	1.00	0.95	1.05	1.00	0.95	0.99	0.04	3.80
001159	100	1.00	1.00	1.05	1.00	1.00	1.00	1.01	0.02	2.02
001160	102	1.00	1.00	1.00	1.00	1.00	0.90	0.98	0.04	4.15
001168	120	0.95	1.00	0.95	0.90	1.00	0.95	0.96	0.04	3.93
001169	102	0.95	1.00	1.00	1.05	0.95	0.95	0.98	0.04	4.15
001170	101	1.00	1.00	0.95	1.00	1.00	1.00	0.99	0.02	2.06
001171	101	1.05	1.00	1.05	1.00	1.00	1.00	1.02	0.03	2.54
001172	106	1.00	0.95	1.00	1.00	1.00	1.00	0.99	0.02	2.06
001174	120	1.05	0.95	1.00	0.95	1.00	0.90	0.98	0.05	5.38
001177	94	0.95	1.00	0.95	0.90	1.00	0.95	0.96	0.04	3.93
001178	94	0.95	1.05	1.05	1.00	1.00	0.90	0.99	0.06	5.89
001179	184	1.00	1.00	1.00	0.95	1.00	1.00	0.99	0.02	2.06
001185	105	1.00	1.00	0.95	1.00	1.00	0.95	0.98	0.03	2.63
001190	166	1.00	0.90	0.95	0.95	1.00	0.90	0.95	0.04	4.71
001193	110	1.00	0.95	0.95	0.95	0.95	0.95	0.96	0.02	2.13
001196	185	1.05	1.00	0.95	0.95	0.95	1.00	0.98	0.04	4.15
001197	90	1.00	0.95	0.95	0.95	0.95	0.95	0.96	0.02	2.13
001200	250	1.00	0.95	0.95	1.00	1.00	0.95	0.98	0.03	2.81
001201	185	1.00	0.95	0.95	0.95	0.90	1.00	0.96	0.04	3.93
001205	100	0.95	1.00	0.90	1.00	0.95	1.00	0.97	0.04	4.22
001321	105	0.90	1.00	1.00	0.95	1.05	1.05	0.99	0.06	5.89
001383	163	0.95	0.95	0.95	0.95	0.95	0.95	0.95	0.00	0.00
001387	105	0.95	0.95	1.00	0.95	0.90	0.95	0.95	0.03	3.33
001388	108	1.00	1.05	1.00	0.95	0.95	0.95	0.98	0.04	4.15
001393	112	0.95	0.95	1.00	1.00	0.95	0.95	0.97	0.03	2.67
001394	275	0.95	1.00	1.00	1.00	0.90	1.00	0.98	0.04	4.29
001427	108	0.95	1.05	1.05	0.95	1.00	0.95	0.99	0.05	4.96
001428	103	1.00	1.00	1.00	1.00	0.90	1.00	0.98	0.04	4.15
001441	104	1.00	1.00	0.95	1.00	1.00	0.95	0.98	0.03	2.63
001464	115	0.95	0.95	0.95	0.95	0.95	0.90	0.94	0.02	2.17
001479	95	1.00	1.05	0.95	0.95	0.95	1.00	0.98	0.04	4.15
001482	136	1.00	0.95	1.00	0.95	0.90	0.95	0.96	0.04	3.93

Bundle ID	Count	Head W1	Head W2	Head W3	Head W4	Head W5	Head W6	Mean	Std Dev	CV
001483	95	0.95	0.95	0.95	0.95	0.95	0.95	0.95	0.00	0.00
001484	183	0.95	1.00	0.95	0.95	1.00	0.95	0.97	0.03	2.67
001486	96	1.00	1.00	0.90	1.00	1.00	0.95	0.98	0.04	4.29
001490	96	1.00	1.00	1.00	0.95	1.00	1.00	0.99	0.02	2.06
001491	101	0.95	0.95	0.95	0.95	1.00	0.95	0.96	0.02	2.13
001495	96	1.00	0.95	1.00	0.95	0.95	1.00	0.98	0.03	2.81
001496	96	0.95	0.95	0.95	0.95	0.95	0.95	0.95	0.00	0.00
001932	93	1.00	1.00	1.00	1.05	1.00	1.00	1.01	0.02	2.02
001934	101	0.95	0.95	0.95	0.95	0.95	0.95	0.95	0.00	0.00
001935	97	0.95	0.95	0.95	0.95	0.95	0.95	0.95	0.00	0.00
001936	112	1.00	1.00	0.95	1.00	1.00	1.00	0.99	0.02	2.06
001939	95	1.00	1.00	0.95	1.00	1.00	0.95	0.98	0.03	2.63
001940	99	1.00	1.00	0.95	0.95	1.00	1.00	0.98	0.03	2.63
001941	100	0.95	1.00	1.00	0.95	1.00	0.95	0.98	0.03	2.81
001943	100	0.95	0.95	0.95	0.95	1.00	0.95	0.96	0.02	2.13
001944	96	0.95	0.95	0.95	0.95	0.95	0.95	0.95	0.00	0.00
001945	99	0.95	0.95	0.95	0.95	0.90	0.90	0.93	0.03	2.77
001946	92	1.00	1.00	0.95	0.95	1.00	1.00	0.98	0.03	2.63
001947	97	1.00	0.95	0.95	0.95	0.95	1.00	0.97	0.03	2.67
001948	102	0.95	1.00	1.00	1.00	0.90	0.95	0.97	0.04	4.22
001949	100	0.95	0.95	1.00	0.95	0.95	0.95	0.96	0.02	2.13
001950	100	1.05	1.05	1.05	0.95	1.00	0.90	1.00	0.06	6.32
001951	100	0.95	1.00	0.95	0.95	0.95	0.95	0.96	0.02	2.13
001952	100	1.00	0.95	0.95	1.00	0.90	0.95	0.96	0.04	3.93
001953	103	0.95	0.95	0.95	0.90	0.95	0.95	0.94	0.02	2.17
001954	100	1.05	1.00	1.00	1.00	0.95	0.95	0.99	0.04	3.80
001955	95	1.00	1.05	1.00	1.05	0.95	0.95	1.00	0.04	4.47
001956	91	1.00	1.00	0.95	0.95	1.00	0.95	0.98	0.03	2.81
001957	97	1.00	1.00	0.95	1.00	0.95	0.95	0.98	0.03	2.81
001958	94	1.00	1.00	1.00	1.00	0.95	0.95	0.98	0.03	2.63
001959	98	0.95	0.95	1.00	1.00	1.00	1.00	0.98	0.03	2.63
001960	100	1.00	1.00	1.00	1.00	1.00	0.95	0.99	0.02	2.06
001962	98	1.00	1.00	1.00	1.00	1.00	0.95	0.99	0.02	2.06
001963	99	0.95	0.95	0.95	0.95	0.95	0.95	0.95	0.00	0.00
001964	96	1.00	1.00	1.00	1.00	1.00	1.00	1.00	0.00	0.00
001965	98	0.95	0.95	0.95	0.95	0.95	0.95	0.95	0.00	0.00
001966	101	1.00	0.95	1.00	1.00	1.00	1.00	0.99	0.02	2.06

Bundle ID	Count	Head W1	Head W2	Head W3	Head W4	Head W5	Head W6	Mean	Std Dev	CV
001967	98	1.00	0.95	1.00	0.95	0.95	0.95	0.97	0.03	2.67
001968	100	1.00	0.95	1.00	1.00	0.95	0.95	0.98	0.03	2.81
001969	100	1.00	0.95	1.00	0.95	0.95	0.90	0.96	0.04	3.93
001970	98	1.00	0.95	1.00	0.95	0.95	1.00	0.98	0.03	2.81
001971	95	1.00	1.00	0.95	0.95	0.95	0.95	0.97	0.03	2.67
001979	100	1.05	1.05	1.00	0.95	0.95	0.95	0.99	0.05	4.96
001980	100	0.95	1.00	1.00	1.05	0.90	1.00	0.98	0.05	5.25
001981	101	1.00	1.00	1.00	1.00	1.00	1.00	1.00	0.00	0.00
001982	101	1.00	0.95	0.95	1.00	0.95	0.95	0.97	0.03	2.67
001983	101	1.00	1.00	0.95	0.95	1.00	0.95	0.98	0.03	2.81
001984	101	0.95	0.90	0.90	0.95	0.95	1.00	0.94	0.04	4.00
001985	100	1.00	1.00	0.95	0.95	0.95	0.95	0.97	0.03	2.67
001989	100	1.00	1.05	1.00	1.00	1.00	1.00	1.01	0.02	2.02
001994	100	0.95	1.00	1.00	1.00	0.95	1.00	0.98	0.03	2.63
001995	100	1.00	1.00	1.00	0.95	1.00	0.90	0.98	0.04	4.29
001998	101	1.00	1.00	1.00	0.95	1.00	0.90	0.98	0.04	4.29
001999	97	1.00	1.00	0.95	0.95	0.95	0.95	0.97	0.03	2.67
002000	100	1.00	1.00	1.00	1.00	0.95	1.00	0.99	0.02	2.06
002003	100	1.00	1.00	0.95	0.95	1.00	0.95	0.98	0.03	2.81
002004	101	1.05	0.95	1.00	0.90	1.00	0.95	0.98	0.05	5.38
002009	100	1.00	0.95	1.00	1.00	0.95	1.00	0.98	0.03	2.63
002010	96	1.00	1.00	0.95	0.95	0.95	1.00	0.98	0.03	2.81
002012	100	0.90	0.95	0.95	0.90	0.95	0.95	0.93	0.03	2.77
002013	100	0.95	1.05	1.00	1.00	1.00	0.95	0.99	0.04	3.80
002014	97	0.95	0.95	0.95	1.00	1.00	0.90	0.96	0.04	3.93
002016	100	1.05	1.05	1.05	0.95	1.00	1.00	1.02	0.04	4.02
002018	103	1.00	1.00	1.00	1.00	0.95	0.95	0.98	0.03	2.63
002020	98	0.95	0.95	1.00	1.00	1.00	0.90	0.97	0.04	4.22
002022	100	0.95	1.00	0.95	0.95	0.95	0.90	0.95	0.03	3.33
002023	99	1.00	1.00	1.00	1.00	1.00	1.00	1.00	0.00	0.00
002025	100	1.05	1.05	1.05	0.95	1.00	1.00	1.02	0.04	4.02
002029	97	1.00	0.95	1.05	1.00	0.95	1.00	0.99	0.04	3.80
002030	100	1.00	1.00	1.05	1.00	1.00	1.05	1.02	0.03	2.54
002031	96	1.00	1.05	0.95	1.00	0.90	0.95	0.98	0.05	5.38
002034	100	1.00	0.95	0.95	0.95	0.95	0.95	0.96	0.02	2.13
002035	98	1.05	1.05	1.05	1.00	1.05	1.00	1.03	0.03	2.50
002036	98	0.95	1.00	0.95	0.95	1.00	0.95	0.97	0.03	2.67

Bundle ID	Count	Head W1	Head W2	Head W3	Head W4	Head W5	Head W6	Mean	Std Dev	CV
002037	99	1.00	0.95	1.00	1.00	0.95	0.95	0.98	0.03	2.81
002038	99	0.95	0.95	0.95	0.95	0.95	0.95	0.95	0.00	0.00
002041	99	1.00	1.00	0.95	1.00	1.00	1.00	0.99	0.02	2.06
002042	99	0.95	0.95	0.95	0.95	0.95	0.95	0.95	0.00	0.00
002044	102	1.00	1.00	1.05	1.00	1.05	0.95	1.01	0.04	3.73
002050	100	1.00	1.05	1.05	1.00	1.00	1.00	1.02	0.03	2.54
002051	98	1.00	1.00	0.95	0.95	0.95	0.95	0.97	0.03	2.67
002052	104	1.05	1.05	1.05	1.05	1.00	0.90	1.02	0.06	5.96
002053	98	1.05	1.00	1.00	0.95	0.95	0.95	0.98	0.04	4.15
002054	100	1.05	1.00	0.95	1.00	1.00	0.95	0.99	0.04	3.80
002055	109	1.00	0.95	1.00	0.95	0.95	0.95	0.97	0.03	2.67
002056	100	1.00	1.00	1.00	1.00	1.00	1.00	1.00	0.00	0.00
002057	100	1.00	1.00	1.00	1.00	0.95	1.00	0.99	0.02	2.06
002059	96	1.00	0.95	1.00	1.00	0.95	0.90	0.97	0.04	4.22
002061	92	0.95	1.00	0.95	0.95	0.95	0.95	0.96	0.02	2.13
002062	96	1.05	0.90	1.00	1.00	1.00	0.95	0.98	0.05	5.25
002063	100	1.05	0.95	1.00	0.95	0.95	0.95	0.98	0.04	4.29
002064	100	0.95	1.05	1.05	1.00	1.00	0.95	1.00	0.04	4.47
002065	100	1.00	1.05	1.00	1.05	0.90	0.90	0.98	0.07	6.95
002066	98	1.00	1.05	1.00	1.05	1.00	1.05	1.03	0.03	2.67
002067	99	1.00	0.95	0.95	0.95	1.00	0.95	0.97	0.03	2.67
002068	100	1.00	1.00	1.00	0.95	0.95	1.00	0.98	0.03	2.63
002069	99	0.95	0.95	1.00	0.95	0.90	0.95	0.95	0.03	3.33
002073	95	1.00	0.95	1.05	1.00	1.00	1.00	1.00	0.03	3.16
002075	100	1.05	1.00	1.05	1.05	1.00	0.95	1.02	0.04	4.02
002076	97	1.00	0.95	0.90	0.95	0.95	0.95	0.95	0.03	3.33
002077	114	1.00	0.95	0.95	0.95	1.00	0.95	0.97	0.03	2.67
002078	98	0.95	0.95	1.00	0.95	0.95	0.95	0.96	0.02	2.13
002079	97	1.00	0.95	0.95	0.95	0.95	0.95	0.96	0.02	2.13
002081	95	1.00	1.00	0.95	0.95	1.00	0.95	0.98	0.03	2.81
002082	100	1.00	0.95	0.95	0.95	0.95	0.95	0.96	0.02	2.13
002083	100	1.05	1.00	1.00	1.05	1.00	1.00	1.02	0.03	2.54
002086	100	1.05	1.00	1.00	1.00	1.00	1.00	1.01	0.02	2.02
002088	102	1.00	1.05	1.00	0.95	1.00	1.00	1.00	0.03	3.16
002089	93	1.05	1.05	1.05	0.95	1.00	1.00	1.02	0.04	4.02
002090	99	1.00	0.95	0.95	1.00	0.95	0.95	0.97	0.03	2.67
002091	100	1.00	0.95	0.95	0.90	0.95	0.90	0.94	0.04	4.00

Bundle ID	Count	Head W1	Head W2	Head W3	Head W4	Head W5	Head W6	Mean	Std Dev	CV
002093	102	1.00	1.05	1.00	0.95	0.90	1.05	0.99	0.06	5.89
002095	100	1.00	1.00	0.95	0.90	0.90	0.95	0.95	0.04	4.71
002098	100	1.00	1.00	0.95	0.95	0.95	0.95	0.97	0.03	2.67
002099	95	0.95	0.95	0.90	0.90	0.90	0.90	0.92	0.03	2.82
002101	100	1.05	1.00	1.00	0.95	0.95	0.95	0.98	0.04	4.15
002103	105	1.00	0.95	1.00	0.95	0.95	0.95	0.97	0.03	2.67
002104	97	1.05	1.00	1.00	1.00	1.00	1.00	1.01	0.02	2.02
002105	100	0.95	1.00	0.95	0.95	0.95	0.95	0.96	0.02	2.13
002106	98	0.95	0.95	1.00	1.00	0.95	0.95	0.97	0.03	2.67
002107	94	0.95	0.90	0.95	0.95	0.95	0.95	0.94	0.02	2.17
002108	96	0.95	1.00	1.00	0.95	0.90	1.00	0.97	0.04	4.22
002109	100	0.95	0.95	0.95	0.95	1.00	0.95	0.96	0.02	2.13
002112	98	0.95	0.95	0.95	0.95	0.95	0.95	0.95	0.00	0.00
002113	98	0.95	1.00	0.95	1.00	0.95	0.95	0.97	0.03	2.67
002114	95	1.00	1.00	0.95	0.95	0.95	0.95	0.97	0.03	2.67
002116	97	1.00	1.00	1.00	1.05	1.00	1.00	1.01	0.02	2.02
002118	99	1.00	1.00	0.95	0.90	0.95	0.95	0.96	0.04	3.93
002121	103	1.05	1.05	0.95	1.05	1.00	0.95	1.01	0.05	4.88
002124	97	1.00	1.00	1.00	0.95	0.95	1.00	0.98	0.03	2.63
002125	100	1.05	1.00	1.00	1.05	1.00	0.90	1.00	0.05	5.48
002126	105	1.00	0.95	1.00	1.00	1.00	0.95	0.98	0.03	2.63
002127	94	0.95	1.00	1.00	1.00	0.95	0.95	0.98	0.03	2.81
002136	100	1.00	1.00	1.00	1.00	0.95	0.95	0.98	0.03	2.63
002137	97	0.95	1.00	0.95	1.00	1.00	0.95	0.98	0.03	2.81
002138	100	1.00	0.95	0.95	0.95	0.95	0.90	0.95	0.03	3.33
002139	92	0.95	1.00	1.00	1.00	1.00	0.90	0.98	0.04	4.29
002140	99	1.00	1.00	1.00	0.95	1.00	1.00	0.99	0.02	2.06
002141	99	1.00	0.90	0.90	0.90	1.00	0.95	0.94	0.05	5.22
002144	100	1.00	1.00	1.00	1.00	1.00	1.00	1.00	0.00	0.00
002145	100	0.95	1.00	1.00	0.95	1.05	0.85	0.97	0.07	7.07
002146	100	1.10	1.00	1.00	0.95	0.95	1.00	1.00	0.05	5.48
002147	100	1.10	1.10	1.05	1.05	1.10	1.00	1.07	0.04	3.83
002148	94	1.00	1.00	1.00	1.00	0.95	1.00	0.99	0.02	2.06
002154	100	1.00	1.00	1.00	1.00	0.95	1.00	0.99	0.02	2.06
002155	103	0.95	0.90	0.90	0.95	1.00	0.90	0.93	0.04	4.37
002156	100	1.05	1.00	1.00	1.00	1.00	0.95	1.00	0.03	3.16
002157	92	0.95	0.95	1.00	1.00	1.00	1.00	0.98	0.03	2.63

Bundle ID	Count	Head W1	Head W2	Head W3	Head W4	Head W5	Head W6	Mean	Std Dev	CV
002158	100	1.00	1.00	0.95	0.95	0.90	0.90	0.95	0.04	4.71
002159	100	1.00	1.00	1.00	1.00	0.95	0.95	0.98	0.03	2.63
002160	98	1.00	1.00	0.95	1.00	1.00	1.00	0.99	0.02	2.06
002164	96	1.00	1.05	1.05	1.05	0.95	0.95	1.01	0.05	4.88
002165	101	0.95	0.95	0.95	0.95	0.95	1.00	0.96	0.02	2.13
002166	100	0.95	0.95	0.95	1.00	1.00	0.95	0.97	0.03	2.67
002167	100	1.00	1.00	1.00	1.00	0.95	1.00	0.99	0.02	2.06
002170	100	1.00	0.95	1.00	1.00	0.90	0.95	0.97	0.04	4.22
002171	94	1.00	1.00	1.00	1.00	1.00	1.00	1.00	0.00	0.00
002172	102	1.10	1.10	1.05	1.10	1.05	1.05	1.08	0.03	2.55
002176	102	1.00	1.05	1.00	1.00	0.95	1.00	1.00	0.03	3.16
002177	101	0.95	1.00	0.95	0.95	0.95	0.95	0.96	0.02	2.13
002178	127	1.00	0.95	1.00	1.00	1.00	1.00	0.99	0.02	2.06
002179	94	0.95	1.00	0.95	0.95	0.95	1.00	0.97	0.03	2.67
002181	92	0.95	1.00	1.00	1.05	1.00	1.00	1.00	0.03	3.16
002185	100	1.05	1.10	1.10	1.00	1.00	1.00	1.04	0.05	4.72
002189	100	1.05	1.00	1.00	1.00	0.90	1.00	0.99	0.05	4.96
002191	96	0.95	1.00	0.95	1.00	0.95	0.95	0.97	0.03	2.67
002192	100	1.00	0.95	0.95	0.95	1.00	1.00	0.98	0.03	2.81
002194	100	1.00	0.95	0.95	1.00	0.95	0.95	0.97	0.03	2.67
002195	100	1.00	0.95	1.05	1.00	0.95	0.95	0.98	0.04	4.15
002208	100	1.00	1.00	1.00	1.00	1.00	1.00	1.00	0.00	0.00
002209	100	1.00	1.00	1.00	0.95	1.00	1.00	0.99	0.02	2.06
002210	100	1.00	1.00	0.95	0.95	0.95	0.95	0.97	0.03	2.67
002211	100	1.00	0.95	1.00	1.00	0.95	0.95	0.98	0.03	2.81
002214	103	1.00	1.05	1.00	1.00	1.00	0.95	1.00	0.03	3.16
002215	100	1.10	1.10	1.10	1.10	1.10	1.10	1.10	0.00	0.00
002216	97	0.95	1.00	1.00	0.95	1.00	1.00	0.98	0.03	2.63
002218	100	0.95	1.00	1.00	1.00	1.00	1.00	0.99	0.02	2.06
002219	100	0.95	0.95	0.95	1.00	1.00	1.00	0.98	0.03	2.81
002220	100	1.00	0.95	1.00	1.00	0.95	0.95	0.98	0.03	2.81
002221	95	1.00	1.05	1.05	0.95	0.90	0.90	0.98	0.07	7.07
002222	94	0.95	1.00	1.00	1.00	1.00	0.95	0.98	0.03	2.63
002223	91	1.00	1.05	0.95	0.95	1.10	0.90	0.99	0.07	7.42
002224	100	1.00	0.95	0.95	0.95	0.95	0.90	0.95	0.03	3.33
002226	95	0.95	0.95	1.05	0.90	1.05	0.95	0.98	0.06	6.28
002227	105	0.95	0.95	0.95	1.00	0.95	0.95	0.96	0.02	2.13

Bundle ID	Count	Head W1	Head W2	Head W3	Head W4	Head W5	Head W6	Mean	Std Dev	CV
002229	93	1.00	1.00	1.00	1.00	0.95	1.00	0.99	0.02	2.06
002230	100	1.00	1.05	1.00	1.05	1.00	1.00	1.02	0.03	2.54
003531	100	0.95	1.00	1.00	1.00	0.95	0.95	0.98	0.03	2.81
003716	101	1.00	0.95	1.00	0.95	1.00	1.05	0.99	0.04	3.80
003718	198	0.95	1.00	1.00	0.95	0.95	0.95	0.97	0.03	2.67
003719	105	0.95	1.00	1.00	1.00	1.00	0.95	0.98	0.03	2.63
003759	90	1.00	0.95	1.00	0.95	0.95	0.90	0.96	0.04	3.93
003823	106	0.95	1.00	1.05	1.10	1.10	1.05	1.04	0.06	5.61
003824	125	1.00	0.95	1.00	0.95	1.00	1.00	0.98	0.03	2.63
004029	512	0.80	0.85	0.85	0.85	0.85	0.85	0.84	0.02	2.43
004056	105	1.00	1.05	1.05	1.00	1.00	1.00	1.02	0.03	2.54
004092	229	0.80	0.85	0.85	0.80	0.85	0.85	0.83	0.03	3.10
004093	92	0.95	1.00	1.00	1.00	0.95	1.00	0.98	0.03	2.63
004094	99	1.05	1.00	1.00	1.00	1.00	0.95	1.00	0.03	3.16
004107	100	0.95	0.95	0.95	0.95	1.00	0.90	0.95	0.03	3.33
005611	105	0.95	1.00	1.00	1.00	1.00	0.90	0.98	0.04	4.29

7b

Bundle ID	Count	Head L1	Head L2	Head L3	Head L4	Head L5	Head L6	Mean	Std Dev	CV
001002	120	2.70	2.70	2.70	2.70	2.70	2.70	2.70	0.00	0.00
001003	101	2.90	2.70	2.70	2.70	2.70	2.60	2.72	0.10	3.62
001004	103	2.80	2.60	2.85	2.50	2.70	2.70	2.69	0.13	4.76
001005	100	2.60	2.70	2.70	2.70	2.70	2.60	2.67	0.05	1.94
001009	160	2.70	2.70	2.70	2.70	2.70	2.75	2.71	0.02	0.75
001017	93	2.75	2.60	2.70	2.70	2.80	2.70	2.71	0.07	2.45
001018	96	2.70	2.70	2.70	2.65	2.70	2.80	2.71	0.05	1.82
001020	106	2.80	2.70	2.60	2.70	2.80	2.75	2.73	0.08	2.78
001021	99	2.50	2.60	2.60	2.70	2.70	2.75	2.64	0.09	3.47
001022	102	2.70	2.80	2.75	2.60	2.75	2.60	2.70	0.08	3.10
001026	96	2.85	2.75	2.80	2.80	2.85	2.75	2.80	0.04	1.60
001027	102	2.70	2.75	2.80	2.60	2.70	2.65	2.70	0.07	2.62
001028	94	2.70	2.65	2.70	2.70	2.65	2.75	2.69	0.04	1.40
001030	98	2.80	2.75	2.75	2.90	2.80	2.80	2.80	0.05	1.96
001042	109	2.70	2.60	2.65	2.80	2.65	2.70	2.68	0.07	2.55
001043	105	2.70	2.75	2.65	2.70	2.75	2.70	2.71	0.04	1.39
001050	95	2.70	2.70	2.70	2.70	2.65	2.70	2.69	0.02	0.76

Bundle ID	Count	Head L1	Head L2	Head L3	Head L4	Head L5	Head L6	Mean	Std Dev	CV
001055	90	2.70	2.75	2.70	2.70	2.75	2.75	2.73	0.03	1.00
001056	116	2.80	2.90	2.70	2.80	2.80	2.70	2.78	0.08	2.70
001062	92	2.75	2.75	2.70	2.75	2.75	2.70	2.73	0.03	0.94
001106	143	2.75	2.70	2.75	2.70	2.70	2.75	2.73	0.03	1.00
001107	101	2.60	2.65	2.80	2.65	2.65		2.67	0.08	2.84
001110	119	2.80	2.60	2.75	2.80	2.70	2.60	2.71	0.09	3.39
001112	99	2.75	2.60	2.75	2.60	2.70	2.90	2.72	0.11	4.14
001113	99	2.60	2.70	2.80	2.70	2.70	2.70	2.70	0.06	2.34
001114	122	2.70	2.80	2.75	2.65	2.70	2.80	2.73	0.06	2.22
001116	100	2.60	2.55	2.60	2.70	2.60	2.75	2.63	0.08	2.86
001117	101	2.65	2.75	2.65	2.70	2.70	2.70	2.69	0.04	1.40
001118	100	2.65	2.70	2.70	2.85	2.85	2.70	2.74	0.09	3.14
001119	146	2.70	2.60	2.75	2.70	2.65	2.70	2.68	0.05	1.92
001120	160	2.60	2.60	2.65	2.60	2.70	2.75	2.65	0.06	2.39
001136	154	2.70	2.60	2.65	2.85	2.70	2.75	2.71	0.09	3.18
001139	150	2.65	2.80	2.70	2.75	2.75	2.70	2.73	0.05	1.92
001140	150	2.75	2.75	2.65	2.65	2.65	2.65	2.68	0.05	1.92
001149	125	2.70	2.70	2.70	2.80	2.70		2.72	0.04	1.64
001151	100	2.90	2.90	2.70	2.85	2.70	2.80	2.81	0.09	3.27
001152	92	2.80	2.85	2.80	2.75	2.90	2.80	2.82	0.05	1.83
001155	200	2.75	2.75	2.75	2.80	2.70	2.70	2.74	0.04	1.37
001159	100	2.70	2.90	2.75	2.75	2.70	2.75	2.76	0.07	2.67
001160	102	2.65	2.70	2.85	2.75	2.70	2.80	2.74	0.07	2.68
001168	120	2.70	2.70	2.70	2.65	2.85	2.70	2.72	0.07	2.51
001169	102	2.85	2.70	2.70	2.75	2.65	2.70	2.73	0.07	2.53
001170	101	2.75	2.75	2.70	2.75	2.70	2.70	2.73	0.03	1.00
001171	101	2.75	2.75	2.75	2.80	2.80	2.70	2.76	0.04	1.36
001172	106	2.70	2.70	2.75	2.75	2.80	2.75	2.74	0.04	1.37
001174	120	2.70	2.75	2.60	2.80	2.80	2.80	2.74	0.08	2.92
001177	94	2.70	2.75	2.70	2.75	2.75	2.70	2.73	0.03	1.00
001178	94	2.80	2.75	2.70	2.70	2.80	2.70	2.74	0.05	1.79
001179	184	2.75	2.75	2.85	2.80	2.70	2.85	2.78	0.06	2.18
001185	105	2.75	2.75	2.75	2.70	2.70	2.70	2.73	0.03	1.00
001190	166	2.60	2.65	2.65	2.65	2.65	2.60	2.63	0.03	0.98
001193	110	2.75	2.70	2.70	2.70	2.70	2.75	2.72	0.03	0.95
001196	185	2.80	2.90	2.65	2.65	2.65	2.80	2.74	0.11	3.90
001197	90	2.85	2.70	2.65	2.85	2.85	2.75	2.78	0.09	3.17

Bundle ID	Count	Head L1	Head L2	Head L3	Head L4	Head L5	Head L6	Mean	Std Dev	CV
001200	250	2.65	2.65	2.65	2.70	2.75	2.60	2.67	0.05	1.94
001201	185	2.65	2.65	2.70	2.60	2.60	2.85	2.68	0.09	3.50
001205	100	2.70	2.70	2.65	2.65	2.70	2.75	2.69	0.04	1.40
001321	105	2.70	2.75	2.75	2.75	2.75	2.75	2.74	0.02	0.74
001383	163	2.70	2.75	2.70	2.75	2.75	2.70	2.73	0.03	1.00
001387	105	2.65	2.65	2.80	2.65	2.60	2.70	2.68	0.07	2.58
001388	108	2.70	2.75	2.60	2.70	2.70	2.70	2.69	0.05	1.83
001393	112	2.65	2.65	2.70	2.70	2.65	2.75	2.68	0.04	1.52
001394	275	2.70	2.70	2.70	2.70	2.60	2.70	2.68	0.04	1.52
001427	108	2.65	2.70	2.70	2.70	2.75	2.60	2.68	0.05	1.92
001428	103	2.85	2.75	2.80	2.80	2.60	2.80	2.77	0.09	3.16
001441	104	2.75	2.75	2.75	2.70	2.70	2.75	2.73	0.03	0.94
001464	115	2.70	2.80	2.75	2.65	2.70	2.60	2.70	0.07	2.62
001479	95	2.75	2.85	2.75	2.70	2.70	2.75	2.75	0.05	1.99
001482	136	2.70	2.70	2.70	2.70	2.70	2.60	2.68	0.04	1.52
001483	95	2.60	2.65	2.60	2.60	2.60	2.65	2.62	0.03	0.99
001484	183	2.70	2.80	2.75	2.80	2.75	2.70	2.75	0.04	1.63
001486	96	2.85	2.70	2.75	2.75	2.65	2.70	2.73	0.07	2.50
001490	96	2.70	2.80	2.75	2.75	2.75	2.75	2.75	0.03	1.15
001491	101	2.75	2.70	2.75	2.70	2.70	2.70	2.72	0.03	0.95
001495	96	2.75	2.65	2.75	2.70	2.75	2.70	2.72	0.04	1.50
001496	96	2.70	2.70	2.70	2.70	2.70	2.70	2.70	0.00	0.00
001932	93	2.90	2.85	2.85	2.90	2.85	2.85	2.87	0.03	0.90
001934	101	2.65	2.60	2.65	2.65	2.60	2.65	2.63	0.03	0.98
001935	97	2.70	2.75	2.75	2.65	2.70	2.70	2.71	0.04	1.39
001936	112	2.75	2.70	2.75	2.75	2.70	2.70	2.73	0.03	1.00
001939	95	2.70	2.75	2.75	2.75	2.70	2.60	2.71	0.06	2.16
001940	99	2.40	2.55	2.55	2.65	2.70	2.50	2.56	0.11	4.18
001941	100	2.75	2.65	2.65	2.60	2.80	2.50	2.66	0.11	4.02
001943	100	2.70	2.70	2.75	2.60	2.80	2.50	2.68	0.11	4.05
001944	96	2.60	2.60	2.60	2.60	2.65	2.60	2.61	0.02	0.78
001945	99	2.60	2.60	2.60	2.60	2.60	2.65	2.61	0.02	0.78
001946	92	2.70	2.65	2.75	2.50	2.65	2.65	2.65	0.08	3.16
001947	97	2.75	2.65	2.65	2.65	2.60	2.60	2.65	0.05	2.07
001948	102	2.60	2.75	2.60	2.85	2.50	2.45	2.63	0.15	5.75
001949	100	2.70	2.75	2.75	2.70	2.70	2.65	2.71	0.04	1.39
001950	100	2.90	2.75	2.75	2.55	2.90	2.70	2.76	0.13	4.78

Bundle ID	Count	Head L1	Head L2	Head L3	Head L4	Head L5	Head L6	Mean	Std Dev	CV
001951	100	2.60	2.70	2.65	2.60	2.60	2.65	2.63	0.04	1.55
001952	100	2.60	2.55	2.60	2.70	2.55	2.60	2.60	0.05	2.11
001953	103	2.70	2.65	2.70	2.60	2.60	2.70	2.66	0.05	1.85
001954	100	2.70	2.75	2.65	2.75	2.80	2.80	2.74	0.06	2.13
001955	95	2.70	2.70	2.70	2.70	2.70	2.70	2.70	0.00	0.00
001956	91	2.75	2.70	2.65	2.65	2.75	2.65	2.69	0.05	1.83
001957	97	2.70	2.70	2.75	2.65	2.75	2.75	2.72	0.04	1.50
001958	94	2.70	2.75	2.70	2.70	2.80	2.70	2.73	0.04	1.54
001959	98	2.70	2.70	2.70	2.75	2.75	2.70	2.72	0.03	0.95
001960	100	2.75	2.75	2.75	2.70	2.75	2.70	2.73	0.03	0.94
001962	98	2.55	2.55	2.70	2.60	2.80	2.70	2.65	0.10	3.77
001963	99	2.65	2.70	2.65	2.70	2.65	2.60	2.66	0.04	1.42
001964	96	2.70	2.70	2.70	2.70	2.70	2.70	2.70	0.00	0.00
001965	98	2.65	2.60	2.60	2.60	2.65	2.65	2.63	0.03	1.04
001966	101	2.70	2.70	2.70	2.70	2.70	2.70	2.70	0.00	0.00
001967	98	2.65	2.65	2.75	2.60	2.65	2.55	2.64	0.07	2.52
001968	100	2.70	2.65	2.70	2.65	2.65	2.65	2.67	0.03	0.97
001969	100	2.70	2.70	2.70	2.60	2.70	2.65	2.68	0.04	1.56
001970	98	2.75	2.70	2.75	2.65	2.70	2.65	2.70	0.04	1.66
001971	95	2.80	2.80	2.75	2.60	2.80	2.75	2.75	0.08	2.82
001979	100	2.80	2.80	2.75	2.80	2.75	2.80	2.78	0.03	0.93
001980	100	2.70	2.75	2.60	2.80	2.50	2.60	2.66	0.11	4.19
001981	101	2.75	2.75	2.80	2.75	2.75	2.75	2.76	0.02	0.74
001982	101	2.70	2.65	2.65	2.70	2.70	2.70	2.68	0.03	0.96
001983	101	2.75	2.80	2.70	2.60	2.70	2.65	2.70	0.07	2.62
001984	101	2.90	2.70	2.65	2.65	2.60	2.70	2.70	0.10	3.88
001985	100	2.80	2.70	2.75	2.75	2.70	2.70	2.73	0.04	1.49
001989	100	2.80	2.85	2.85	2.80	2.85	2.85	2.83	0.03	0.91
001994	100	2.75	2.70	2.65	2.65	2.65	2.65	2.68	0.04	1.56
001995	100	2.75	2.75	2.75	2.75	2.75	2.75	2.75	0.00	0.00
001998	101	2.65	2.70	2.70	2.65	2.75	2.65	2.68	0.04	1.52
001999	97	2.80	2.80	2.70	2.85	2.70	2.75	2.77	0.06	2.19
002000	100	2.75	2.75	2.75	2.70	2.70	2.75	2.73	0.03	0.94
002003	100	2.70	2.75	2.65	2.60	2.70	2.70	2.68	0.05	1.92
002004	101	2.80	2.75	2.75	2.65	2.70	2.70	2.73	0.05	1.92
002009	100	2.75	2.75	2.75	2.75	2.75	2.75	2.75	0.00	0.00
002010	96	2.80	2.80	2.80	2.85	2.65	2.65	2.76	0.09	3.12

Bundle ID	Count	Head L1	Head L2	Head L3	Head L4	Head L5	Head L6	Mean	Std Dev	CV
002012	100	2.70	2.65	2.75	2.60	2.70	2.65	2.68	0.05	1.96
002013	100	2.65	2.75	2.60	2.70	2.70	2.80	2.70	0.07	2.62
002014	97	2.70	2.75	2.70	2.75	2.70	2.70	2.72	0.03	0.95
002016	100	2.70	2.80	2.80	2.70	2.70	2.80	2.75	0.05	1.99
002018	103	2.75	2.75		2.65	2.80	2.70	2.73	0.06	2.09
002020	98	2.75	2.55	2.80	2.65	2.90	2.80	2.74	0.12	4.53
002022	100	2.70	2.60	2.60	2.70	2.60	2.75	2.66	0.07	2.50
002023	99	2.75	2.70	2.80	2.75	2.75	2.75	2.75	0.03	1.15
002025	100	2.80	2.80	2.75	2.80	2.75	2.75	2.78	0.03	0.99
002029	97	2.90	2.70	2.85	2.75	2.75	2.75	2.78	0.08	2.70
002030	100	2.60	2.75	2.65	2.65	2.65	2.75	2.68	0.06	2.29
002031	96	2.75	2.75	2.65	2.70	2.70	2.70	2.71	0.04	1.39
002034	100	2.70	2.70	2.65	2.65	2.70	2.70	2.68	0.03	0.96
002035	98	2.70	2.70	2.75	2.70	2.75	2.75	2.73	0.03	1.00
002036	98	2.70	2.65	2.70	2.65	2.60	2.70	2.67	0.04	1.53
002037	99	2.70	2.70	2.60	2.70	2.65	2.70	2.68	0.04	1.56
002038	99	2.60	2.65	2.65	2.65	2.65	2.55	2.63	0.04	1.59
002041	99	2.75	2.70	2.70	2.75	2.70	2.75	2.73	0.03	1.00
002042	99	2.60	2.70	2.60	2.70	2.70	2.65	2.66	0.05	1.85
002044	102	2.75	2.80	2.85	2.75	2.75	2.70	2.77	0.05	1.87
002050	100	2.70	2.75	2.80	2.75	2.80	2.70	2.75	0.04	1.63
002051	98	2.65	2.65	2.65	2.60	2.65	2.65	2.64	0.02	0.77
002052	104	2.80	2.80	2.80	2.80	2.60	2.75	2.76	0.08	2.90
002053	98	2.75	2.75	2.75	2.75	2.55	2.75	2.72	0.08	3.01
002054	100	2.75	2.75	2.65	2.70	2.65	2.75	2.71	0.05	1.82
002055	109	2.70	2.80	2.65	2.80	2.60	2.80	2.73	0.09	3.23
002056	100	2.70	2.70	2.70	2.75	2.70	2.70	2.71	0.02	0.75
002057	100	2.75	2.90	2.70	2.75	2.75	2.70	2.76	0.07	2.67
002059	96	2.85	2.65	2.50	2.75	2.60	2.65	2.67	0.12	4.54
002061	92	2.90	2.90	2.70	2.80	2.80	2.60	2.78	0.12	4.20
002062	96	2.90	2.70	2.65	2.70	2.75	2.70	2.73	0.09	3.20
002063	100	2.80	2.80	2.80	2.75	2.80	2.80	2.79	0.02	0.73
002064	100	2.75	2.75	2.75	2.75	2.75	2.70	2.74	0.02	0.74
002065	100	2.80	2.70	2.75	2.75	2.75	2.65	2.73	0.05	1.89
002066	98	2.75	2.65	2.75	2.70	2.70	2.80	2.73	0.05	1.92
002067	99	2.75	2.75	2.75	2.75	2.70	2.75	2.74	0.02	0.74
002068	100	2.80	2.80	2.70	2.75	2.75	2.75	2.76	0.04	1.36

Bundle ID	Count	Head L1	Head L2	Head L3	Head L4	Head L5	Head L6	Mean	Std Dev	CV
002069	99	2.80	2.75	2.80	2.80	2.80	2.70	2.78	0.04	1.51
002073	95	2.70	2.65	2.40	2.75	2.75	2.70	2.66	0.13	4.96
002075	100	2.80	2.80	2.85	2.70	2.75	2.65	2.76	0.07	2.67
002076	97	2.80	2.90	2.80	2.75	2.70	2.70	2.78	0.08	2.73
002077	114	2.80	2.70	2.80	2.70	2.80	2.80	2.77	0.05	1.87
002078	98	2.80	2.80	2.75	2.80	2.80	2.80	2.79	0.02	0.73
002079	97	2.70	2.75	2.75	2.75	2.75	2.75	2.74	0.02	0.74
002081	95	2.70	2.65	2.70	2.70	2.65	2.65	2.68	0.03	1.02
002082	100	2.75	2.70	2.75	2.75	2.60	2.75	2.72	0.06	2.23
002083	100	2.75	2.65	2.70	2.70	2.65	2.65	2.68	0.04	1.52
002086	100	2.75	2.75	2.70	2.75	2.75	2.70	2.73	0.03	0.94
002088	102	2.70	2.70	2.80	2.70	3.10	2.90	2.82	0.16	5.69
002089	93	2.70	2.80	2.75	2.75	2.70	2.75	2.74	0.04	1.37
002090	99	2.75	2.75	2.75	2.75	2.80	2.75	2.76	0.02	0.74
002091	100	2.75	2.80	2.80	2.80	2.75	2.85	2.79	0.04	1.35
002093	102	2.80	2.80	2.65	2.70	2.50	2.80	2.71	0.12	4.43
002095	100	2.80	2.80	2.75	2.80	2.50	2.70	2.73	0.12	4.30
002098	100	2.60	2.60	2.65	2.65	2.65	2.60	2.63	0.03	1.04
002099	95	2.65	2.65	2.65	2.70	2.60	2.55	2.63	0.05	1.96
002101	100	2.70	2.70	2.75	2.70	2.65	2.75	2.71	0.04	1.39
002103	105	2.85	2.70	2.70	2.70	2.70	2.65	2.72	0.07	2.51
002104	97	2.75	2.75	2.70	2.75	2.75	2.75	2.74	0.02	0.74
002105	100	2.75	2.70	2.70	2.70	2.60	2.70	2.69	0.05	1.83
002106	98	2.70	2.80	2.75	2.70	2.70	2.60	2.71	0.07	2.45
002107	94	2.60	2.70	2.70	2.60	2.70	2.60	2.65	0.05	2.07
002108	96	2.65	2.70	2.70	2.70	2.65	2.65	2.68	0.03	1.02
002109	100	2.70	2.65	2.70	2.65	2.70	2.70	2.68	0.03	0.96
002112	98	2.50	2.65	2.65	2.60	2.70	2.55	2.61	0.07	2.82
002113	98	2.65	2.60	2.50	2.60	2.70	2.50	2.59	0.08	3.09
002114	95	2.70	2.65	2.65	2.55	2.60	2.65	2.63	0.05	1.96
002116	97	2.80	2.75	2.65	2.80	2.70	2.80	2.75	0.06	2.30
002118	99	2.70	2.75	2.80	2.70	2.60	2.60	2.69	0.08	2.98
002121	103	2.70	2.70	2.80	2.75	2.75	2.70	2.73	0.04	1.49
002124	97	2.70	2.70	2.65	2.60	2.65	2.65	2.66	0.04	1.42
002125	100	2.70	2.70	2.80	2.80	2.70	2.60	2.72	0.08	2.77
002126	105	2.70	2.70	2.65	2.70	2.70	2.60	2.68	0.04	1.56
002127	94	2.65	2.70	2.65	2.75	2.65	2.75	2.69	0.05	1.83

Bundle ID	Count	Head L1	Head L2	Head L3	Head L4	Head L5	Head L6	Mean	Std Dev	CV
002136	100	2.75	2.70	2.70	2.70	2.70	2.70	2.71	0.02	0.75
002137	97	2.65	2.75	2.75	2.70	2.90	2.60	2.73	0.10	3.80
002138	100	2.65	2.65	2.65	2.65	2.55	2.70	2.64	0.05	1.86
002139	92	2.75	2.75	2.80	2.80	2.75	2.80	2.78	0.03	0.99
002140	99	2.70	2.70	2.65	2.70	2.70	2.60	2.68	0.04	1.56
002141	99	2.70	2.45	2.60	2.60	2.60	2.60	2.59	0.08	3.09
002144	100	2.65	2.65	2.65	2.70	2.65	2.55	2.64	0.05	1.86
002145	100	2.50	2.80	2.75	2.60	2.75	2.80	2.70	0.12	4.54
002146	100	2.75	2.75	2.75	2.70	2.60	2.75	2.72	0.06	2.23
002147	100	2.80	2.75	2.75	2.70	2.80	2.75	2.76	0.04	1.36
002148	94	2.75	2.75	2.75	2.75	2.75	2.70	2.74	0.02	0.74
002154	100	2.70	2.75	2.75	2.80	2.80	2.70	2.75	0.04	1.63
002155	103	2.60	2.60	2.50	2.70	2.80	2.60	2.63	0.10	3.92
002156	100	2.75	2.70	2.75	2.80	2.75	2.75	2.75	0.03	1.15
002157	92	2.75	2.70	2.70	2.70	2.70	2.70	2.71	0.02	0.75
002158	100	2.70	2.70	2.65	2.70	2.65	2.60	2.67	0.04	1.53
002159	100	2.80	2.70	2.75	2.75	2.75	2.70	2.74	0.04	1.37
002160	98	2.85	2.85	2.85	2.90	2.75	3.00	2.87	0.08	2.85
002164	96	2.70	2.80	2.75	2.70	2.70	2.75	2.73	0.04	1.49
002165	101	2.90	2.70	2.60	2.70	2.80	2.75	2.74	0.10	3.72
002166	100	2.60	2.65	2.80	2.75	2.80	2.75	2.73	0.08	3.01
002167	100	2.70	2.80	2.60	2.70	2.75	2.80	2.73	0.08	2.78
002170	100	2.70	2.70	2.75	2.70	2.70	2.60	2.69	0.05	1.83
002171	94	2.80	2.70	2.80	2.75	2.70	2.75	2.75	0.04	1.63
002172	102	2.75	2.80	2.75	2.75	2.75	2.70	2.75	0.03	1.15
002176	102	2.60	2.90	2.80	2.75	2.75	2.50	2.72	0.14	5.29
002177	101	2.80	2.90	2.90	2.70	2.75	2.75	2.80	0.08	2.99
002178	127	2.70	2.75	2.75	2.75	2.75	2.65	2.73	0.04	1.54
002179	94	2.85	2.90	2.85	2.60	2.70	2.90	2.80	0.12	4.37
002181	92	2.70	2.75	2.70	2.70	2.70	2.70	2.71	0.02	0.75
002185	100	2.70	2.75	2.75	2.75	2.60	2.70	2.71	0.06	2.16
002189	100	2.70	2.70	2.70	2.70	2.70	2.75	2.71	0.02	0.75
002191	96	2.70	2.60	2.70	2.65	2.65	2.45	2.63	0.09	3.56
002192	100	2.70	2.60	2.70	2.60	2.65	2.70	2.66	0.05	1.85
002194	100	2.70	2.70	2.60	2.65	2.60	2.65	2.65	0.04	1.69
002195	100	2.70	2.70	2.70	2.70	2.70	2.70	2.70	0.00	0.00
002208	100	2.70	2.75	2.70	2.65	2.70	2.60	2.68	0.05	1.92

Bundle ID	Count	Head L1	Head L2	Head L3	Head L4	Head L5	Head L6	Mean	Std Dev	CV
002209	100	2.75	2.75	2.60	2.70	2.70	2.75	2.71	0.06	2.16
002210	100	2.70	2.70	2.60	2.75	2.75	2.65	2.69	0.06	2.17
002211	100	2.70	2.70	2.70	2.75	2.75	2.75	2.73	0.03	1.00
002214	103	2.70	2.75	2.70	2.70	2.70	2.70	2.71	0.02	0.75
002215	100	2.85	2.80	2.80	2.75	2.80	2.80	2.80	0.03	1.13
002216	97	2.65	2.65	2.65	2.65	2.70	2.80	2.68	0.06	2.26
002218	100	2.70	2.70	2.70	2.70	2.70	2.75	2.71	0.02	0.75
002219	100	2.80	2.65	2.65	2.55	2.80	2.70	2.69	0.10	3.61
002220	100	2.90	2.70	2.60	2.90	2.75	2.65	2.75	0.13	4.60
002221	95	2.60	2.60	2.75	2.70	2.70	2.70	2.68	0.06	2.29
002222	94	2.70	2.85	2.85	2.85	2.70	2.80	2.79	0.07	2.64
002223	91	2.85	2.85	2.75	2.80	2.90	2.75	2.82	0.06	2.15
002224	100	2.70	2.80	2.80	2.75	2.75	2.75	2.76	0.04	1.36
002226	95	2.60	2.65	2.80	2.80	2.80	2.80	2.74	0.09	3.35
002227	105	2.70	2.75	2.75	2.70	2.75	2.70	2.73	0.03	1.00
002229	93	2.80	2.70	2.65	2.70	2.60	2.70	2.69	0.07	2.47
002230	100	2.75	2.70	2.80	2.70	2.70	2.80	2.74	0.05	1.79
003531	100	2.70	2.80	2.70	2.70	2.70	2.65	2.71	0.05	1.82
003716	101	2.70	2.70	2.75	2.75	2.75	2.75	2.73	0.03	0.94
003718	198	2.60	2.75	2.70	2.60	2.65	2.65	2.66	0.06	2.20
003719	105	2.60	2.80	2.65	2.70	2.60	2.60	2.66	0.08	3.01
003759	90	2.70	2.60	2.70	2.75	2.60	2.60	2.66	0.07	2.50
003823	106	2.75	2.75	3.00	2.90	2.75	2.75	2.82	0.11	3.83
003824	125	2.70	2.65	2.70	2.60	2.65	2.70	2.67	0.04	1.53
004029	512	3.20	3.35	3.30	3.15	3.25	3.30	3.26	0.07	2.26
004056	105	2.80	2.90	2.80	2.80	2.75	2.80	2.81	0.05	1.75
004092	229	3.25	3.25	3.40	3.25	3.25	3.40	3.30	0.08	2.35
004093	92	2.75	2.75	2.65	2.70	2.70	2.70	2.71	0.04	1.39
004094	99	2.70	2.70	2.80	2.70	2.70	2.70	2.72	0.04	1.50
004107	100	2.65	2.60	2.65	2.70	2.75	2.65	2.67	0.05	1.94
005611	105	2.80	2.75	2.80	2.75	2.80	2.70	2.77	0.04	1.48

7c

Bundle ID	Count	Tang 1	Tang 2	Tang 3	Tang 4	Tang 5	Tang 6	Mean	Std Dev	CV	Notes
001002	120	13.6	14.2	12.3	15.4	15.6	12.1	13.9	1.5	10.8	
001003	101	12.9	13.1	13.4	12.2	13.9	0.0	10.9	5.4	49.3	
001004	103	13.1	13.3	12.8	12.3	11.8	9.1	12.1	1.6	12.9	
001005	100	13.7	15.5	16.1	14.8	14.9	14.1	14.9	0.9	5.9	longer
001009	160	14.6	15.6	11.7	15.6	6.8	15.6	13.3	3.5	26.5	
001017	93	14.3	16.0	14.6	14.2	14.0	16.2	14.9	1.0	6.5	
001018	96	14.5	12.3	13.4	13.7	14.4	12.7	13.5	0.9	6.6	
001020	106	10.7	10.1	12.2	11.7	11.3	11.9	11.3	0.8	7.0	
001021	99	14.7	13.8	12.4	14.0	14.2	14.4	13.9	0.8	5.8	short
001022	102	7.6	8.3	6.7	6.5	7.4	8.5	7.5	0.8	10.8	
001026	96	12.3	11.0	11.4	12.5	12.4	12.7	12.1	0.7	5.7	
001027	102	16.9	15.8	13.9	14.7	12.0	15.0	14.7	1.7	11.4	
001028	94	11.2	12.5	12.0	14.3	12.1	13.9	12.7	1.2	9.4	thin
001030	98	15.9	13.3	13.8	14.1	12.6	8.3	13.0	2.6	19.6	thin
001042	109	10.3	15.2	10.7	11.4	16.2	13.6	12.9	2.5	19.2	
001043	105	14.5	13.0	14.2	15.3	10.3	12.8	13.4	1.8	13.2	
001050	95	12.5	12.1	12.2	9.6	12.8	12.2	11.9	1.2	9.7	mix long and short
001055	90	8.2	7.7	12.4	10.7	15.8	14.2	11.5	3.2	28.2	mix long and short
001056	116	11.5	13.5	9.7	10.3	13.2	12.5	11.8	1.6	13.2	
001062	92	14.8	13.4	11.7	14.2	15.2	16.3	14.3	1.6	11.1	
001106	143	14.7	11.8	14.5	14.5	13.7	10.9	13.4	1.6	12.1	
001107	101	13.6	14.3	12.2	12.1	0.0	0.0	8.7	6.8	78.1	
001110	119	12.0	11.3	12.9	9.5	14.6	11.3	11.9	1.7	14.4	
001112	99	11.5	13.7	11.3	12.2	13.8	9.7	12.0	1.6	13.0	
001113	99	12.6	14.0	14.5	15.3	15.4	10.7	13.8	1.8	13.2	thin
001114	122	12.7	12.3	15.1	11.8	9.7	12.9	12.4	1.7	14.1	
001116	100	9.2	11.8	15.3	11.4	11.6	13.9	12.2	2.1	17.4	
001117	101	12.7	13.6	11.6	14.2	13.2	15.1	13.4	1.2	9.0	
001118	100	13.9	15.2	12.8	14.7	12.8	10.1	13.3	1.8	13.8	
001119	146	11.1	8.7	11.4	11.1	14.1	12.4	11.5	1.8	15.5	
001120	160	13.8	13.3	13.0	13.0	13.3	7.8	12.4	2.3	18.2	
001136	154	12.0	11.5	10.9	15.5	13.0	13.8	12.8	1.7	13.2	
001139	150	10.5	9.5	13.3	12.6	12.3	16.8	12.5	2.5	20.3	short
001140	150	14.0	15.5	17.4	12.9	11.2	11.7	13.8	2.4	17.1	
001149	125	6.9	8.1	7.3	7.5	6.7	7.0	7.3	0.5	7.0	thin, bent

Bundle ID	Count	Tang 1	Tang 2	Tang 3	Tang 4	Tang 5	Tang 6	Mean	Std Dev	CV	Notes
001151	100	12.2	11.9	10.0	10.0	10.7	10.4	10.9	1.0	8.8	
001152	92	11.7	11.2	11.7	8.0	12.2	11.0	11.0	1.5	13.8	short
001155	200	11.0	11.1	10.8	11.1	11.6	10.4	11.0	0.4	3.6	
001159	100	8.6	8.8	7.2	7.7	7.0	8.0	7.9	0.7	9.2	
001160	102	14.7	11.8	14.6	14.5	13.7	11.0	13.4	1.6	11.9	
001168	120	13.8	9.9	12.1	14.6	15.3	12.2	13.0	2.0	15.2	
001169	102	10.9	11.2	12.7	12.4	11.7	9.9	11.5	1.0	9.0	short
001170	101	12.6	13.1	15.0	12.5	12.3	13.4	13.2	1.0	7.6	
001171	101	8.6	8.1	9.8	7.4	7.3	7.2	8.1	1.0	12.5	
001172	106	15.4	13.3	13.4	14.9	13.5	13.0	13.9	1.0	7.1	
001174	120	11.4	9.3	10.9	12.3	12.5	10.6	11.2	1.2	10.6	broken
001177	94	14.1	12.7	12.8	11.2	13.3	12.3	12.7	1.0	7.6	mixed
001178	94	8.0	11.2	9.6	14.2	6.2	7.6	9.5	2.9	30.5	
001179	184	5.2	6.5	12.6	14.6	14.2	10.9	10.7	4.0	37.3	
001185	105	14.0	13.7	16.5	15.3	16.5	12.0	14.7	1.8	12.0	
001190	166	12.6	12.1	15.1	13.9	15.1	16.9	14.3	1.8	12.5	
001193	110	12.4	12.1	10.8	13.4	13.9	13.6	12.7	1.2	9.2	
001196	185	11.4	14.5	13.5	12.0	11.2	11.7	12.4	1.3	10.7	mixed long and short
001197	90	12.6	11.1	8.7	11.4	11.0	14.2	11.5	1.8	15.9	
001200	250	14.8	14.9	13.8	10.3	7.0	7.4	11.4	3.6	32.0	
001201	185	12.2	8.2	15.2	15.4	11.7	13.4	12.7	2.7	21.0	
001205	100	13.7	13.9	14.3	14.2	14.9	13.0	14.0	0.6	4.6	mixed thin and thick
001321	105	10.4	15.7	11.4	13.8	11.9	13.3	12.8	1.9	15.0	
001383	163	11.8	12.5	13.7	9.0	10.1	9.4	11.1	1.9	16.9	
001387	105	13.0	15.9	12.3	13.2	15.0	15.2	14.1	1.5	10.3	
001388	108	15.9	14.5	17.3	15.3	15.4	15.9	15.7	0.9	5.9	
001393	112	13.0	14.7	15.5	15.4	11.8	11.4	13.6	1.8	13.3	
001394	275	14.1	16.7	16.1	15.5	15.0	13.1	15.1	1.3	8.8	
001427	108	12.5	12.6	13.2	12.0	13.2	14.3	13.0	0.8	6.1	
001428	103	11.3	12.6	13.6	13.4	13.8	11.9	12.8	1.0	7.9	
001441	104	10.0	16.2	14.1	13.1	13.8	14.7	13.7	2.1	15.2	thin
001464	115	12.2	13.4	12.2	13.2	14.2	12.8	13.0	0.8	5.9	
001479	95	12.0	16.7	11.6	10.8	7.2	12.9	11.9	3.1	26.0	
001482	136	13.7	11.8	15.3	13.2	12.1	14.7	13.5	1.4	10.3	
001483	95	13.6	14.8	16.6	15.4	14.7	11.5	14.4	1.7	12.1	
001484	183	14.6	14.7	11.8	11.9	12.6	14.0	13.3	1.3	10.0	

Bundle ID	Count	Tang 1	Tang 2	Tang 3	Tang 4	Tang 5	Tang 6	Mean	Std Dev	CV	Notes
001486	96	13.7	13.5	13.2	13.3	13.5	11.1	13.1	1.0	7.4	
001490	96	15.2	14.0	14.7	15.6	12.6	13.8	14.3	1.1	7.6	
001491	101	14.4	14.0	12.5	15.4	13.7	14.4	14.1	1.0	6.8	
001495	96	12.1	12.3	12.5	12.5	11.6	11.6	12.1	0.4	3.4	
001496	96	13.9	12.8	13.5	10.9	10.8	9.6	11.9	1.7	14.5	
001932	93	12.5	10.7	11.6	10.6	10.8	11.6	11.3	0.7	6.5	
001934	101	14.4	16.1	11.0	13.8	14.0	15.2	14.1	1.7	12.3	
001935	97	14.7	13.5	12.5	12.4	13.9	13.8	13.5	0.9	6.6	thin
001936	112	13.5	12.7	14.0	13.4	13.3	14.3	13.5	0.6	4.1	thin
001939	95	11.9	11.4	13.6	13.5	14.1	10.3	12.5	1.5	12.0	thin
001940	99	12.9	13.0	13.4	11.5	10.2	13.3	12.4	1.3	10.3	
001941	100	12.4	13.8	12.6	10.4	11.4	12.8	12.2	1.2	9.7	
001943	100	12.5	11.8	12.8	8.7	13.9	10.6	11.7	1.8	15.7	
001944	96	11.9	14.2	16.1	14.1	13.5	13.5	13.9	1.4	9.8	
001945	99	15.5	16.6	15.0	13.3	12.5	15.5	14.7	1.5	10.4	
001946	92	16.0	15.0	12.4	13.1	15.3	13.4	14.2	1.4	10.0	
001947	97	12.0	13.6	14.3	14.1	13.2	13.8	13.5	0.8	6.1	
001948	102	15.4	10.5	11.6	11.4	11.4	15.1	12.6	2.1	16.8	
001949	100	15.5	14.3	15.5	14.5	15.0	14.9	15.0	0.5	3.3	
001950	100	13.6	11.6	12.5	13.6	11.5	15.0	13.0	1.4	10.4	
001951	100	13.8	13.1	15.1	14.5	13.3	14.9	14.1	0.8	6.0	
001952	100	15.3	17.2	14.9	17.2	14.0	13.5	15.4	1.6	10.2	
001953	103	13.4	14.3	12.8	15.6	16.3	13.3	14.3	1.4	9.8	
001954	100	14.6	14.0	15.7	12.6	12.1	11.3	13.4	1.7	12.4	
001955	95	14.5	13.5	15.2	14.9	13.7	13.4	14.2	0.8	5.4	
001956	91	14.6	15.7	13.5	13.9	13.6	15.7	14.5	1.0	6.9	
001957	97	15.7	12.7	14.4	14.3	15.2	12.6	14.2	1.3	9.0	
001958	94	13.4	14.8	14.2	14.1	12.2	14.1	13.8	0.9	6.5	
001959	98	13.5	15.1	13.9	16.2	14.5	10.5	14.0	1.9	13.9	
001960	100	14.4	15.0	15.1	13.6	13.3	13.2	14.1	0.8	6.0	
001962	98	12.9	12.2	11.0	14.2	13.3	12.8	12.7	1.1	8.5	
001963	99	14.9	13.5	13.8	13.1	13.7	11.6	13.4	1.1	8.0	
001964	96	15.9	15.5	12.1	15.0	14.6	14.1	14.5	1.4	9.3	
001965	98	13.3	15.3	13.2	15.2	16.3	12.4	14.3	1.5	10.7	
001966	101	14.5	12.7	15.0	11.9	12.6	15.7	13.7	1.5	11.2	
001967	98	11.5	12.0	12.8	10.8	12.3	12.8	12.0	0.8	6.5	
001968	100	13.7	13.6	16.1	14.7	15.5	15.3	14.8	1.0	6.8	

Bundle ID	Count	Tang 1	Tang 2	Tang 3	Tang 4	Tang 5	Tang 6	Mean	Std Dev	CV	Notes
001969	100	17.1	15.5	12.5	14.5	13.1	13.4	14.4	1.7	12.0	
001970	98	15.3	15.7	10.5	11.7	11.8	15.3	13.4	2.3	17.2	
001971	95	12.8	11.0	11.0	11.8	9.9	10.3	11.1	1.0	9.4	
001979	100	12.6	10.6	11.9	13.2	12.1	10.6	11.8	1.1	8.9	
001980	100	17.1	17.2	15.9	13.4	15.1	7.7	14.4	3.6	24.8	
001981	101	11.4	14.7	11.9	12.9	12.8	10.9	12.4	1.4	10.9	
001982	101	12.2	11.2	12.8	13.8	13.4	13.3	12.8	1.0	7.4	
001983	101	15.7	10.5	14.5	10.8	12.5	12.2	12.7	2.0	16.1	
001984	101	12.2	13.4	12.6	12.5	11.5	14.0	12.7	0.9	7.0	
001985	100	12.3	12.9	13.1	10.6	13.7	12.7	12.6	1.1	8.5	
001989	100	8.9	12.4	12.7	12.5	11.2	12.5	11.7	1.5	12.6	
001994	100	13.2	13.3	14.3	15.2	11.9	13.7	13.6	1.1	8.2	
001995	100	13.4	13.8	13.3	14.2	14.2	14.3	13.9	0.4	3.1	
001998	101	13.6	12.5	15.7	13.0	11.0	14.1	13.3	1.6	11.9	
001999	97	13.3	12.6	12.3	14.2	13.1	11.5	12.8	0.9	7.2	
002000	100	16.1	14.7	14.5	15.6	12.3	16.7	15.0	1.6	10.4	
002003	100	14.4	12.2	13.6	12.7	13.3	14.0	13.4	0.8	6.1	
002004	101	14.6	14.7	13.4	14.7	16.3	15.3	14.8	1.0	6.4	
002009	100	14.7	13.0	14.4	13.9	14.6	12.7	13.9	0.9	6.1	
002010	96	11.7	12.1	10.5	10.8	10.9	12.4	11.4	0.8	6.8	
002012	100	9.4	12.7	15.7	15.6	13.7	15.3	13.7	2.4	17.7	
002013	100	12.4	11.4	13.0	12.2	11.8	11.2	12.0	0.7	5.6	
002014	97	15.3	15.6	11.6	11.3	13.5	15.1	13.7	1.9	13.9	
002016	100	14.8	14.7	13.2	13.3	15.8	14.8	14.4	1.0	6.9	
002018	103	14.7	12.1	13.9	12.4	12.1	13.5	13.1	1.1	8.2	
002020	98	12.9	12.3	11.3	11.6	12.5	12.9	12.3	0.7	5.5	
002022	100	14.3	13.0	15.8	14.1	14.7	14.3	14.4	0.9	6.3	
002023	99	16.3	15.3	13.0	15.9	13.5	15.2	14.9	1.3	8.9	
002025	100	11.8	14.0	15.8	13.0	12.9	15.3	13.8	1.5	11.1	
002029	97	12.3	9.7	11.6	11.6	12.5	10.5	11.4	1.1	9.5	
002030	100	13.9	14.1	10.3	14.4	13.5	13.8	13.3	1.5	11.4	
002031	96	14.0	13.8	12.5	13.4	13.9	12.3	13.3	0.7	5.6	
002034	100	17.5	15.4	14.3	13.3	12.6	13.8	14.5	1.8	12.1	
002035	98	14.5	13.4	14.9	13.4	14.1	11.9	13.7	1.1	7.8	
002036	98	17.3	14.5	14.2	13.8	13.3	13.1	14.4	1.5	10.7	
002037	99	15.2	14.4	15.2	15.2	13.8	14.9	14.8	0.6	3.9	
002038	99	13.7	14.4	14.1	15.0	16.8	14.1	14.7	1.1	7.6	

Bundle ID	Count	Tang 1	Tang 2	Tang 3	Tang 4	Tang 5	Tang 6	Mean	Std Dev	CV	Notes
002041	99	12.5	13.2	13.0	13.1	12.4	12.9	12.9	0.3	2.5	
002042	99	14.8	14.1	15.5	12.8	13.8	14.8	14.3	0.9	6.6	broken
002044	102	12.2	13.7	15.0	14.3	14.1	15.3	14.1	1.1	7.8	
002050	100	13.7	16.1	12.3	11.5	15.7	12.0	13.6	2.0	14.5	
002051	98	14.0	13.8	13.6	13.8	13.5	13.0	13.6	0.3	2.6	
002052	104	12.5	15.2	9.9	8.9	8.7	5.3	10.1	3.4	33.8	
002053	98	12.7	13.0	12.3	12.5	14.4	8.7	12.3	1.9	15.5	
002054	100	15.1	15.7	14.8	14.4	15.3	14.9	15.0	0.4	3.0	
002055	109	13.2	13.1	12.2	11.6	13.2	10.5	12.3	1.1	8.9	thin
002056	100	16.6	15.0	16.4	16.2	16.7	12.0	15.5	1.8	11.7	mix long and short
002057	100	17.0	16.5	16.9	15.6	16.4	17.1	16.6	0.6	3.4	
002059	96	13.5	13.9	13.7	13.7	14.1	12.1	13.5	0.7	5.3	
002061	92	8.5	10.2	12.4	10.3	10.0	10.8	10.4	1.3	12.2	mix long and short
002062	96	12.6	15.5	8.3	7.8	9.9	12.6	11.1	3.0	26.7	
002063	100	14.2	13.7	10.5	11.0	10.9	11.8	12.0	1.6	13.0	
002064	100	11.4	11.0	11.3	11.2	12.8	11.5	11.5	0.6	5.6	
002065	100	14.1	9.2	8.0	7.3	8.5	7.2	9.1	2.6	28.6	thin
002066	98	16.0	12.4	15.9	13.4	15.3	12.0	14.2	1.8	12.6	
002067	99	14.6	15.8	15.9	14.9	16.5	15.8	15.6	0.7	4.5	
002068	100	12.3	14.2	15.3	16.0	11.9	12.1	13.6	1.8	13.1	thin
002069	99	11.9	13.1	9.7	10.3	11.9	12.2	11.5	1.3	11.0	
002073	95	8.8	8.1	8.7	12.4	11.6	13.5	10.5	2.3	21.6	
002075	100	13.0	13.5	12.6	12.5	14.2	11.5	12.9	0.9	7.2	
002076	97	11.3	11.6	11.1	11.1	11.7	12.2	11.5	0.4	3.7	
002077	114	9.3	11.0	14.1	13.9	12.8	10.5	11.9	2.0	16.4	
002078	98	11.8	10.5	11.0	9.4	11.3	12.5	11.1	1.1	9.7	
002079	97	13.5	13.2	13.9	13.0	13.2	13.7	13.4	0.3	2.6	
002081	95	14.0	15.3	17.8	15.5	15.3	13.8	15.3	1.4	9.4	
002082	100	14.1	14.2	14.3	16.7	14.3	14.3	14.7	1.0	6.9	
002083	100	12.4	11.4	10.7	11.1	11.3	12.4	11.6	0.7	6.1	
002086	100	15.0	15.3	17.3	13.5	15.3	12.2	14.8	1.7	11.8	thin
002088	102	11.5	11.4	11.4	12.5	8.1	7.9	10.5	2.0	18.7	thin
002089	93	14.9	14.3	13.7	11.3	10.4	13.6	13.0	1.8	13.6	
002090	99	11.4	11.7	12.2	12.2	12.2	9.2	11.5	1.2	10.2	
002091	100	11.2	11.3	11.2	9.9	12.2	13.2	11.5	1.1	9.7	
002093	102	11.9	12.8	11.8	10.6	12.7	10.5	11.7	1.0	8.5	

Bundle ID	Count	Tang 1	Tang 2	Tang 3	Tang 4	Tang 5	Tang 6	Mean	Std Dev	CV	Notes
002095	100	11.3	11.0	9.1	7.6	12.1	10.3	10.2	1.6	16.0	
002098	100	15.4	13.3	14.2	13.1	13.7	13.8	13.9	0.8	5.9	
002099	95	14.5	13.3	13.6	14.6	15.8	14.2	14.3	0.9	6.1	
002101	100	13.0	14.6	13.1	14.7	14.2	15.8	14.2	1.1	7.4	
002103	105	13.7	12.2	12.8	13.2	12.2	12.0	12.7	0.7	5.3	
002104	97	8.0	8.0	7.3	7.9	8.2	8.5	8.0	0.4	5.0	
002105	100	12.9	13.5	13.7	12.7	14.6	14.7	13.7	0.8	6.1	
002106	98	8.3	8.0	7.8	6.5	8.6	6.3	7.6	1.0	12.6	mixed long and short
002107	94	12.5	13.3	15.6	14.7	13.6	13.2	13.8	1.1	8.2	
002108	96	13.8	13.8	13.8	13.5	13.2	13.7	13.6	0.2	1.8	
002109	100	15.0	15.3	14.5	13.2	13.3	13.7	14.2	0.9	6.3	
002112	98	15.2	13.5	12.8	15.0	14.7	16.2	14.6	1.2	8.4	thin
002113	98	14.9	10.8	13.7	13.8	13.5	12.3	13.2	1.4	10.8	
002114	95	13.2	14.4	10.7	14.8	13.1	14.9	13.5	1.6	11.7	
002116	97	8.6	10.6	8.6	9.0	7.9	12.1	9.5	1.6	16.6	
002118	99	12.6	14.3	16.2	14.0	15.4	17.7	15.0	1.8	11.9	
002121	103	14.2	14.2	12.8	13.0	10.9	12.3	12.9	1.2	9.7	
002124	97	12.6	11.0	14.3	13.6	13.7	13.7	13.2	1.2	9.0	
002125	100	7.5	9.7	11.3	12.2	8.8	13.3	10.5	2.2	20.9	
002126	105	12.4	15.3	12.3	14.6	15.6	14.2	14.1	1.4	10.1	
002127	94	13.8	14.3	14.0	14.0	13.3	14.2	13.9	0.4	2.6	
002136	100	15.3	13.1	14.5	12.0	12.7	10.6	13.0	1.7	13.0	thin
002137	97	13.0	14.0	14.0	15.1	12.8	10.7	13.3	1.5	11.3	
002138	100	15.2	15.4	15.3	13.7	12.8	14.0	14.4	1.1	7.4	
002139	92	12.3	14.2	13.2	14.1	14.4	13.6	13.6	0.8	5.8	
002140	99	17.0	17.0	13.5	15.9	12.3	14.4	15.0	1.9	12.9	thin
002141	99	14.5	13.8	14.1	14.0	12.3	12.5	13.5	0.9	6.7	
002144	100	15.2	14.5	14.3	15.1	15.0	14.1	14.7	0.5	3.1	
002145	100	11.9	11.6	8.9	9.7	12.5	11.7	11.1	1.4	12.8	
002146	100	11.6	14.1	12.4	12.1	12.2	14.3	12.8	1.1	8.8	
002147	100	14.9	15.2	11.7	11.8	14.2	13.2	13.5	1.5	11.3	
002148	94	13.6	13.9	14.0	13.0	13.8	14.6	13.8	0.5	3.8	thin
002154	100	14.4	11.7	11.2	14.0	13.2	13.2	13.0	1.3	9.7	short
002155	103	14.7	14.0	16.8	14.2	13.4	15.2	14.7	1.2	8.1	thin
002156	100	12.2	12.4	12.2	9.8	15.4	12.0	12.3	1.8	14.5	
002157	92	13.8	12.5	12.9	12.8	11.0	12.2	12.5	0.9	7.4	
002158	100	13.5	13.8	13.0	13.8	13.9	14.1	13.7	0.4	2.8	

Bundle ID	Count	Tang 1	Tang 2	Tang 3	Tang 4	Tang 5	Tang 6	Mean	Std Dev	CV	Notes
002159	100	14.1	12.7	15.0	13.1	13.6	15.5	14.0	1.1	7.8	
002160	98	10.8	10.4	11.7	12.2	10.2	12.5	11.3	1.0	8.6	
002164	96	8.4	8.8	6.0	7.6	7.2	7.8	7.6	1.0	12.9	mixed long and short
002165	101	12.7	10.6	12.5	13.1	10.3	12.8	12.0	1.2	10.2	thin
002166	100	10.3	13.9	13.7	12.0	14.6	15.2	13.3	1.8	13.7	
002167	100	13.5	13.5	12.0	12.8	12.2	12.0	12.7	0.7	5.6	
002170	100	13.6	13.8	13.6	14.6	13.7	13.6	13.8	0.4	2.8	
002171	94	14.5	13.8	13.7	14.2	11.9	11.6	13.3	1.2	9.2	
002172	102	14.2	11.3	15.5	15.4	16.8	14.3	14.6	1.9	12.8	
002176	102	8.1	7.5	8.5	8.4	15.5	12.6	10.1	3.2	31.8	
002177	101	12.7	10.4	13.0	12.7	10.1	13.3	12.0	1.4	11.7	
002178	127	15.4	14.5	17.6	13.2	15.8	15.0	15.3	1.5	9.6	
002179	94	12.2	12.0	12.0	10.5	10.9	12.4	11.7	0.8	6.6	
002181	92	15.8	11.3	14.8	14.8	11.7	14.6	13.8	1.9	13.4	
002185	100	14.7	15.6	16.8	14.3	16.2	16.5	15.7	1.0	6.4	
002189	100	15.4	9.5	11.2	14.8	13.8	9.8	12.4	2.6	20.8	
002191	96	14.7	13.7	14.7	14.2	10.3	13.9	13.6	1.7	12.2	
002192	100	14.1	15.2	13.3	15.0	13.4	10.2	13.5	1.8	13.4	
002194	100	13.6	13.6	14.1	13.6	14.0	14.0	13.8	0.2	1.7	
002195	100	12.6	14.0	10.3	12.5	14.0	11.3	12.5	1.5	11.8	
002208	100	14.2	13.7	14.1	15.7	13.3	15.4	14.4	1.0	6.6	
002209	100	15.9	15.3	15.3	15.7	16.1	14.9	15.5	0.4	2.9	short
002210	100	13.8	13.0	13.5	13.6	15.4	12.1	13.6	1.1	8.0	broken
002211	100	14.3	14.2	14.9	12.5	15.0	14.6	14.3	0.9	6.4	
002214	103	15.6	13.3	13.7	14.2	16.0	12.0	14.1	1.5	10.5	thin
002215	100	16.0	13.5	13.5	16.3	13.4	13.6	14.4	1.4	9.5	thin
002216	97	12.2	12.7	13.2	11.5	12.4	11.6	12.3	0.7	5.3	thin
002218	100	15.1	14.2	14.9	15.7	14.3	16.0	15.0	0.7	4.8	thin
002219	100	8.4	5.8	6.0	7.0	5.8	7.5	6.8	1.1	15.8	thin
002220	100	10.8	11.0	8.7	6.3	3.6	7.2	7.9	2.8	35.7	thick
002221	95	6.6	8.3	7.1	6.4	7.7	7.3	7.2	0.7	9.7	short
002222	94	12.0	13.7	9.2	9.1	11.8	6.9	10.5	2.5	23.7	
002223	91	12.4	12.5	10.7	9.1	11.8	10.4	11.2	1.3	11.9	short
002224	100	10.9	10.6	12.5	12.3	8.9	12.2	11.2	1.4	12.4	
002226	95	12.7	10.4	9.7	10.6	10.3	11.6	10.9	1.1	10.0	
002227	105	12.5	10.4	11.6	10.8	11.0	12.4	11.5	0.9	7.6	
002229	93	17.0	15.2	13.8	12.8	12.1	12.9	14.0	1.8	13.1	broken

Bundle ID	Count	Tang 1	Tang 2	Tang 3	Tang 4	Tang 5	Tang 6	Mean	Std Dev	CV	Notes
002230	100	6.7	6.3	7.8	8.0	6.7	7.4	7.2	0.7	9.6	
003531	100	12.1	11.9	13.3	11.4	15.3	15.8	13.3	1.9	14.0	longer
003716	101	7.1	7.3	7.9	8.0	7.1	7.9	7.6	0.4	5.7	
003718	198	15.1	10.8	14.6	11.7	11.4	10.1	12.3	2.1	16.8	
003719	105	13.5	12.3	13.0	13.6	11.5	12.7	12.8	0.8	6.2	
003759	90	13.4	13.3	9.4	9.6	12.9	9.8	11.4	2.0	17.4	
003823	106	7.4	5.5	6.6	5.6	7.5	11.2	7.3	2.1	28.7	mixed thin and thick
003824	125	12.4	12.6	14.0	15.8	14.5	13.9	13.9	1.3	9.1	
004029	512	29.0	28.6	28.8	28.5	27.1	29.0	28.4	0.7	2.5	
004056	105	12.1	11.8	12.0	12.1	10.2	10.8	11.5	0.8	7.0	
004092	229	29.8	29.8	28.3	27.6	29.8	27.8	28.9	1.1	3.7	
004093	92	15.4	13.4	13.5	12.3	15.3	15.3	14.2	1.3	9.2	
004094	99	14.2	15.7	13.0	14.7	12.2	11.1	13.5	1.7	12.6	
004107	100	11.6	14.8	16.4	14.8	13.7	14.0	14.2	1.6	11.2	
005611	105	15.9	11.5	10.0	10.8	10.4	11.3	11.7	2.2	18.5	

Appendix 8

Spatial and metric data on ferrules

Ferrule ID	Spatial ID	X	Y	Generic location	Location error	Type	Inscrip-tion?	Length	Diameter T1	Diameter T2	Diameter B1	Diameter T2	Thick-ness	Upper length
001675	27	30.36332	-21.988289	Pit1main	exact	I	0	11.32	4.09	3.29	3.73	3.02	0.21	4.53
001676	29	21.678343	-22.002189	Pit1main	exact	III	0	3.50	3.77	3.19	3.76	3.18	0.35	
001677	28	29.633117	-22.166095	Pit1main	exact	III	0	3.48	3.81	3.19	3.78	3.23	0.20	
001678	24	26.544154	-9.318164	Pit1main	exact	I	0	11.22	4.05	3.46	3.69	3.09	0.30	4.57
001679	79	15.669819	-14.075431	Pit1main	exact	III	0	3.49	3.79	3.14	3.79	3.19	0.20	
001680	33	11.667551	-15.747676	Pit1main	exact	III	1	3.43	3.77	3.36	3.80	3.38	0.26	
001681	36	9.653734	-14.896122	Pit1main	exact	III	1	3.52	3.83	3.68	3.85	3.66	0.25	
001682	42	7.26974	-14.992024	Pit1main	exact	III	0	3.42	3.80	3.18	3.79	3.21	0.26	
001683	32	16.06179	-15.851278	Pit1main	exact	III	0	3.47	3.80	3.18	3.78	3.22	0.23	
001684	37	13.935687	-20.672958	Pit1main	exact	III	1	3.48	3.82	3.22	3.76	3.20	0.30	
001685	38	15.401094	-20.732604	Pit1main	exact	I	0	11.21	4.01	3.45	3.63	3.21	0.20	4.52
001686	18	33.897212	9.545137	Pit1main	exact	II	0	10.07	3.27	3.19	3.13	3.12	0.27	
001687	25	30.719264	-4.065418	Pit1main	exact	I	0	11.28	4.08	3.42	3.72	3.03	0.19	4.49
001688	23	29.749301	-4.789183	Pit1main	exact	I	0	11.25	4.15	3.37	3.82	3.01	0.18	4.49
001689	22	30.733698	1.342202	Pit1main	exact	I	0	11.27	4.09	3.23	3.87	3.07	0.13	4.54
001690	65	23.325236	8.436779	Pit1main	exact	III	0	3.15	2.83	2.79	2.83	2.79	0.23	
001691	61	14.624853	24.198183	Pit1main	exact	III	0	3.51	3.87	3.42	3.83	3.40	0.21	
001692	62	17.384535	23.128079	Pit1main	exact	III	0	3.51	3.80	3.25	3.76	3.23	0.21	
001693	74	24.332277	17.010839	Pit1main	exact	I	0	11.22	3.97	3.40	3.60	3.13	0.16	4.53
001694	60	17.929902	13.499618	Pit1main	exact	I	0	11.53	4.29	3.42	3.85	3.19	0.16	4.85
001695	15	31.762395	22.241842	Pit1main	exact	I	0	11.19	4.01	3.49	3.62	3.14	0.16	4.58
001696	71	29.45671	23.054828	Pit1main	exact	I	0	11.19	3.96	3.33	3.63	3.19	0.21	4.52
001697	14	17.435182	22.497996	Pit1main	exact	I	0	11.55	4.15	3.29	4.13	3.23	0.17	
001698	72	27.032034	22.217045	Pit1main	exact	II	0	10.25	3.31	3.26	3.22	3.18	0.24	
001699	19	32.194966	8.21549	Pit1main	exact	I	0	11.21	4.00	3.43	3.55	3.06	0.14	4.43
001700	21	32.394268	5.378035	Pit1main	exact	I	0	11.19	4.00	3.36	3.59	3.10	0.16	4.55

Ferrule ID	Spatial ID	X	Y	Generic location	Location error	Type	Inscrip-tion?	Length	Diameter T1	Diameter T2	Diameter B1	Diameter T2	Thick-ness	Upper length
001701	17	32.927437	12.904796	Pit1main	exact	I	0	11.21	3.95	3.39	3.66	3.24	0.13	4.49
001702	67	12.710556	19.050626	Pit1main	exact	III	1	3.53	3.90	3.66	3.88	3.67	0.20	
001703	1	2.035892	18.712645	Pit1main	exact	III	1	3.52	3.87	3.67	3.84	3.66	0.25	
001704	3	2.185588	17.545321	Pit1main	exact	III	0	3.49	3.81	3.24	3.81	3.22	0.20	
001705	2	2.012032	18.070084	Pit1main	exact	III	1	3.48	3.82	3.43	3.82	3.40	0.25	
001706	7	8.889727	18.400762	Pit1main	exact	III	1	3.43	3.80	3.35	3.79	3.35	0.24	
001707	5	0.439067	17.953937	Pit1main	exact	III	1	3.44	3.81	3.32	3.82	3.36	0.20	
001708	11	14.247334	17.667337	Pit1main	exact	I	0	11.24	4.00	3.39	3.66	3.09	0.19	4.52
001709	4	1.133928	17.746426	Pit1main	exact	III	0	3.46	3.84	3.21	3.83	3.23	0.22	
001710	8	7.108188	24.862539	Pit1main	exact	III	0	3.46	3.85	3.16	3.82	3.28	0.16	
001711	13	19.829236	28.580815	Pit1main	poor	I	0	11.45	3.88	3.24	3.83	2.99	0.18	4.77
001712	16	32.845686	27.46743	Pit1main	poor	I	0	11.18	3.98	3.35	3.61	3.01	0.22	4.45
002361	73	2.761638	-25.586641	Pit1main	exact	I	0	11.30	4.10	3.36	3.83	3.12	0.18	4.50
002363	40	10.899877	-22.016127	Pit1main	exact	III	1	3.47	3.85	3.16	3.86	3.25	0.22	
002364	41	9.482286	-21.28394	Pit1main	exact	III	1	3.46	3.85	3.25	3.83	3.24	0.11	
002365	31	18.445393	-22.757078	Pit1main	exact	III	0	3.50	3.84	3.21	3.84	3.24	0.24	
002366	51	7.44323	-20.678943	Pit1main	exact	III	1	3.52	3.80	3.23	3.85	3.24	0.18	
002367	57	8.022527	-20.74168	Pit1main	exact	I	0	11.55	4.14	3.45	3.88	3.17	0.19	5.04
002368	50	7.086615	-20.889377	Pit1main	exact	I	0	11.20	4.08	3.39	3.68	3.06	0.18	4.53
002369	52	6.921574	-21.394079	Pit1main	exact	III	0	3.51	3.81	3.25	3.84	3.23	0.20	
002370	49	8.492875	-22.124172	Pit1main	exact	III	1	3.50	3.90	3.66	3.89	3.69	0.25	
002371	55	7.769361	-21.55276	Pit1main	exact	III	1	3.51	3.78	3.19	3.81	3.24	0.17	
002372	54	7.437248	-22.723216	Pit1main	exact	I	0	11.46	3.94	3.16	3.86	3.02	0.13	4.87
002373	53	6.856912	-22.654856	Pit1main	exact	III	0	3.48	3.87	3.22	3.85	3.24	0.16	
002374	56	2.233985	-22.63557	Pit1main	exact	II	0	10.06	3.20	3.19	3.13	3.11	0.19	
002375	35	12.391512	-16.30439	Pit1main	exact	III	0	3.50	3.81	3.21	3.81	3.25	0.17	

Ferrule ID	Spatial ID	X	Y	Generic location	Location error	Type	Inscrip-tion?	Length	Diameter T1	Diameter T2	Diameter B1	Diameter T2	Thick-ness	Upper length
002376	75	8.422856	-15.540867	Pit1main	exact	III	0	3.51	3.84	3.24	3.82	3.27	0.19	
002377	80	7.572583	-15.64771	Pit1main	exact	III	1	3.46	3.81	3.41	3.87	3.41	0.28	
002378	76	7.328495	-14.068083	Pit1main	exact	III	0	3.44	3.83	3.28	3.84	3.25	0.28	
002379	77	6.581876	-14.34423	Pit1main	exact	III	0	3.51	3.83	3.21	3.82	3.22	0.22	
002380	46	5.586146	-15.713276	Pit1main	exact	III	1	3.53	3.83	3.61	3.84	3.64	0.18	
002381	43	5.228233	-14.695826	Pit1main	exact	III	1	3.52	3.87	3.58	3.87	3.64	0.16	
002382	44	4.209489	-14.853052	Pit1main	exact	III	0	3.49	3.81	3.25	3.86	3.25	0.24	
002383	34	10.71324	-16.405878	Pit1main	exact	III	0	3.50	3.87	3.63	3.90	3.66	0.21	
002384	45	4.254981	-15.422654	Pit1main	exact	III	1	3.55	3.86	3.61	3.86	3.64	0.24	
002385	48	3.399724	-14.148695	Pit1main	exact	III	0	3.42	3.82	3.19	3.84	3.24	0.17	
002386	47	2.43995	-15.440479	Pit1main	exact	III	1	3.50	3.78	3.20	3.78	3.20	0.20	
002387	78	0.015713	-15.837469	Pit1main	exact	III	1	3.45	3.80	3.41	3.81	3.40	0.22	
002388	26	24.811109	-5.654508	Pit1main	exact	I	0	11.19	4.00	3.37	3.69	3.05	0.15	4.65
002390	64	24.760509	11.320852	Pit1main	exact	I	0	11.22	4.03	3.41	3.67	3.12	0.19	4.46
002391	63	27.474209	11.514237	Pit1main	exact	I	0	11.27	3.90	3.51	3.62	3.09	0.17	4.56
002392	6	7.486247	19.486179	Pit1main	exact	III	1	3.52	3.86	3.64	3.86	3.73	0.18	
002393	58	10.214415	16.852817	Pit1main	exact	III	1	3.49	3.83	3.23	3.82	3.28	0.15	
002394	70	6.719265	17.771521	Pit1main	exact	III	0	3.49			3.81	3.34	0.20	
002395	59	9.607143	17.001072	Pit1main	exact	III	1	3.51	3.91	3.68	3.87	3.68	0.16	
002396	9	7.907952	24.332307	Pit1main	exact	III	1	3.53	3.82	3.71	3.90	3.70	0.19	
002397	10	7.785266	24.074517	Pit1main	exact	III	1	3.50	3.87	3.65	3.87	3.69	0.18	
002398	69	13.408053	22.364595	Pit1main	exact	III	1	3.46	3.77	3.37	3.81	3.41	0.19	
002431	12	10.17646	24.964964	Pit1main	exact	III	0	3.50	3.83	3.20	3.83	3.21	0.25	
002432	66	4.507266	23.819083	Pit1main	exact	I	0	11.22	3.94	3.41	3.59	3.06	0.13	4.59
002434	68	14.206073	22.43175	Pit1main	exact	III	0	3.52	3.80	3.26	3.81	3.22	0.16	
003806	30	29.859227	-22.51681	Pit1main	exact	I	0	11.18	3.97	3.38	3.64	3.16	0.19	4.53

www.ingramcontent.com/pod-product-compliance
Ingram Content Group UK Ltd.
Pitfield, Milton Keynes, MK11 3LW, UK
UKHW061847041026
15591UKWH00015B/954

9 781407 316901